T4-AQM-422

High Frequency Financial Econometrics
Recent Developments

Luc Bauwens · Winfried Pohlmeier
David Veredas (Eds.)

High Frequency Financial Econometrics

Recent Developments

With 57 Figures and 64 Tables

Physica-Verlag
A Springer Company

Prof. Luc Bauwens
CORE
Voie du Roman Pays
1348 Louvain-la-Neuve
Belgium
bauwens@ucl.ac.be

Prof. Winfried Pohlmeier
Department of Economics
University of Konstanz
78457 Konstanz
Germany
winfried.pohlmeier@uni-konstanz.de

Prof. David Veredas
ECARES
Université Libre des Bruxelles
30, Avenue Roosevelt
1050 Brussels
Belgium
dveredas@ulb.ac.be

Parts of the papers have been first published in
"Empirical Economics," Vol. 30, No. 4, 2006

Library of Congress Control Number: 2007933836

ISBN 978-3-7908-1991-5 Physica-Verlag Heidelberg New York

Physica-Verlag is a part of Springer Science+Business Media

springer.com

Typesetting by the author and SPi using a Springer LaTeX macro package
Cover-design: WMX design GmbH, Heidelberg

Printed on acid-free paper SPIN: 12107759 88/SPi 5 4 3 2 1 0

Contents

Luc Bauwens · Winfried Pohlmeier · David Veredas

Editor's introduction: recent developments in high frequency financial econometrics

> *"But there are several aspects of the quantitative approach to **finance**, and no single one of these aspects, taken by itself, should be confounded with **financial econometrics**. Thus, **financial econometrics** is by no means the same as **finance** statistics. Nor is it identical with what we call general **financial** theory...Not should be **financial econometrics** a synonymous with the application of mathematics to **finance**. Experience has shown that each of these three view-points, that of statistics, **financial** theory, and mathematics, is a necessary, but not sufficient, condition for a real understanding of the quantitative relations of modern **financial** life. It is the unification of all the three that is powerful. And it is this unification that constitutes **financial econometrics**."*

This paragraph is a virtual copy of the one in p. 2 of Frisch's Editor Note on *Econometrica* Vol. 1, No. 1. The only difference is that economics has been replaced by finance, economic by financial, econometrics by financial econometrics.

It was written 74 years ago but it fully reflects the spirit of this special issue. High frequency finance is an archetypical example of Ragnar Frisch's words. It represents a unification of (1) financial theory, in particular market microstructure, (2) mathematical finance, exemplified in derivative markets, and (3) statistics, for instance the theory of point processes. It is the intersection of these three components that yields an incredibly active research area, with contributions that enhance the understanding of today's complex intra-daily financial world.

> *"Theory, in formulating its abstract quantitative notions, must be inspired to a larger extent by the technique of observations. And fresh statistical and other factual studies must be the healthy element of disturbance that constantly threatens and disquiets theorists and prevents them from coming to rest on some inherited, obsolete set of assumptions."*

Here again high frequency finance is fully reflected in Ragnar Frisch's words. Its *modus vivendi* is a perfect combination of observed real facts, market microstructure theory, and statistics and they all form a system in which each component nicely dovetails with the others. Market microstructure theory deals with models explaining price and agent's behavior in a market governed by certain rules. These markets have different ways to operate (with/without market makers, with/without order books), opening and closing hours, maximum price variations, minimum traded volume, etc. On the other hand, empirical analysis deals with the study of market behavior using real data. For example, what are the relations between traded volume, price variations, and liquidity? What are the potential problems? Last, statistically speaking, high frequency data are realizations of so-called point processes, that is, the arrival of the observations is random. This, jointly with the fact that financial data has pathological and unique features (long memory, strong skewness, and kurtosis) implies that new methods and new econometric models are needed.

The econometric analysis of high frequency data permits us to answer to questions that are of great interest for policy markers. For instance, how much information should regulators disclose to market participants? Or, how do extreme movements in the book affect market liquidity? Or is a market maker really necessary?

On the other side, the practitioners, the traders that participate in the market every day, also have a growing interest in the understanding of financial markets that operate at high frequency. For instance, trading rules may be constructed based on the markets conditions that, in turn, may be explained with financial econometrics.

This volume presents some advanced research in this area. In order to document the potential of high frequency finance, it is our goal to select a wide range of papers, including studies of the order book dynamics, the role of news events, and the measurement of market risks as well as new econometric approaches to the analysis of market microstructures.

Bauwens, Rime, and Sucarrat shed new light on the mixture of distribution hypothesis by means of a study of the weekly exchange rate volatility of the Norwegian krone. They find that the impact of information arrival on exchange rate volatility is positive and statistically significant, and that the hypothesis that an increase in the number of traders reduces exchange rate volatility is not supported. Moreover, they document that the positive impact of information arrival on volatility is relatively stable across three different exchange rate regimes, and in that the impact is relatively similar for both weekly volatility and weekly realised volatility.

Despite its rather weak theoretical and statistical foundation, chart analysis is still a frequently used tool among financial analysts. *Omrane and van Oppens* investigate the existence of chart patterns in the Euro/Dollar intra-daily foreign exchange market at the high frequency level. Checking 12 types of chart patterns, they study the detected patterns through two criteria: predictability and profitability and find an apparent existence of some chart patterns in the currency market. More than one half of detected charts present a significant predictability. But only two chart patterns imply a significant profitability which is, however, too small to cover the transaction costs.

Tick data is, by market structure, discrete. Prices move by multiples of the tick, the minimum price variation. Two approaches can be taken to account for price discreteness. One, which stems from the realized variance literature, is to consider tick changes as market microstructure noise. The other is to consider price discreteness as structural information. *Bien, Nolte, and Pohlmeier* pursue this second line of research and propose a model for multivariate discrete variables. Econometric models for univariate discrete price processes have been suggested recently but a multivariate version of it was still missing. The multivariate integer count hurdle model (MICH) proposed by *Bien, Nolte, and Pohlmeier* can be viewed as a combination of the copula approach by Cameron et al. (2004) with the integer count hurdle (ICH) model of Liesenfeld, Nolte and, Pohlmeier (2006), which allows the dynamic specification of a univariate conditional distribution with discrete support. They illustrate the usefulness of the model for estimating the joint distribution of the EUR/GBP and the EUR/USD exchange rate changes at the 1-min level. Their approach leaves the door open to other applications such as the measurement of multivariate conditional volatilities, the quantification of intradaily liquidity and value-at-risk applications, and the joint analysis of several marks of the trading process (volumes, price and volume durations, discrete quote changes).

Escribano and Pascual propose a new approach of jointly modeling the trading process and the revisions of market quotes. This method accommodates asymmetries in the dynamics of ask and bid quotes after trade-related shocks. The empirical specification is a vector error correction (VEC) model for ask and bid quotes, with the spread as the co-integrating vector, and with an endogenous trading process. Contrary to some hypothesis implied from market microstructure theory, they provide evidence against several symmetry assumptions and report asymmetric adjustments of ask and bid prices to trade-related shocks, and asymmetric impacts of buyer and seller-initiated trades. In general, buys are more informative than sells.

Frey and Grammig analyze adverse selection costs and liquidity supply in a pure open limit order book market using the Glosten/Sandas modeling framework. Relaxing some assumptions of Sandas' (2001) basic model, they show that their revised methodology delivers improved empirical results.[1] They find empirical support for one of the main hypothesis put forth by the theory of limit order book markets, which states that liquidity supply and adverse selection costs are inversely related. Furthermore, adverse selection cost estimates based on the structural model and those obtained using popular model-free methods are strongly correlated.

In the mid-1990s, financial institutions started implementing VaR type measures to meet the 1988 and 1996 Basel Accords' capital requirements to cover their market risk. Based on an internal model, they compute the "Value-at-Risk," which represents the loss they can incur over 10 trading days at a 1% confidence level. However, most of these models do not account for the liquidity risk that has been widely documented in the microstructure literature. Due to the price impact of trades, which relies on trade size, there may indeed be a difference between the market value of a portfolio, computed over "no-trade returns," and its liquidation value. *Giot and Grammig* propose an original way to shed light on the liquidity discount that should be part of the evaluation of market risk borne by financial institutions. They quantify the liquidity risk premiums over different time horizons, for portfolios of different sizes, composed of three stocks traded on Xetra. This paper thus not only contributes to the existing literature on market liquidity,

[1] Sandås, P. (2001), "Adverse Selection and Competitive Market Making: Empirical Evidence from a Limit Order Market", *Review of Financial Studies*, 14, 705–734.

but provides also an answer to practitioners' concerns relative to the measurement of market risk.

Hall and Hautsch study the determinants of order aggressiveness and traders' order submission strategy in an open limit order book market. Applying an order classification scheme, they model the most aggressive market orders, limit orders as well as cancellations on both sides of the market employing a six-dimensional autoregressive conditional intensity model. Using order book data from the Australian Stock Exchange, they find that market depth, the queued volume, the bid-ask spread, recent volatility, as well as recent changes in both the order flow and the price play an important role in explaining the determinants of order aggressiveness. Overall, their empirical results broadly confirm theoretical predictions on limit order book trading.

Liesenfeld, Nolte, and Pohlmeier develop a dynamic model to capture the fundamental properties of financial prices at the transaction level. They decompose the price in discrete components—direction and size of price changes—and, using autoregressive multinomial models, they show that the model is well suited to test some theoretical implications of market microstructure theory on the relationship between price movements and other marks of the trading process.

Intradaily financial data is characterized by its dynamic behavior as well by deterministic seasonal patterns that are due to the market structure. Volatility is known to be larger at the opening and closing than during the lunch time. Similarly for financial durations: they are shorter at the opening and closing, indicating higher activity at these times of the day. Any econometric model should therefore incorporate these features. *Rodriguez-Poo, Veredas, and Espasa* propose a semiparametric model for financial durations. The dynamics are specified parametrically, with an ACD type of model, while seasonality is left unspecified and hence nonparametric. Estimation rests on generalized profile likelihood, which allows for joint estimation of the parametric—an ACD type of model—and nonparametric components, providing consistent and asymptotically normal estimators. It is possible to derive the explicit form for the nonparametric estimator, simplifying estimation to a standard maximum likelihood problem.

Tay and Ting carry out an empirical analysis using high frequency data and more specifically estimate the distribution of price changes conditional on trade volume and duration between trades. Their main empirical finding is that even when controlling for the trade volume level, duration has an effect on the distribution of price changes, and the higher the conditioning volume level, the higher the impact of duration on price changes. The authors find significant positive (negative) skewness in the distribution of price changes in buyer (respectively seller)—initiated trades, and see this finding as support of the Diamond and Verrecchia (1987) analysis of the probability of large price falls with high levels of duration.[2] The analysis is carried out using up-to-date techniques for the nonparametric estimation of conditional distributions, and outlines a descriptive procedure that can be useful in choosing the specification of the relationship between duration, volume, and prices when performing a parametric investigation.

[2] Diamond, D.W. and Verrecchia, R.E. (1987), "Constraints on Short-Selling and Asset Price Adjustment to Private Information", *Journal of Financial Economics*, 18, 277–311.

News is the driving force of price movements in financial markets. *Veredas* analyses the effect of macroeconomic news on the price of the USA 10-year treasury bond future. Considering 15 fundamentals, he investigates the effect of their forecasting errors conditional upon their sign and the momentum of the business cycle. The results show that traders react when the forecasting error differs from zero. The reaction to a positive or negative forecasting error is different and, most importantly, the reaction varies significantly depending on the momentum of the economic cycle. Moreover, the time of the release matters: the closer it is to the covering period, the more effect it has on the bond future.

Modeling and forecasting the covariance of a large number of financial return series has always been a challenge due to the so-called "curse of dimensionality." For example, the multivariate GARCH models are heavily parameterized or the dynamics of conditional variances and covariances must be restricted to reduce the number of parameters, e.g., through factor structures. As an alternative, the sample covariance matrix has often been used, based on rolling windows, e.g., a monthly covariance is estimated from monthly returns of the last 5 years. *Voev* compare this approach, and variants of it, with others that use higher frequency data (daily data), the so-called realized covariance matrix. In each approach, there are different ways to define forecasts. For example, the realized covariance matrix for month t may serve to predict the covariance matrix of next month. A more sophisticated forecast is obtained by taking a convex combination of the realized covariance of month t and an equicorrelated covariance matrix, a technique known as "shrinkage." The previous forecasts are static. Another method consists in modeling the different elements of the realized covariance matrix by using separate univariate time series models to construct forecasts. This raises the difficulty to obtain always a positive definite forecast. *Voev* measures the deviation of the forecast as a matrix from its target by using the Frobenius norm, where the target or "true" covariance matrix is the realized covariance matrix (observed ex post), and by Diebold–Mariano statistical tests. His main conclusion is that the dynamic models result in the smallest errors in the covariance matrix forecasts for most of the analyzed data series.

Luc Bauwens · Dagfinn Rime · Genaro Sucarrat

Exchange rate volatility and the mixture of distribution hypothesis

Abstract This study sheds new light on the mixture of distribution hypothesis by means of a study of the weekly exchange rate volatility of the Norwegian krone. In line with other studies we find that the impact of information arrival on exchange rate volatility is positive and statistically significant, and that the hypothesis that an increase in the number of traders reduces exchange rate volatility is not supported. The novelties of our study consist in documenting that the positive impact of information arrival on volatility is relatively stable across three different exchange rate regimes, and in that the impact is relatively similar for both weekly volatility and weekly realised volatility. It is not given that the former should be the case since exchange rate stabilisation was actively pursued by the central bank in parts of the study period. We also report a case in which undesirable residual properties attained within traditional frameworks are easily removed by applying the log-transformation on volatilities.

Keywords Exchange rate volatility · Mixture of distribution hypothesis

JEL Classification F31

1 Introduction

If exchange rates walk randomly and if the number of steps depends positively on the number of information events, then exchange rate volatility over a given period

---

L. Bauwens · G. Sucarrat (✉)
CORE and Department of Economics, Université catholique de Louvain,
Louvain-la-Nueve, Belgium
E-mail: sucarrat@core.ucl.ac.be, bauwens@core.ucl.ac.be,
URL: http://www.core.ucl.ac.be/~sucarrat/index.html

D. Rime
Norges Bank, Oslo, Norway
E-mail: dagfinn.rime@norges-bank.no

should increase with the number of information events in that period. This chain of reasoning is the essence of the so-called "mixture of distribution hypothesis" (MDH) associated with Clark (1973) and others. Several versions of the MDH have been put forward, including one that suggests the size of the steps depends negatively on the number of traders, see for example Tauchen and Pitts (1983). In other words, an increase in the number of traders, a measure of liquidity, should decrease the size of the steps and thus volatility. Exchange rate volatility may of course depend on other factors too, including country-specific institutional factors, market conditions and economic fundamentals. Bringing such factors together in a general framework and trying to disentangle their distinct effects on exchange rate volatility leads to economic or explanatory volatility modelling as opposed to "pure" forecast modelling, which may remain silent about the economic reasons for variation in volatility.

When Karpoff (1987) surveyed the relationship between financial volatility and trading volume (a measure of information intensity) during the mid-eighties, only one out of the 19 studies he cited was on exchange rates. The increased availability of data brought by the nineties has changed this, and currently we are aware of ten studies that directly or indirectly investigate the relationship between exchange rate volatility and information intensity. The ten studies are summarised in Table 1 and our study of Norwegian weekly exchange rate volatility from 1993 to 2003 adds to this literature in several ways. First, our study spans more than a decade covering three different exchange rate regimes. Second, not only do we find that the impact of changes in the number of information events on exchange rate volatility is positive and statistically significant, recursive parameter analysis suggests the impact is relatively stable across the different exchange rate regimes. Finally, our results do not support the hypothesis that an increase in the number of traders reduces exchange rate volatility.

Another contribution of our study concerns the economic modelling of exchange rate volatility as such. We report a case in which undesirable residual properties are easily removed by applying the logarithmic transformation on volatilities. In particular, we show that OLS-regressions of the *logarithm* of volatility on its own lags and on several economic variables can produce uncorrelated and homoscedastic residuals. Moreover, in the log of realised volatility case the residuals are also normal. When Geweke (1986), Pantula (1986) and Nelson (1991) proposed that volatilities should be analysed in logs it was first and foremost in order to ensure non-negativity. In our case the motivation stems from unsatisfactory residual properties and fragile inference results. Without the log-transformation we do not generally produce uncorrelated residuals, and when we do the results are very sensitive to small changes in specification.

The rest of this paper contains three sections. In Section 2, we review the link between exchange rate volatility and the MDH hypothesis, and discuss measurement issues. We also present our data and other economic variables that we believe may impact on the volatility of the Norwegian exchange rate. In Section 3, we present the models we use and the empirical results. We conclude in the last section, whereas an Appendix provides the details of the data sources and transformations.

Table 1 Summary of empirical studies that investigate the impact of information intensity on exchange rate volatility

Publication	Data	Period	Supportive of MDH?
Grammatikos and Saunders (1986)	Daily currency futures contracts (DEM, CHF, GBP, CAD and JPY) denominated in USD	1978–1983	Yes
Goodhart (1991)	Intradaily quotes (USD against GBP, DEM, CHF, JPY, FRF, NLG, ITL, ECU) and Reuters' news-headline page	14/9–15/9 1987	No
Goodhart (2000)	Intradaily quotes (USD against GBP, DEM, JPY, FRF, AUD) and Reuters' news-headline pages	9/4-19/6 1989	No
Bollerslev and Domowitz (1993)	Intradaily USD/DEM quotes and quoting frequency	9/4–30/6 1989	No
Demos and Goodhart (1996)	Intradaily DEM/USD and JPY/USD quotes and quoting frequency	5 weeks in 1989	Yes
Jorion (1996)	Daily DEM/USD futures and options	Jan. 1985– Feb. 1992	Yes
Melvin and Xixi (2000)	Intradaily DEM/USD and JPY/USD quotes, quoting frequency and Reuters' headline-news screen	1/12 1993– 26/4 1995	Yes
Galati (2003)	Daily quotes (USD against JPY and seven emerging market currencies) and trading volume	1/1 1998– 30/6 1999	Yes
Bauwens et al. (2005)	Intradaily EUR/USD quotes, quoting frequency and Reuters' news-alert screens	15/5 2001– 14/11 2001	Yes
Bjønnes et al. (2005)	Daily SEK/EUR quotes and transaction volume	1995–2002	Yes

2 Exchange rate volatility and economic determinants

The purpose of this section is to motivate and describe our exchange rate volatility measures, and the economic determinants that we use in our empirical study. In Subsection 2.1, we define our volatility measures and present the Norwegian exchange rate data. We make a distinction between period volatility on the one hand and within or intra-period volatility on the other, arguing that analysis of both is desirable since level-expectations may have an impact. In Subsection 2.2, we review the link between volatility and the MDH, and after presenting our quote frequency data we explain how we use them to construct the explanatory variables we include in our volatility equations. In Subsection 2.3, we motivate and describe the other economic determinants of volatility which we include as explanatory variables in the empirical part.

2.1 Period vs. intra-period volatility measures

Conceptually we may distinguish between period volatility on the one hand and within or intra-period volatility on the other. If $\{S_0,S_1,...,S_n,..., S_{N-1},S_N\}$ denotes a sequence of exchange rates between two currencies at times $\{0,1,...,N\}$, then the squared (period) return $[\log(S_N/S_0)]^2$ is an example of a period measure of observable volatility, and realised volatility $\sum_{n=1}^{N}[\log(S_n/S_{n-1})]^2$ is an example of a within-period measure of volatility. (Another example of a within-period measure of volatility is high–low.) It has been showed that realised volatility is an unbiased and consistent measure of integrated volatility under certain assumptions, see Andersen et al. (2001). The reader should be aware though that nowhere do we rely on such assumptions. Rather, our focus is on the *formula* of realised volatility. The main difference between period volatility and realised volatility is that in addition to time 0 to time N variation the latter is also capable of capturing variation between 0 and N. For example, if S_n fluctuates considerably between 0 and N but ends up close to S_0 at N, then the two measures may produce substantially different results. Essentially this can be due to one of two reasons. If the random walk model provides a decent description of how exchange rates behave, then it is due to chance. On the other hand, if there are strong level-effects present among market participants, then the return back to the level of S_0 might be due to market expectations rather than chance. Although market participants' views on exchange rate level clearly matter, we believe most observers would agree that such level-effects are relatively small or infrequent on a day-to-day basis for most exchange rates. Differently put, at very short horizons the random walk model provides a reasonably good description of exchange rate increments. However, the two measures are still qualitatively different, so that any eventual differences in their relation with (say) the rate of information arrival should be investigated, in particular for weekly data where level-expectations is more likely to play a role.

Our period measure will be referred to as "weekly volatility" whereas our within-period measure will be referred to as "within-weekly volatility" or "realised volatility." Weekly volatility is just the squared return from the end of one week to the end of the subsequent week. More precisely, if $S_{N(t)}$ denotes the closing value in the last day of trading in week t and $S_{N(t-1)}$ denotes the closing value in the last day of trading in the previous week, then weekly volatility recorded in week t is denoted by V_t^w and defined as

$$V_t^w = \left[\log\left(S_{N(t)}/S_{N(t-1)}\right)\right]^2. \tag{1}$$

On the other hand, realised volatility in week t, denoted by V_t^r, is the sum of squared returns of the sequence $\{S_{N(t-1)}, S_{1(t)}, S_{2(t)},.., S_{N(t)}\}$, that is,

$$V_t^r = \sum_{n=1(t)}^{N(t)} [\log(S_n/S_{n-1})]^2, \tag{2}$$

where $1(t)-1:=N(t-1)$. It should be noted though that we use only a small sub-set of the within-week observations in the construction of realised volatility (typically ten observations per week).

In order to distinguish between volatilities and logs of volatilities we use lower and upper case letters. So $v_t^w=\log V_t^w$ and $v_t^r=\log V_t^r$. Our data set span the period from 8 January 1993 to 26 December 2003, a total of 573 observations, and before 1 January 1999 we use the BID NOK/DEM exchange rate converted to euro-equivalents with the official conversion rate 1.95583 DEM=1 EURO. After 1 January 1999 we use the BID NOK/EUR rate.

The main characteristics of the two measures are contained in Table 2 and in Fig. 1. At least three attributes of the graphs should be noted. First, although the two measures of volatility are similar level-wise, that is, if plotted in the same diagram they would be "on top of each other," the sample correlation between the log of weekly volatility and the log of realised volatility is only 0.55. In other words, the two measures differ considerably and one of the differences is that the realised volatility measure is less variable. Second, sustained increases in volatility around 1 January 1999 and 29 March 2001 are absent or at least seemingly so. On the first date the current central bank governor assumed the job and reinterpreted the guidelines, which in practice entailed a switch from exchange rate stabilisation to "partial" inflation targeting. On the second date the Norwegian central bank was instructed by the Ministry of Finance to pursue an inflation target of 2.5% as main policy objective. One might have expected that both of these changes would have resulted in shifts upwards in volatility. However, if this is the case then this is not evident by just looking at the graphs. Alternatively, the apparent absence of shifts in volatility might be due to the fact that the markets had expected these changes and already adapted to them. A third interesting feature is that there is a marked and lasting increase in volatility around late 1996 or in the beginning of 1997. This is partly in line with Giot (2003) whose study supports the view that the Asian crisis in the second half of 1997 brought about a sustained increase in the volatility of financial markets in general. In the case of Norwegian exchange rate volatility, however, the shift upwards seems to have taken place earlier, namely towards the end of 1996 or in the beginning of 1997. This may be attributed to the appreciatory pressure on the Norwegian krone in late 1996 and early 1997.

Table 2 Descriptive statistics of selected variables

	S_t	Δs_t	$\lvert\Delta s_t\rvert$	V_t^w	v_t^w	V_t^r	v_t^r	q_t
Mean	8.208	0.000	0.005	0.578	−2.489	0.488	−1.659	7.5115
Median	8.224	0.000	0.003	0.120	−2.121	0.209	−1.567	7.5192
Max.	9.063	0.044	0.044	19.365	2.963	16.033	2.775	9.1363
Min.	7.244	−0.035	0.000	0.000	−10.757	0.004	−5.497	5.6131
St. dev.	0.352	0.008	0.006	1.679	2.447	1.035	1.413	0.5739
Skew.	0.025	0.774	2.886	7.036	−0.845	8.418	−0.102	−0.3287
Kurt.	2.174	9.399	14.991	65.150	3.884	107.193	2.793	3.4512
Obs.	573	572	572	573	573	573	573	573

	Δq_t	$\lvert\Delta q_t\rvert$	M_t	Δm_t	$\lvert\Delta m_t\rvert$	m_t^w	f_t^a	f_t^b
Mean	0.004	0.226	1.115	0.000	0.011	−0.556	0.007	0.010
Median	−0.003	0.145	1.124	−0.001	0.009	−0.190	0.000	0.000
Max.	2.141	2.141	1.429	0.047	0.053	3.325	0.500	1.000
Min.	−1.278	0.000	0.838	−0.053	0.000	−9.150	0.000	0.000
St.dev.	0.339	0.252	0.150	0.014	0.009	2.018	0.057	0.081
Skew.	0.530	2.722	0.025	0.099	1.228	−1.015	8.271	9.136
Kurt.	9.008	14.119	2.174	3.377	5.023	4.181	70.632	94.602
Obs.	572	572	573	572	572	573	573	573

Some zero-values are due to rounding, and the variables are explained in Subsection 2.3 and in the Appendix

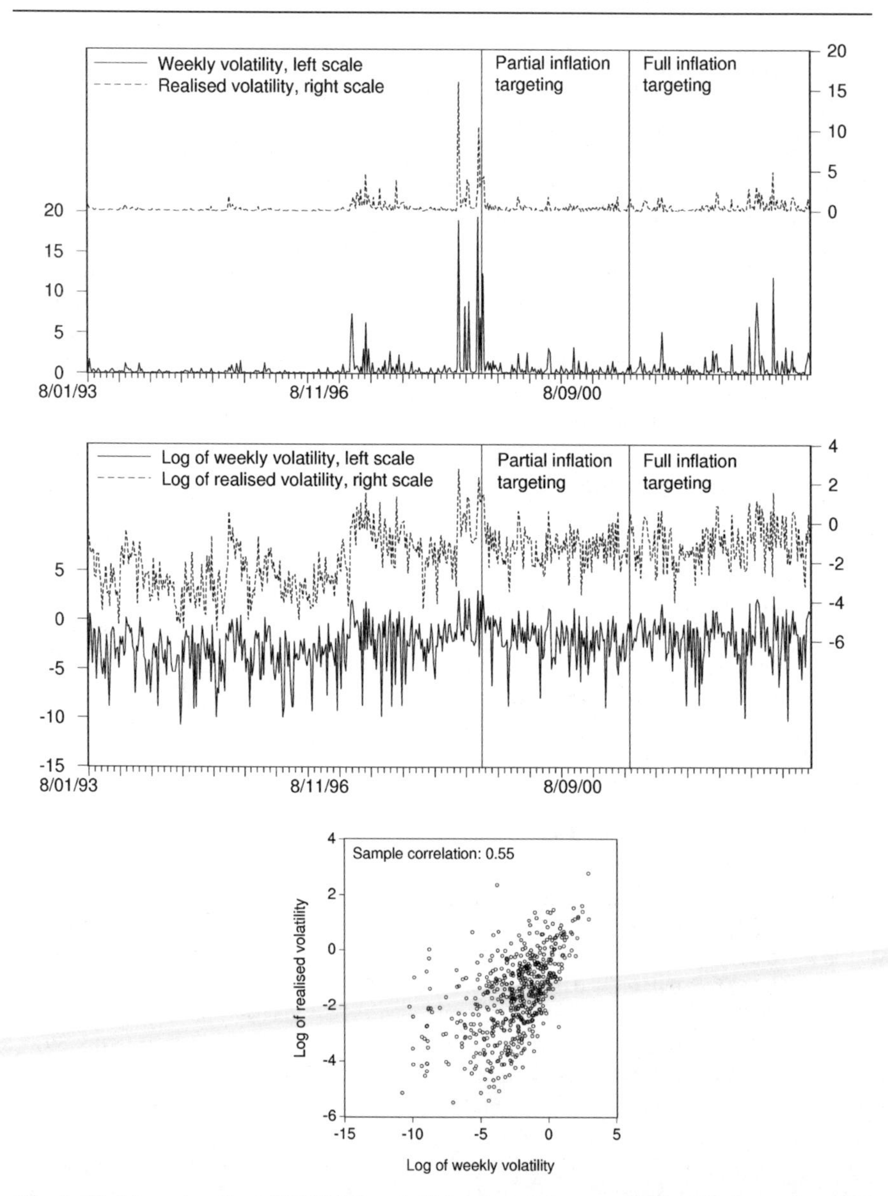

Fig. 1 Weekly and realised NOK/EUR volatilities from 8 January 1993 to 26 December 2003 (NOK/DEM before 1 January 1999) in the upper graph, log of weekly and realised volatilities in the middle graph, and a scatter plot of the log-volatilities

2.2 MDH and quote frequency

If exchange rates follow a random walk and if the number of steps depends positively on the number of information events, then exchange rate volatility over a given period should increase with the number of information events in that period. This chain of reasoning is the essence of the MDH, an acronym which is due to the statistical setup used by Clark (1973). Formally, focusing on the economic content of the hypothesis, the MDH can also be formulated as

$$\Delta s_t = \sum_{n=1}^{N(t)} \Delta s_n, \quad n = 1, \ldots, N(t), \quad s_0 = s_{N(t-1)}, \tag{3}$$

$$\{\Delta s_n\}\text{IID}, \quad \Delta s_n \sim N(0, 1), \tag{4}$$

$$\frac{\partial E[N(t)|\nu_t]}{\partial \nu_t} > 0. \tag{5}$$

where s_t=log S_t. The first line Eq. (3) states that the price increment of period t is equal to the sum of the intra-period increments, Eq. (4) is a random walk hypothesis (any "random walk" hypothesis would do), and Eq. (5) states that the mean of the number of intra-period increments $N(t)$ conditioned on the number of information events ν_t in period t is strictly increasing in ν_t. Several variations of the MDH have been formulated, but for our purposes it is the economic content of Tauchen and Pitts (1983) that is of most relevance. In a nutshell, they argue that an increase in the number of traders reduces the size of the intra-period increments. Here this is akin to replacing Eq. (4) with (say)

$$\Delta s_n = \sigma_n(\eta_n) z_n, \quad \sigma_n' < 0, \quad \{z_n\}\text{IID}, \quad z_n \sim N(0, 1), \tag{6}$$

where η_n denotes the number of traders at time n and where σ_n' is the derivative. But markets differ and theoretical models thus have to be adjusted accordingly. In particular, in a comparatively small currency market like the Norwegian an increase in the number of currency traders is also likely to increase substantially the number of increments per period, that is, $N(t)$, resulting in two counteracting effects. One effect would tend to reduce period-volatility through the negative impact on the size of the intra-period increments, whereas the other effect would tend to increase period-volatility by increasing the number of increments. So it is not known beforehand what the overall effect will be. Replacing Eq. (5) with

$$\frac{\partial E[N(t)|\nu_t, \eta_t]}{\partial \nu_t} > 0, \quad \frac{\partial E[N(t)|\nu_t, \eta_t]}{\partial \eta_t} > 0, \tag{7}$$

means the conditional mean of the number of increments $N(t)$ is strictly increasing in both the number of information events ν_t and the number of traders η_t. Taking Eq. (7) together with Eqs. (3) and (6) as our starting point we may formulate our null hypotheses as

$$\frac{\partial \text{Var}(\Delta s_t|\nu_t \eta_t)}{\partial \nu_t} > 0 \tag{8}$$

$$\frac{\partial \text{Var}(\Delta s_t|\nu_t, \eta_t)}{\partial \eta_t} < 0. \tag{9}$$

In words, the first hypothesis states that an increase in the number of information events given the number of traders increases period volatility, whereas the second holds that an increase in the number of traders without changes in the information intensity reduces volatility. That Eq. (8) is the case is generally suggested by Table 1, whereas Eq. (9) is suggested by Tauchen and Pitts (1983). However, it should be noted that the empirical results of Jorion (1996) and Bjønnes et al. (2005) do not support the hypothesis that an increase in the number of traders reduces volatility.

The most commonly used indicators of information arrival are selected samples from the news-screens of Reuters or Telerate, quoting frequency, the number of transacted contracts and transaction volume. The former is laborious to construct and at any rate not exhaustive with respect to the range of information events that might induce price revision, and the latter two are not readily available in foreign exchange markets. So quote frequency is our indicator of information arrival. More precisely, before 1 January 1999 our quote series consists of the number of BID NOK/DEM quotes per week, and after 1 January 1999 it consists of the number of BID NOK/EUR quotes per week. We denote the log of the number of quotes in week t by q_t, but it should be noted that we have adjusted the series for two changes in the underlying data collection methodology, see the data Appendix for details. Graphs of q_t and Δq_t are contained in Fig. 2. In empirical analysis it is common to distinguish between "expected" and "unexpected" activity, see amongst others Bessembinder and Seguin (1992), Jorion (1996) and Bjønnes et al. (2005). Expected activity is supposed to reflect "normal" or "everyday" quoting or trading activity by traders, and should thus be negatively associated with volatility according to Eq. (9) since this essentially reflects the number of active traders. Unexpected activity on the other hand refers to changes in the rate at which relevant information arrives to the market and should increase volatility. The strategy that is used in order to obtain the expected and unexpected components is to interpret the fitted values of an ARMA–GARCH model as the expected component and the residual as the unexpected. In our case an ARMA(1,1) specification of Δq_t with a GARCH(1,1) structure on the error terms suffices in order to obtain uncorrelated standardised residuals and uncorrelated squared standardised residuals. The model and estimation output is contained in Table 3. The expected values are then computed by generating fitted values of q_t (not of Δq_t) and are denoted $\widehat{q}_t$. The unexpected values are defined as $q_t - \widehat{q}_t$. It has been argued that such a strategy might result in a so-called "generated regressor bias"-see for example Pagan (1984), so we opt for an alternative strategy which yields virtually identical results. As it turns out using q_t directly instead of $\widehat{q}_t$, and Δq_t instead of the residual, has virtually no effect on the estimates in Section 3. The reason can be deduced by looking at the bottom graph of Fig. 2. For statistical purposes q_t is virtually identical to $\widehat{q}_t$, and Δq_t is virtually identical to the residual (the sample correlations are 0.85 and 0.94, respectively). Summarised, then, we use q_t as our measure of the number of active traders and Δq_t as our measure of changes in the rate at which information arrives to the market. Both variables serve as explanatory variables in the modelling of volatility in Section 3.

2.3 Other impact variables

Other economic variables may also influence the level of volatility and should be controlled for in empirical models. In line with the conventions introduced above

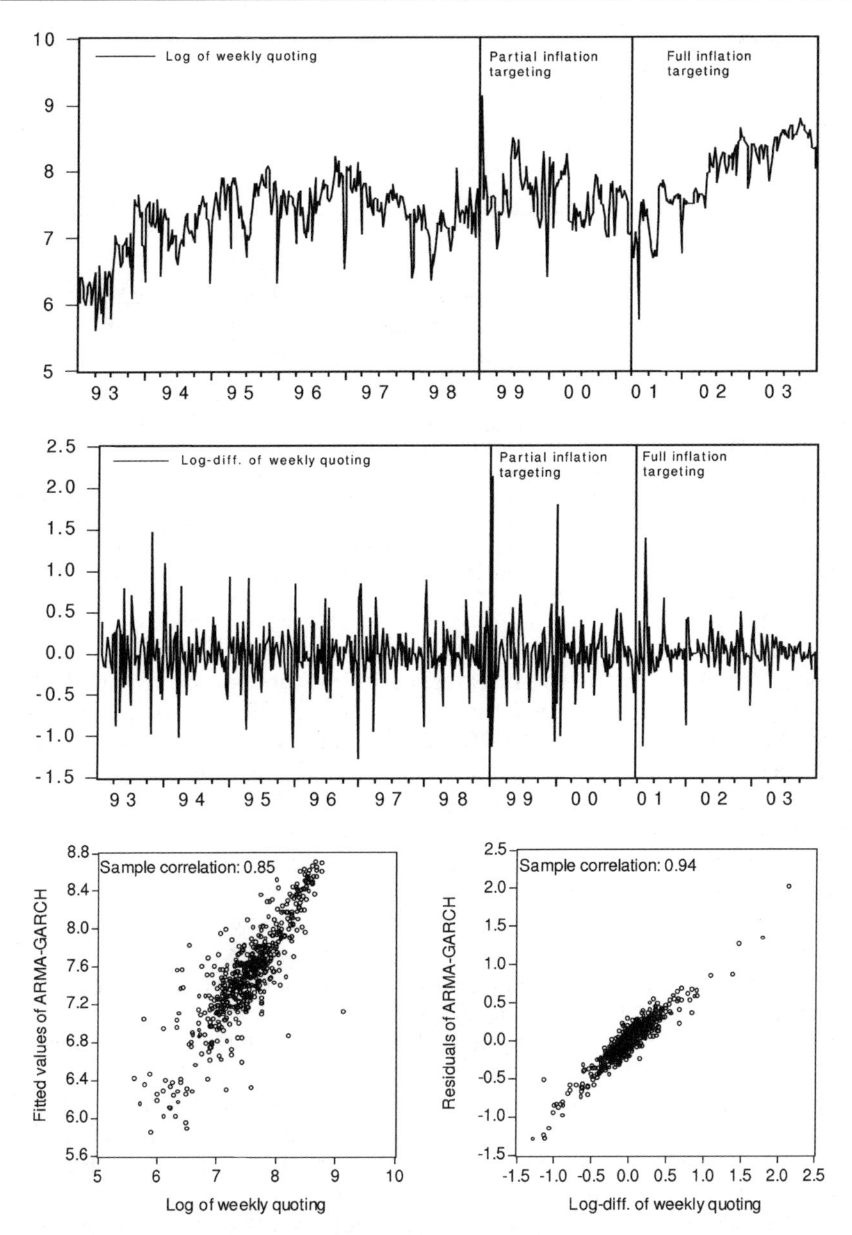

Fig. 2 The log of weekly number of BID NOK/EUR quotes (BID NOK/DEM before 1999) in the upper graph, the log-difference of weekly quoting in the middle graph, and scatter plots of q_t vs. $\widehat{q}_t$ and Δq_t vs. residual in the bottom graph

lower-case means the log-transformation is applied, and upper-case means it is not. The only exceptions are the interest-rate variables, a Russian moratorium dummy id_t equal to 1 in one of the weeks following the Russian moratorium (the week containing Friday 28 August 1998 to be more precise) and 0 elsewhere, and a step dummy sd_t equal to 0 before 1997 and 1 after.

Table 3 ARMA–GARCH model of Δq_t:
$\Delta q_t = b_0 + b_1 \Delta q_{t-1} + b_2 e_{t-1} + e_t$,
$e_t = \sigma_t z_t, \sigma_t^2 = \alpha_0 + \alpha_1 e_{t-1}^2 + \beta_1 \sigma_{t-1}^2$

Parameter			Diagnostics		
	Est.	Pval.		Est.	Pval.
b_0	0.004	0.13	R^2	0.19	
b_1	0.569	0.00	*Log L*	-103.27	
b_2	−0.910	0.00	*Q*(10)	11.03	0.20
α_0	0.034	0.02	$ARCH_{1-10}$	0.29	0.98
α_1	0.299	0.00	JB	691.45	0.00
β_1	0.368	0.06	Obs.	571	

Computations are in EViews 5.1 and estimates are ML with heteroscedasticity consistent standard errors of the Bollerslev and Wooldridge (1992) type. *Pval* stands for *p*-value and corresponds to a two-sided test with zero as null, *Log L* stands for log-likelihood, AR_{1-10} is the Ljung and Box (1979) test for serial correlation in the standardised residuals up to lag 10, $ARCH_{1-10}$ is the *F*-form of the Lagrange-mulitplier test for serial correlation in the squared standardised residuals up to lag 10, *Skew.* is the skewness of the standardised residuals, *Kurt.* is the kurtosis of the standardised residuals, and *JB* is the Jarque and Bera (1980) test for non-normality of the standardised residuals

The first economic variable is a measure of general currency market turbulence and is measured through EUR/USD-volatility. If m_t=log (EUR/USD)$_t$, then Δm_t denotes the weekly return of EUR/USD, M_t^w stands for weekly volatility, m_t^w is its log-counterpart, M_t^r is realised volatility and m_t^r is its log-counterpart. The petroleum sector plays a major role in the Norwegian economy, so it makes sense to also include a measure of oilprice volatility. If the log of the oilprice is denoted o_t, then the weekly return is Δo_t, weekly volatility is O_t^w with o_t^w as its log-counterpart, and realised volatilities are denoted O_t^r and o_t^r, respectively. We proceed similarly for the Norwegian and US stock market variables. If x_t denotes the log of the main index of the Oslo stock exchange, then the associated variables are Δx_t, X_t^w, x_t^w, X_t^r and x_t^r. In the US case u_t is the log of the New York stock exchange (NYSE) index and the associated variables are Δu_t, U_t^w, u_t^w, U_t^r and u_t^r.

The interest-rate variables that are included are constructed using the main policy interest rate variable of the Norwegian central bank. We do not use market interest-rates because this produces interest-rate based measures that are substantially intercorrelated with q_t and sd_t, with the consequence that inference results are affected. The interest-rate variables reflect two important regime changes that took place over the period in question. As the current central bank governor assumed the position in 1999, the bank switched from exchange rate stabilisation to "partial" inflation targeting. However, a full mandate to target inflation was not given before 29 march 2001, when the Ministry of Finance instructed the bank to target an inflation of 2.5%. So an interesting question is whether policy interest rate changes contributed differently to exchange rate volatility in the partial and full inflation targeting periods, respectively.[1] This motivates the construction of our interest rate variables. Let F_t denote the main policy interest rate in percentages and

[1] Prior to 1999 central bank interest rates were very stable, at least from late 1993 until late 1996, and it was less clear to the market what role the interest rate actually had.

let ΔF_t denote the change from the end of 1 week to the end of the next. Furthermore, let I_a denote an indicator function equal to 1 in the period 1 January 1999–Friday 30 March 2001 and 0 otherwise, and let I_b denote an indicator function equal to 1 after 30 March 2001 and 0 before. Then $\Delta F_t^a = \Delta F_t \times I_a$ and $\Delta F_t^b = \Delta F_t \times I_b$, respectively, and f_t^a and f_t^b stand for $|\Delta F_t^a|$ and $|\Delta F_t^b|$, respectively.

3 Models and empirical results

In this section, we present the econometric models of volatility and their estimated versions, together with interpretations. In Subsection 3.1 we use linear regression models for the log of our volatility measures defined in Subsection 2.1, hence the expression "log–linear analysis." In Subsection 3.2 we use EGARCH models. Of these two our main focus is on the results of the log–linear analysis, and the motivation for the EGARCH analysis is that it serves as a point of comparison since both frameworks model volatility in logs.

3.1 Log–linear analysis

In this part we report the estimates of six specifications:

$$v_t^w = b_0 + b_1 v_{t-1}^w + b_2 v_{t-2}^w + b_3 v_{t-3}^w + b_{14} id_t + b_{15} sd_t + e_t \tag{10}$$

$$\begin{aligned} v_t^w = {} & b_0 + b_1 v_{t-1}^w + b_2 v_{t-2}^w + b_3 v_{t-3}^w + b_6 q_t + b_7 \Delta q_t \\ & + b_{14} id_t + b_{15} sd_t + e_t \end{aligned} \tag{11}$$

$$\begin{aligned} v_t^w = {} & b_0 + b_1 v_{t-1}^w + b_2 v_{t-2}^w + b_3 v_{t-3}^w + b_6 q_t + b_7 \Delta q_t + b_8 m_t^w + b_9 o_t^w \\ & + b_{10} x_t^w + b_{11} u_t^w + b_{12} f_t^a + b_{13} f_t^b + b_{14} id_t + b_{15} sd_t + e_t \end{aligned} \tag{12}$$

$$\begin{aligned} v_t^r = {} & b_0 + b_1 v_{t-1}^r + b_2 v_{t-2}^r + b_3 v_{t-3}^r + b_4 v_{t-4}^r + b_5 v_{t-5}^r \\ & + b_{14} id_t + b_{15} sd_t + b_{16} e_{t-1} + e_t \end{aligned} \tag{13}$$

$$\begin{aligned} v_t^r = {} & b_0 + b_1 v_{t-1}^r + b_2 v_{t-2}^r + b_3 v_{t-3}^r + b_4 v_{t-4}^r + b_5 v_{t-5}^r + b_6 q_t \\ & + b_7 \Delta q_t + b_{14} id_t + b_{15} sd_t + b_{16} e_{t-1} + e_t \end{aligned} \tag{14}$$

$$\begin{aligned} v_t^r = {} & b_0 + b_1 v_{t-1}^r + b_2 v_{t-2}^r + b_3 v_{t-3}^r + b_4 v_{t-4}^r + b_5 v_{t-5}^r + b_6 q_t + b_7 \Delta q_t \\ & + b_8 m_t^r + b_9 o_t^r + b_{10} x_t^r + b_{11} u_t^r + b_{12} f_t^a + b_{13} f_t^b + b_{14} id_t \\ & + b_{15} sd_t + b_{16} e_{t-1} + e_t. \end{aligned} \tag{15}$$

The first three have log of weekly volatility v_t^w as left-side variable and the latter three have log of realised volatility v_t^r as left-side variable. In each triple the first specification consists of an autoregression augmented with the Russian moratorium dummy id_t and the step dummy sd_t for the lasting shift upwards in volatility in 1997. In the realised case a moving average (MA) term e_{t-1} is also added for reasons to be explained below. The second specification in each triple consists of the first together with the quote variables, and the third specification is an autoregression augmented by all the economic variables. The estimates of the first triple is contained in Table 4, whereas the estimates of the second triple is contained in Table 5. The results can be summarised in five points.

1. *Information arrival.* The estimated impacts of changes in the rate at which information arrives to the market Δq_t carry the hypothesised positive sign and are significant at all conventional levels. In the weekly case the estimates are virtually identical and equal to about 1, whereas in the realised case the coefficient drops from 0.88 to 0.73 as other variables are added. Summarised, then, the results

Table 4 Regressions of log of weekly NOK/EUR volatility

	(10)		(11)		(12)	
	Est.	Pval.	Est.	Pval.	Est.	Pval.
Const.	−2.917	0.00	−3.887	0.01	−0.660	0.67
v_{t-1}^w	0.019	0.64	0.023	0.59	0.007	0.87
v_{t-2}^w	0.077	0.04	0.078	0.04	0.076	0.05
v_{t-3}^w	0.096	0.03	0.105	0.02	0.099	0.02
q_t			0.141	0.46	0.029	0.88
Δq_t			0.995	0.00	0.986	0.00
m_t^w					0.139	0.00
o_t^w					0.015	0.74
x_t^w					0.123	0.01
u_t^w					0.112	0.01
f_t^a					−0.116	0.92
f_t^b					3.545	0.00
id	4.745	0.00	4.400	0.00	3.563	0.00
sd_t	1.396	0.00	1.306	0.00	1.037	0.00
R^2	0.14		0.16		0.21	
AR_{1-10}	0.34	0.97	0.81	0.62	0.32	0.98
$ARCH_{1\text{-}10}$	0.99	0.45	0.78	0.64	0.56	0.84
Het.	9.42	0.31	13.40	0.34	24.81	0.42
Hetero.	21.89	0.08	45.40	0.01	79.22	0.63
JB	120.94	0.00	117.12	0.00	146.16	0.00
Obs.	570		570		570	

Computations are in EViews 5.1 and estimates are OLS with heteroscedasticity consistent standard errors of the White (1980) type. *Pval* stands for *p*-value and corresponds to a two-sided test with zero as null, AR_{1-10} is the *F*-form of the Lagrange-multiplier test for serially correlated residuals up to lag 10, $ARCH_{1-10}$ is the *F*-form of the Lagrange-multiplier test for serially correlated squared residuals up to lag 10, *Het.* and *Hetero.* are White's (1980) heteroscedasticity tests without and with cross products, respectively, *Skew.* is the skewness of the residuals, *Kurt.* is the kurtosis of the residuals, and *JB* is the Jarque and Bera (1980) test for non-normality in the residuals

Table 5 Regressions of log of realised NOK/EUR volatility

	(3)		(14)		(15)	
	Est.	Pval.	Est.	Pval.	Est.	Pval.
Const.	−1.012	0.00	−1.690	0.00	−1.916	0.02
v^r_{t-1}	0.405	0.05	0.643	0.00	0.483	0.00
v^r_{t-2}	0.078	0.29	0.014	0.81	0.047	0.36
v^r_{t-3}	0.104	0.04	0.086	0.07	0.085	0.05
v^r_{t-4}	−0.059	0.25	−0.065	0.16	−0.050	0.28
v^r_{t-5}	0.122	0.00	0.087	0.03	0.069	0.08
q_t			0.139	0.03	0.173	0.02
Δq_t			0.876	0.00	0.725	0.00
m^r_t					0.194	0.00
o^r_t					−0.021	0.62
x^r_t					0.070	0.08
u^r_t					−0.007	0.85
f^a_t					−0.256	0.63
f^b_t					1.403	0.00
id	4.275	0.00	3.777	0.00	3.985	0.00
sd_t	0.659	0.00	0.382	0.00	0.532	0.00
e_{t-1}	−0.130	0.53	−0.380	0.00	−0.238	0.03
R^2	0.53		0.57		0.60	
AR_{1-10}	0.86	0.57	1.17	0.31	0.81	0.62
$ARCH_{1-10}$	0.44	0.93	1.34	0.20	1.15	0.32
Het.	6.16	0.91	10.91	0.82	33.69	0.21
Hetero.	30.94	0.27	50.15	0.24	128.63	0.10
JB	3.15	0.52	0.44	0.80	0.47	0.79
Obs.	568		568		568	

See Table 4 for details

support the idea that exchange rate variability increases with the number of information events, and the results suggest the impact is higher for weekly than for realised volatility. There might be a small caveat in the realised case though. The MA(1) term e_{t-1} is needed in Eqs. (14) and (15) in order to account for residual serial correlation at lag 1 induced by the inclusion of Δq_t. We have been unsuccessful so far in identifying why Δq_t induces this serial correlation, and excluding Δq_t from (15) also removes the signs of heteroscedasticity indicated by White's (1980) test with cross products in the sense that the p-value increases from 10% to 24%.

2. *Number of traders*. The hypothesised effect of an increase in the number of traders as measured by q_t is negative, but in all the four specifications in which it is included it comes out positive. Moreover, it is significantly positive at 5% in both realised specifications. Figure 3 aims at throwing light on why we obtain these unanticipated results and contains recursive OLS estimates of the impact of q_t with approximate 95% confidence bands. In the weekly case the value starts out negative, but then turns positive and stays so for the rest of the sample. However, it descends steadily towards the end. In the realised case, the value is positive all the

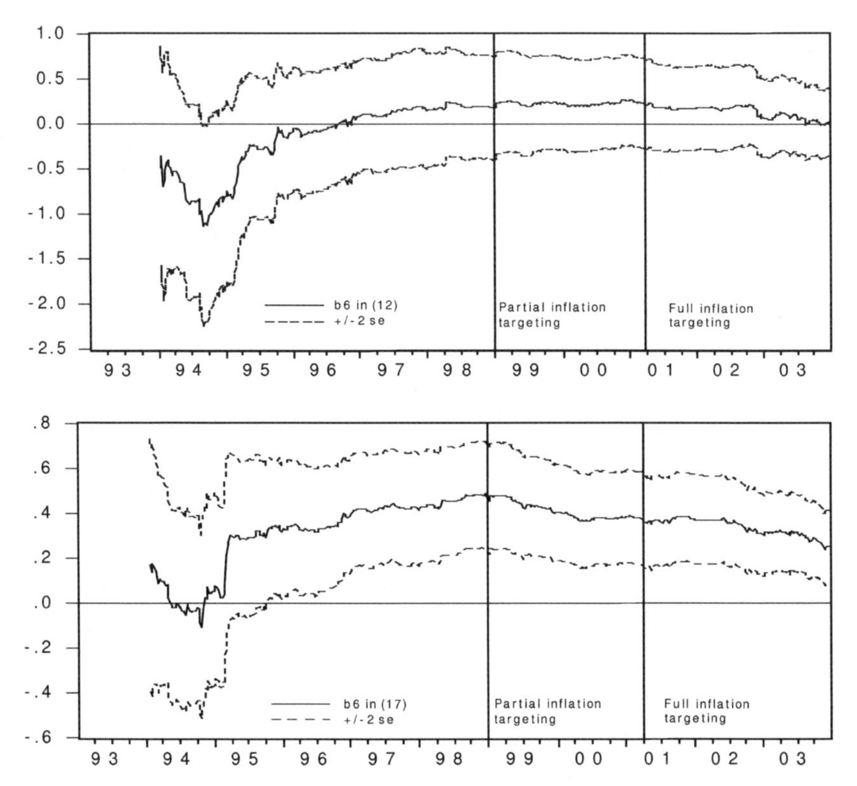

Fig. 3 Stability analysis of the impact of q_t in the general unrestricted specification (12) and in the parsimonious specification (17). Computations in PcGive 10.4 with OLS and initialisation at observation number 50

time but for a short interval in the beginning, and exhibits the same downwards tendency towards the end as in the weekly case. The recursive estimates are more stable here though than in the weekly case. All in all, then, the recursive graphs suggest the impact of q_t over the sample is positive rather than negative, and this may be explained in one of two ways: Either our measure of number of traders is faulty, or the impact of number of traders is positive rather than negative.

3. *Volatility persistence*. The autoregressions Eqs. (10) and (13) were constructed according to a simple-to-general philosophy. The starting equation was volatility regressed on a constant, volatility lagged once, the step dummy sd_t and the impulse dummy id_t, and then lags of volatility were added until two properties were satisfied in the following order of importance: (1) Residuals and squared residuals were serially uncorrelated, and (2) the coefficient in question was significantly different from zero at 5%. Interestingly such simple autoregressions are capable of producing uncorrelated and almost homoscedastic residuals in the weekly case, and uncorrelated, homoscedastic and normal residuals in the realised case. One might suggest that normality in the log-realised specifications comes as no surprise since Andersen et al. (2001) have shown that taking the log of realised exchange rate volatility produces variables close to the normal. In our data, however, the Russian moratorium dummy id_t is necessary for residual normality. The

step dummy sd_t is necessary for uncorrelatedness in all six specifications, but not the impulse dummy id_t. The MA(1) term in Eq. (13) is not needed for any of the residual properties but is included for comparison with Eqs. (14) and (15). However, it does influence the coefficient estimates and the inference results of the lag-structure in all three specifications. Most importantly v_{t-2}^r would be significant if the MA(1) term were not included. Finally, when the lag coefficients are significant at the 10% level, then they are relatively similar across the specifications in both the weekly and realised cases. The only possible exception is the coefficient of the first lag in the realised case, which ranges from 0.41 to 0.64 across the three specifications.

4. *Policy interest rate changes*. One would expect that policy interest rate changes in the full inflation targeting period – as measured by f_t^b – increase contemporaneous volatility, whereas the hypothesised contemporaneous effect in the partial inflation period–as measured by f_t^a – is lower or at least uncertain. The results in both Eqs. (12) and (15) support this since they suggest a negative but insignificant contemporaneous impact in the partial inflation targeting period, and a positive, significant and substantially larger contemporaneous impact (in absolute value) in the full inflation targeting period.

5. *Other*. The effect of general currency market volatility, as measured by m_t^w and m_t^r, is positive as expected, significant in both Eqs. (12) and (15), but a little bit higher in the latter specification. The effect of oilprice volatility, as measured by o_t^w and o_t^r, is estimated to be positive in the first case and negative in the second, but the coefficients are not significant in either specification. This might come as a surprise since Norway is a major oil-exporting economy, currently third after Saudi-Arabia and Russia, and since the petroleum sector plays a big part in the Norwegian economy. A possible reason for this is that the impact of oilprice volatility is non-linear in ways not captured by our measure, see Akram (2000). With respect to the effects of Norwegian and US stock market volatility the two equations differ noteworthy. In the weekly case both x_t^w and u_t^w are estimated to have an almost identical, positive impact on volatility, and both are significant at 1%. In the realised case on the other hand everything differs. Norwegian stock market volatility x_t^r is estimated to have a positive and significant (at 10%) impact albeit somewhat smaller than in the weekly case, whereas US stock market volatility u_t^r is estimated to have an insignificant negative impact.

In order to study the evolution of the impact of Δq_t free from any influence of (statistically) redundant regressors, we employ a general-to-specific (GETS) approach to derive more parsimonious specifications. In this way we reduce the possible reasons for changes in the evolution of the estimates. In a nutshell GETS proceeds in three steps. First, formulate a general model. Second, simplify the general model sequentially while tracking the residual properties at each step. Finally, test the resulting model against the general starting model. See Hendry (1995), Hendry and Krolzig (2001), Mizon (1995) and Gilbert (1986) for more extensive and rigorous expositions of the GETS approach. In our case we posited Eqs. (12) and (15) without the MA(1) term as general models, and it should be noted that a GETS "purist" would probably oppose to the use of the second specification as a starting model, since it exhibits residual serial correlation. Then we tested hypotheses regarding the parameters sequentially with a Wald-test (these tests are not reported), where at each step the simpler model was posited as null. In the weekly case we used heteroscedasticity consistent standard errors of the White (1980) type, and in the realised case we used heteroscedasticity and autocorrelation consistent standard errors of the Newey and West (1987) type. Our final models are not rejected in favour of the general starting models when all the restrictions are tested jointly, their estimates are contained in Table 6, and their specifications are

$$\widehat{v}_t^w = b_2\left(v_{t-2}^w + v_{t-3}^w\right) + b_7\Delta q_t + b_8 m_t^w + b_{10}\left(x_t^w + u_t^w\right) + b_{13} f_t^b + b_{14} id_t + b_{15} sd_t \quad (16)$$

$$\widehat{v}_t^r = b_1 v_{t-1}^r + b_2\left(v_{t-2}^r + v_{t-3}^r + v_{t-5}^r\right) + b_7\Delta q_t + b_8 m_t^r + b_{13} f_t^b + b_{14} id_t + b_{15} sd_t. \quad (17)$$

In both cases the estimates of the impact of Δq_t in the parsimonious specifications are close to those of the general starting specifications. In the weekly case the estimates are equal to 0.99 in the general specification and 1.00 in the specific, whereas in the realised case the estimate changes from 0.57 in the general specification Eq. (15) without the MA(1) term (not reported) to 0.56 in the parsimonious specification Eq. (17). Figure 4 contains recursive OLS estimates of the coefficients of Δq_t in the parsimonious specifications. They are relatively stable over the sample, but admittedly we do not test this formally. Also, the estimates seems to be more stable in the realised case than in the weekly, in the sense that the difference between the maximum and minimum values is larger in the weekly case

Table 6 Parsimonious log–linear specifications obtained by GETS analysis

	v_t^w			v_t^r	
	Est.	Pval.		Est.	Pval.
			Const.	−2.526	0.00
$v_{t-2}^w+v_{t-3}^w$	0.095	0.00	3 $v_{t-1}^r+v_{t-2}^r+v_{t-3}^r+v_{t-5}^r$	0.091	0.00
			q_t	0.256	0.00
Δq_t	0.998	0.00	Δq_t	0.561	0.00
m_t^w	0.141	0.00	m_t^r	0.206	0.00
$x_t^w+u_t^w$	0.143	0.00	x_t^r	0.075	0.07
f_t^b	3.529	0.00	f_t^b	1.635	0.00
id_t	3.445	0.00	id_t	3.878	0.00
sd_t	0.951	0.00	sd_t	0.628	0.00
R^2	0.21			0.60	
AR_{1-10}	0.39	0.95		0.81	0.62
$ARCH_{1-10}$	0.56	0.85		1.07	0.38
Het.	12.93	0.37		18.20	0.20
Hetero.	20.23	0.78		50.19	0.04
JB.	143.16	0.00		0.57	0.75
Obs.	570			568	

See Table 4 for details

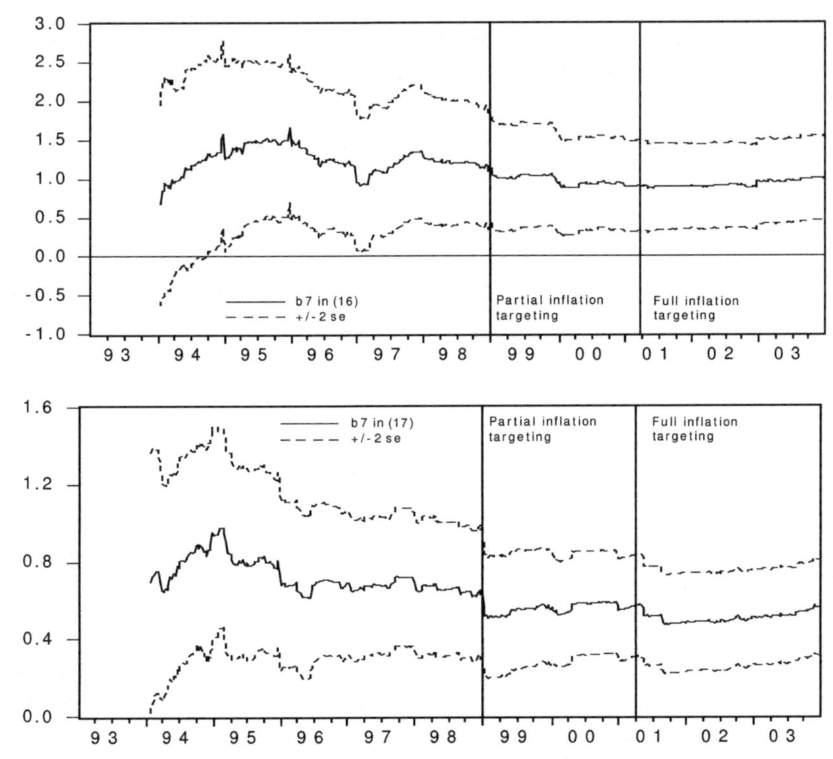

Fig. 4 Recursive estimates of b_7 in the parsimonious specifications Eqs. (16) and (17). Computations in PcGive 10.4 with OLS and initialisation at observation number 50

(1.66−0.67=0.99) than in the realised (0.98−0.47=0.51). Both graphs appear to be trending downward for most of the sample, the exception being towards the end in the weekly case, and in both graphs there seems to be a distinct shift downwards as the change to partial inflation targeting takes place in the beginning of 1999. One should be careful however in attributing the shift to the change in regime without further investigation. Indeed, another possible reason is the transition to the euro, since Δq_t attains both its maximum and minimum in the first weeks of 1999.

3.2 EGARCH analysis

The estimates of the three EGARCH specifications which we report have all equal mean-specification $r_t/\widehat{\sigma}_r = \mu + e_t = \mu + \sigma_t z_t$, where r_t=log(S_t/S_{t-1}) is the weekly return, $\widehat{\sigma}_r = 0.007615$ is the sample standard deviation of the returns, and where $\{z_t\}_{t=1,572}$ is an IID sequence. For exchange rates it is also common to include an AR(1) term in the mean-equation in order to account for the possibility of negative serial correlation in the returns. In our data however there are signs that this term induces serial correlation in either the standardised residuals or in the squared standardised residuals or in both. So we do not include it in the specifications reported here. The three EGARCH specifications can be considered as the ARCH counterparts of the weekly log–linear equations, that is, Eqs. (10)–(12), and their log-variance specifications are

$$\log\sigma_t^2 = \alpha_0 + \alpha_1\left|\frac{e_{t-1}}{\sigma_{t-1}}\right| + \gamma_1\frac{e_{t-1}}{\sigma_{t-1}} + \beta_1\log\sigma_{t-1}^2 + c_{11}id_t + c_{12}sd_t \quad (18)$$

$$\log\sigma_t^2 = \alpha_0 + \alpha_1\left|\frac{e_{t-1}}{\sigma_{t-1}}\right| + \gamma_1\frac{e_{t-1}}{\sigma_{t-1}} + \beta_1\log\sigma_{t-1}^2 + c_1q_t^* + c_2\Delta q_t^* \quad (19)$$
$$+c_{11}\,id_t + c_{12}\,sd_t$$

$$\log\sigma_t^2 = \alpha_0 + \alpha_1\left|\frac{e_{t-1}}{\sigma_{t-1}}\right| + \gamma_1\frac{e_{t-1}}{\sigma_{t-1}} + \beta_1\log\sigma_{t-1}^2 + c_1q_t^* + c_2\Delta q_t^* \quad (20)$$
$$+c_3m_t^{f*} + c_4o_t^{f*} + c_5x_t^{f*} +.c_6u_t^{f*} + c_7f_t^a + c_8f_{t-1}^a + c_9f_t^b$$
$$+c_{10}f_{t-1}^b + c_{11}\,id_t + c_{12}\,sd_t$$

Specification Eq. (18) is an EGARCH(1,1) with the Russian moratorium dummy id_t and the step dummy sd_t as only regressors, Eq. (19) is an EGARCH(1,1) augmented with the quote variables and the dummies, and Eq. (20) is an EGARCH (1,1) with all the economic variables as regressors. Note that * as superscript means the variable has been divided by its sample standard deviation. Specifications Eqs. (18)–(20) are analogous to the ARCH-specifications in Lamoureux and Lastrapes (1990), but note that our results are not directly comparable to theirs since our measure of information intensity Δq_t does not exhibit strong positive serial correlation (in fact, our measure Δq_t exhibits weak negative serial correlation). Strong positive serial correlation is an important assumption for their conclusions.

The estimates of Eqs. (18)–(20) are contained in Table 7 and are relatively similar significance-wise to the results of the weekly log–linear analysis above, that is, to the estimates of Eqs. (10)–(12). Note however that the magnitudes of the coefficient estimates are not directly comparable since the variables are scaled differently. The most important similarity is that the coefficient of Δq_t^* is positive and significant in both Eqs. (19) and (20), and that the coefficient estimates are almost identical in Eqs. (19) and (20). Another important similarity is that the measure of number of traders q_t^* is insignificant in the two EGARCH specifications in which it is included. There are three minor differences in the inference results compared with the weekly log–linear analysis. The first is that the measure of US stock market volatility u_t^* is significant at 9% in the EGARCH specification Eq. (20) containing all the variables, whereas it is significant at 1% in the weekly log–linear counterpart Eq. (12). The second minor difference is that in the EGARCH case the impacts of x_t^{w*} and u_t^{w*} respectively are not so similar as in the weekly case. Finally, the step dummy sd_t is not significant in the EGARCH specification that only contains the dummies as economic variables, whereas it is in its weekly counterpart.

There are also some parameters particular to the EGARCH setup that merit attention. The news term $|e_{t-1}/\sigma_{t-1}|$ is estimated to be positive as expected and

Table 7 EGARCH-analysis of NOK/EUR return volatility

	(18)		(19)		(20)	
	Est.	Pval.	Est.	Pval.	Est.	Pval.
Const. (mean)	−0.025	0.37	−0.054	0.05	−0.065	0.01
Const. (var.)	−0.303	0.11	−0.994	0.13	0.343	0.70
$\lvert e_{t-1}/\sigma_{t-1} \rvert$	0.230	0.09	0.247	0.07	0.169	0.11
e_{t-1}/σ_{t-1}	0.005	0.95	0.085	0.21	0.084	0.18
$\log(\sigma_{t-1}^2)$	0.906	0.00	0.789	0.00	0.587	0.00
q_t^*			0.038	0.38	0.037	0.49
Δq_t^*			0.373	0.00	0.356	0.00
m_t^{w*}					0.148	0.03
o_t^{w*}					−0.057	0.29
x_t^{w*}					0.312	0.00
u_t^{w*}					0.098	0.09
f_t^a					−0.352	0.71
f_t^b					1.611	0.00
id_t	3.002	0.00	2.665	0.00	0.441	0.00
sd_t	0.151	0.19	0.329	0.03	1.552	0.01
Log L.	−710.16		−687.85		−656.38	
Q(10)	11.88	0.29	10.95	0.36	11.91	0.29
$ARCH_{1-10}$	0.88	0.55	11.69	0.31	12.46	0.26
JB	161.00	0.00	119.73	0.00	20.65	0.00
Obs.	572		572		572	

See Table 3 for details

reasonably similar across the three specifications, but its significance is at the borderline since the two-sided p-values range from 7 to 11%. The impact of the asymmetry term e_{t-1}/σ_{t-1} is not significant in any of the equations at conventional significance levels, which suggest no (detectable) asymmetry as is usually found for exchange rate data. Persistence is high as suggested by the estimated impact of the autoregressive term $\log \sigma_{t-1}^2$ since it is 0.91 in Eq. (18), but it drops to 0.79 when the quote variables are included, and then to 0.59 when the rest of the economic variables are included, though it remains quite significant in all cases. Finally, the standardised residuals are substantially closer to the normal distribution in Eq. (20) compared with the other two EGARCH specifications.

4 Conclusions

Our study of weekly Norwegian exchange rate volatility sheds new light on the mixture of distribution hypothesis in several ways. We find that the impact of changes in the number of information events is positive and statistically significant within two different frameworks, that the impact is relatively stable across three different exchange rate regimes for both weekly and realised volatility, and that the estimated impacts are relatively similar in both cases. One might have expected that the effect of changes in the number of information events would increase with a shift in regime from exchange rate stabilisation to partial inflation targeting, and then to full inflation targeting, since the Norwegian central bank actively sought to stabilise the exchange rate previous to the full inflation targeting regime. In our data however there are no clear breaks, shifts upwards nor trends following the points of regime change. Moreover, our results do not support the hypothesis that an increase in the number of traders reduces volatility. Finally, we have shown that simply applying the log of volatility can improve inference and remove undesirable residual properties. In particular, OLS-estimated autoregressions of the log of volatility are capable of producing uncorrelated and (almost) homoscedastic residuals, and in the log of realised volatility case the residuals are also Gaussian.

Our study suggests at least two avenues for future research. First, our results suggest there is no impact of the number of traders on exchange rate volatility, but this might be due to our measure being unsatisfactory. So the first avenue of research is to reconsider the hypothesis with a different approach. The second avenue of future research is to uncover why applying the log works so well. Pantula (1986), Geweke (1986) and Nelson (1991) proposed that volatility should be analysed in logs in order to ensure nonnegativity. In our case the motivation stems from unsatisfactory residual properties and fragile inference-results. Before we switched to the log–linear framework we struggled only to obtain uncorrelated residuals within the ARCH, ARMA and linear frameworks, and when we did attain satisfactory residual properties the results turned out to be very sensitive to small changes in the specification. With the log-transformation, however, results are robust across a number of specifications. So the second avenue of further research consists of understanding better why the log works. Is it due to particularities in our data? For example, is it due to our in financial contexts relatively small sample of 573 observations? Is it due to influential observations? Is it due to both? Or is it just due to the simple fact that applying the log is believed to lead to faster convergence towards the asymptotic theory which our residual tests rely upon? Further application of log–linear analysis is necessary in order to answer these questions, and to verify the possible usefulness of the log–linear framework more generally.

Acknowledgements We are indebted to various people for useful comments and suggestions at different stages, including Farooq Akram, Sébastien Laurent, an anonymous referee, participants at the JAE conference in Venice June 2005, participants at the poster session following the joint CORE-ECARES-KUL seminar in Brussels April 2005, participants at the MICFINMA summer school in Konstanz in June 2004, and participants at the bi-annual doctoral workshop in economics at Université catolique de Louvain (Louvain la Neuve) in May 2004. The usual disclaimer about remaining errors and interpretations being our own applies of course. This work was supported by the European Community's Human Potential Programme under contract HPRN-CT-2002-00232, Microstructure of Financial Markets in Europe, and by the Belgian Program on Interuniversity Poles of Attraction initiated by the Belgian State, Prime Minister's Office, Science Policy Programming. The third author would like to thank Finansmarkedsfondet (the Norwegian Financial Market Fund) and Lånekassen (the Norwegian government's student funding scheme) for financial support at different stages, and the hospitality of the Department of Economics at the University of Oslo and the Norwegian Central Bank in which part of the research was carried out.

Appendix: Data sources and transformations

The data transformations were undertaken in Ox 3.4 and EViews 5.1.

$S_{n(t)}$ — $n(t)$=1(t), 2(t),.., $N(t)$, where $S_{1(t)}$ is the first BID NOK/1EUR opening exchange rate of week t, $S_{2(t)}$ is the first closing rate, $S_{3(t)}$ is the second opening rate, and so on, with $S_{N(t)}$ denoting the last closing rate of week t. Before 1.1.1999 the BID NOK/1EUR rate is obtained by the formula BID NOK/100DEM×0.0195583, where 0.0195583 is the official DEM/1EUR conversion rate 1.95583 DEM=1 EUR divided by 100. The first untransformed observation is the opening value of BID NOK/100DEM on Wednesday 6.1.1993 and the last is the BID NOK/1EUR closing value on Friday 26.12.2003. The source of the BID NOK/100DEM series is Olsen and the source of the BID NOK/1EUR series is Reuters.

S_t — $S_{N(t)}$, the last closing value of week t

r_t — $\log S_t - \log S_{t-1}$

V_t^w — $\{\{\log[S_t+I(S_t=S_{t-1})\times 0.0009]-\log(S_{t-1})\}\times 100\}^2$. $I(S_t=S_{t-1})$ is an indicator function equal to 1 if $S_t=S_{t-1}$ and 0 otherwise, and $S_t=S_{t-1}$ occurs for t=10/6/1994, t=19/8/1994 and t=17/2/2000.

v_t^w — $\log V_t^w$

V_t^r — $\Sigma_n\,[\log(S_n/S_{n-1})\times 100]^2$, where n=1(t), 2(t),..., $N(t)$ and $1(t)-1{:=}N(t-1)$

v_t^r — $\log V_t^r$

$M_{n(t)}$ — $n(t)$=1(t), 2(t),.., $N(t)$, where $M_{1(t)}$ is the first BID USD/EUR opening exchange rate of week t, $M_{2(t)}$ is the first closing rate, $M_{3(t)}$ is the second opening rate, and so on, with $M_{N(t)}$ denoting the last closing rate of week t. Before 1.1.1999 the BID USD/EUR rate is obtained with the formula 1.95583/(BID DEM/USD). The first untransformed observation is the opening value of BID DEM/USD on Wednesday 6.1.1993 and the last is the closing value on Friday 30.12.2003. The source of the BID DEM/USD and BID USD/EUR series is Reuters.

M_t — $M_{N(t)}$, the last closing value of week t

m_t — $\log M_t$

M_t^w — $\{\{\log[M_t+I(M_t=M_{t-1})\times k_t]-\log(M_{t-1})\}\times 100\}^2$. $I(M_t=M_{t-1})$ is an indicator function equal to 1 if $M_t=M_{t-1}$ and 0 otherwise, and k_t is a positive number that ensures the log-transformation is not performed on a zero-value. $M_t=M_{t-1}$ occurs for t=23/2/1996, t=19/12/1997 and t=20/2/1998, and the value of k_t was set on a case to case basis depending on the number of decimals in the original, untransformed data series. Specifically the values of k_t were set to 0.00009, 0.0009 and 0.00009, respectively.

m_t^w — $\log M_t^w$

M_t^r — $\Sigma_n\,[\log(M_n/M_{n-1})\times 100]^2$, where n =1(t), 2(t),.., $N(t)$ and $1(t)-1{:=}N(t-1)$

m_t^r — $\log M_t^r$

Q_t — Weekly number of NOK/EUR quotes (NOK/100DEM before 1.1.1999). The underlying data is a daily series from Olsen Financial Technologies, and the weekly values are obtained by summing the values of the week.

q_t — $\log Q_t$. Note that this series is "synthetic" in that it has been adjusted for changes in the underlying quote-collection methodology at Olsen Financial Technologies. More precisely q_t has been generated under the assumption that Δq_t was equal to zero in the weeks containing Friday 17 August 2001 and Friday 5 September 2003, respectively. In the first week the underlying feed was changed from Reuters to Tenfore, and on the second a feed from Oanda was added.

$O_{n(t)}$ — $n(t)$=2(t), 4(t),.., $N(t)$, where $O_{2(t)}$ is the first closing value of the Brent Blend spot oilprice in USD per barrel in week t, $O_{4(t)}$ is the second closing value of week t, and so on, with $O_{n(t)}$ denoting the last closing value of week t. The untransformed series is Bank of Norway database series D2001712, which is based on Telerate page 8891 at 16.00.

O_t — $O_{N(t)}$, the last closing value in week t

o_t — $\log O_t$

O_t^w — $\{\log[O_t+I(O_t=O_{t-1})\times 0.009]-\log(O_{t-1})\}^2$. $I(O_t=O_{t-1})$ is an indicator function equal to 1 if $O_t=O_{t-1}$ and 0 otherwise, and $O_t=O_{t-1}$ occurs three times, for t=1/7/1994, t=13/10/1995 and t=25/7/1997.

---

o_t^w	$\log O_t^w$
O_t^r	$\Sigma_n [\log(O_n/O_{n-2})]^2$, where n=2(t), 4(t),.., $N(t)$ and 2(t)−2:=$N(t-1)$
o_t^r	$\log O_t^r$
$X_{n(t)}$	$n(t)$=2(t), 4(t),.., $N(t)$, where $X_{2(t)}$ is the first closing value of the main index of the Norwegian Stock Exchange (TOTX) in week t, $X_{4(t)}$ is the second closing value, and so on, with $X_{N(t)}$ denoting the last closing value of week t. The source of the daily untransformed series is EcoWin series ew:nor15565.
X_t	$X_{N(t)}$, the last closing value in week t
x_t	$\log X_t$
X_t^w	$[\log(X_t/X_{t-1})]^2$. $X_t=X_{t-1}$ does not occur for this series.
x_t^w	$\log X_t^w$
X_t^r	$\Sigma_n [\log(X_n/X_{n-2})]^2$, where n=2(t), 4(t),.., $N(t)$ and 2(t)−2:=$N(t-1)$
x_t^r	$\log X_t^r$
$U_{n(t)}$	$n(t)$=2(t), 4(t),.., $N(t)$, where $U_{2(t)}$ is the first closing value in USD of the composite index of the New York Stock Exchange (the NYSE index) in week t, $U_{4(t)}$ is the second closing value, and so on, with $U_{N(t)}$ denoting the last closing value of week t. The source of the daily untransformed series is EcoWin series ew:usa15540.
U_t	$U_{N(t)}$, the last closing value in week t
U_t^w	$[\log(U_t/U_{t-1})]^2$. $U_t=U_{t-1}$ does not occur for this series.
u_t^w	$\log U_t^w$
U_t^r	$\Sigma_n [\log(U_n/U_{n-2})]^2$, where n=2(t), 4(t),.., $N(t)$ and 2(t)−2:=$N(t-1)$
u_t^r	$\log U_t^r$
F_t	The Norwegian central bank's main policy interest-rate, the so-called "folio", at the end of the last trading day of week t. The source of the untransformed daily series is Bank of Norway's web-pages.
f_t^a	$\lvert\Delta F_t\rvert \times I_a$, where I_a is an indicator function equal to 1 in the period 1 January 1999–Friday 30 March 2001 and 0 elsewhere
f_t^b	$\lvert\Delta F_t\rvert \times I_b$, where I_b is an indicator function equal to 1 after Friday 30 March 2001 and 0 before
id_t	Russian moratorium impulse dummy, equal to 1 in the week containing Friday 28 August 1998 and 0 elsewhere.
sd_t	Step dummy, equal to 0 before 1997 and 1 thereafter.

References

Akram QF (2000) When does the oil price affect the Norwegian exchange rate? Working Paper 2000/8. The Central Bank of Norway, Oslo

Andersen TG, Bollerslev T, Diebold FS, Labys P(2001) The distribution of realized exchange rate volatility. J Am Stat Assoc 96:42–55

Bauwens L, Ben Omrane W, Giot P (2005) News announcements, market activity and volatility in the euro/dollar foreign exchange market. J Int Money Financ, Forthcoming

Bessembinder H, Seguin P (1992) Futures-trading activity and stock price volatility. J Financ 47:2015–2034

Bjønnes G, Rime D, Solheim H (2005) Volume and volatility in the FX market: does it matter who you are? In: De Grauwe P (ed) Exchange rate modelling: where do we stand? MIT Press, Cambridge, MA

Bollerslev T, Wooldridge J (1992) Quasi-maximum likelihood estimation and inference in dynamic models with time varying covariances. Econ Rev 11:143–172

Bollerslev T, Domowitz I (1993) Trading patterns and prices in the interbank foreign exchange market. J Financ 4:1421–1443

Clark P (1973) A subordinated stochastic process model with finite variance for speculative prices. Econometrica 41:135–155

Demos A, Goodhart CA (1996) The interaction between the frequency of market quotations, spreads and volatility in the foreign exchange market. Appl Econ 28:377–386

Galati (2003) Trading volume, volatility and spreads in foreign exchange markets: evidence from emerging market countries. BIS Working Paper

Geweke J (1986) Modelling the persistence of conditional variance: a comment. Econ Rev 5:57–61
Gilbert CL (1986) Professor Hendry's econometric methodology. Oxf Bull Econ Stat 48:283–307
Giot P (2003) The Asian financial crisis: the start of a regime switch in volatility. CORE Discussion Paper 2003/78
Goodhart C (1991) Every minute counts in financial markets. J Int Money Financ Mark 10:23–52
Goodhart C (2000) News and the foreign exchange market. In: Goodhart C (ed) The foreign exchange market. MacMillan, London
Grammatikos T, Saunders A (1986) Futures price variability: a test of maturity and volume effects. J Bus 59:319–330
Hendry DF (1995) Dynamic econometrics. Oxford University Press, Oxford
Hendry DF, Krolzig H-M (2001) Automatic econometric model selection using PcGets. Timberlake Consultants, London
Jarque C, Bera A (1980) Efficient tests for normality, homoskedasticity, and serial independence of regression residuals. Econ Lett 6:255–259
Jorion P (1996) Risk and turnover in the foreign exchange market. In: Frankel J et al (ed) The microstructure of foreign exchange markets. University of Chicago Press, Chicago
Karpoff J (1987) The relation between price changes and trading volume: a survey. J Financ Quant Anal 22:109–126
Lamoureux CG, Lastrapes WD (1990) Heteroscedasticity in stock return data: volume versus GARCH Effects. J Financ, pp 221–229
Ljung G, Box G (1979) On a measure of lack of fit in time series models. Biometrika 66:265–270
Melvin M, Xixi Y (2000) Public information arrival, exchange rate volatility, and quote frequency. Econ J 110:644–661
Mizon G (1995) Progressive modeling of macroeconomic time series: the LSE methodology. In: Hoover KD (ed) Macroeconometrics. Developments, tensions and prospects. Kluwer
Nelson DB (1991) Conditional heteroscedasticity in asset returns: a new approach. Econometrica 51:485–505
Newey W, West K (1987) A simple positive semi-definite, heteroskedasticity and autocorrelation consistent covariance matrix. Econometrica 55:703–708
Pagan A (1984) Econometric issues in the analysis of regressions with generated regressors. Int Econ Rev 25:221–247
Pantula S (1986) Modelling the persistence of conditional variance: a comment. Econ Rev 5:71–73
Tauchen G, Pitts M (1983) The price variability–volume relationship on speculative markets. Econometrica 51:485–505
White H (1980) A heteroskedasticity-consistent covariance matrix and a direct test for heteroskedasticity. Econometrica 48:817–838

Katarzyna Bien · Ingmar Nolte · Winfried Pohlmeier

A multivariate integer count hurdle model: theory and application to exchange rate dynamics

Abstract In this paper we propose a model for the conditional multivariate density of integer count variables defined on the set $\mathbb{Z}^n$. Applying the concept of copula functions, we allow for a general form of dependence between the marginal processes, which is able to pick up the complex nonlinear dynamics of multivariate financial time series at high frequencies. We use the model to estimate the conditional bivariate density of the high frequency changes of the EUR/GBP and the EUR/USD exchange rates.

Keywords Integer count hurdle · Copula functions · Discrete multivariate distributions · Foreign exchange market

JEL Classification G10 · F30 · C30

1 Introduction

In this paper we propose a model for the multivariate conditional density of integer count variables. Our modelling framework can be used for a broad set of applications to multivariate processes where the primary characteristics of the variables are: first, their discrete domain spaces, each being the whole space $\mathbb{Z}$; and second, their contemporaneous dependence.

Katarzyna Bien
University of Konstanz, Konstanz, Germany

Ingmar Nolte
University of Konstanz, CoFE, Konstanz, Germany

Winfried Pohlmeier
University of Konstanz, CoFE, ZEW, Konstanz, Germany

Ingmar Nolte (✉)
Department of Economics, University of Konstanz,
Box D124, 78457 Konstanz, Germany
E-mail: Ingmar.Nolte@uni-konstanz.de

Although econometric modelling of univariate processes with a discrete support has been studied extensively, the multivariate counterpart is still underdeveloped. Most of the existing approaches (e.g. Kocherlakota and Kocherlakota (1992) and Johnson et al. (1997)) concentrate on the parametric modelling of multivariate discrete distributions with a non-negative domain and a non-negative contemporaneous dependence only. Alternatively, Cameron et al. (2004) exploit the concept of copula functions to derive a more flexible form of the bivariate distribution for non-negative count variables that allows for both a positive or a negative dependence between the discrete random variables. The multivariate integer count hurdle model (MICH) proposed here can be viewed as an combination of the copula approach by Cameron et al. (2004) with the Integer Count Hurdle (ICH) model of Liesenfeld et al. (2006), which allows for the dynamic specification of a univariate conditional distribution with discrete domain $\mathbb{Z}$.

Quite a number of applications of the MICH model are conceivable in many academic disciplines. Most apparent are applications to high frequent financial data, which are characterized by a set of contemporaneously correlated trade marks, many of them are discrete in nature at high or ultra high frequency. In empirical studies on financial market microstructure, characteristics of the multivariate time-varying conditional densities (moments, ranges, quantiles, etc.) are crucial. For instance, with our model we are able to derive multivariate conditional volatility or liquidity measures. As an application, we propose a model for the bivariate process of exchange rate changes sampled at the 1 min frequency. Other possible applications would be, for example, modelling joint movements of stock transaction prices or the changes of the bid and ask quotes of selected financial instruments.

The discreteness of price changes plays an important role for financial theory and applications. Huang and Stoll (1994), Crack and Ledoit (1996) and Szpiro (1998) among others show that discrete price changes imply a ray shaped pattern in the scatter plot of returns against one period lagged returns, which is referred to as the 'compass rose'. The compass rose can be found for many financial instruments on different markets, such as futures (Lee et al. (1999)), exchange rates (Gleason et al. (2000) and Szpiro (1998)) and stocks (Crack and Ledoit (1996) and Antoniou and Vorlow (2005)).

It has several implications for the dynamics of the data generating process of asset returns which may render naively applied statistical tests such as the Brock et al. (1996) test (Krämer and Runde (1997)), random walk tests or simple autocorrelation estimates (Fang (2002)) invalid. Moreover, GARCH models estimated for such data may be misspecified (Amilon (2003)) and the assumption of a geometric Brownian Motion as the true price process can at least be questioned, which has consequences, for instance, for option pricing (Ball (1988)) and the discrimination between the market microstructure noise and the underlying price process in the realized volatility literature (Andersen et al. (1999), Oomen (2005) and Hansen and Lunde (2006)). Furthermore, Vorlow (2004) analyzes to which extent such patterns can be exploited for forecasting issues. Our approach nicely contributes to this literature since the MICH is able to pick up complex nonlinear structure such as the compass rose in a multivariate setting.

The data used in the application part of the paper are 1 min changes of the EUR/GBP and the EUR/USD midquotes. Figure 1 shows its bivariate histogram. The changes of exchange rates are discrete, since bid or ask quotes of the GBP and the USD against the EUR can jump by a multiple of a fixed tick size of 0.0001

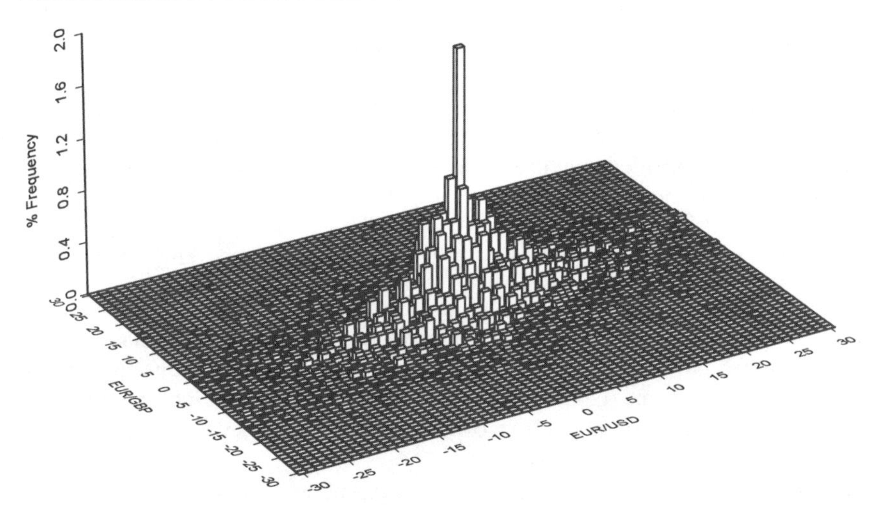

Fig. 1 Bivariate histogram of the tick changes of the EUR/GBP and the EUR/USD exchange rates.

EUR only. The bid quotes (and the ask quotes, analogously) are aggregated to the 1 min level by taking the average of the highest and the lowest best bid within that minute, resulting in a smallest bid quote change of 0.00005 EUR, so that the smallest observable mid quote change amounts to 0.000025 EUR.

Due to the discreteness of the bivariate process, the surface of the histogram is rough, characterized by distinct peaks with the most frequent outcome (0,0) having a sample probability of 2.02%, that corresponds to the simultaneous zero movement of both exchange rates. The discrete changes of the variables are positively correlated, since the positive (negative) movements of the EUR/GBP exchange rate go along with the positive (negative) movements of the EUR/USD exchange rate more frequently.

The sequence of the paper is organized as follows. In Section 2 we describe the general framework of our multivariate modelling approach. The description of the theoretical settings customized with respect to modelling the bivariate density of exchange rate changes follows in Section 3. There, we also present the results of empirical application as well as some statistical inference. Section 4 discusses the results and concludes.

2 The general model

Let $Y_t = (Y_{1t}, \ldots, Y_{nt})' \in \mathbb{Z}^n$, with $t = 1, \ldots, T$, denote the multivariate process of n integer count variables and let $\mathcal{F}_{t-1}$ denote the associated filtration at time $t-1$. Moreover, let $F(y_{1t}, \ldots, y_{nt}|\mathcal{F}_{t-1})$ denote the conditional cumulative density function of Y_t and $f(y_{1t}, \ldots, y_{nt}|\mathcal{F}_{t-1})$ its conditional density. Each marginal process Y_{kt}, $k = 1, \ldots, n$ is assumed to follow the ICH distribution of Liesenfeld et al. (2006) and the dependency between the marginal processes is modelled with a copula function.

2.1 Copula function

The copula concept of Sklar (1959) has been extended by Patton (2001) to conditional distributions. In that framework the marginal distributions and/or the copula function can be specified conditional on $\mathfrak{F}_{t-1}$, so that the conditional multivariate distribution of Y_t can be modelled as

$$F(y_{1t}, \ldots, y_{nt}|\mathcal{F}_{t-1}) = \boldsymbol{C}(F(y_{1t}|\mathcal{F}_{t-1}), \ldots, F(y_{nt}|\mathcal{F}_{t-1})|\mathcal{F}_{t-1}), \qquad (1)$$

where $F(y_{kt}|\mathcal{F}_{t-1})$ denotes the conditional distribution function of the kth component and $\boldsymbol{C}(\cdot|\mathcal{F}_{t-1})$ the conditional copula function defined on the domain $[0, 1]^n$. This approach provides a flexible tool for modelling multivariate distributions as it allows for the decomposition of the multivariate distribution into the marginal distributions, which are bound by a copula function, being solely responsible for their contemporaneous dependence.

If the marginal distribution functions are continuous, the copula function $\boldsymbol{C}$ is unique on its domain $[0, 1]^n$, because the random variables Y_{kt}, $k = 1, \ldots, n$ are mapped through the strictly monotone increasing functions $F(y_{kt}|\mathcal{F}_{t-1})$ onto the entire set $[0, 1]^n$. The joint density function can then be derived by differentiating $\boldsymbol{C}$ with respect to the continuous random variables Y_{kt}, as:

$$f(y_{1t}, \ldots, y_{nt}|\mathcal{F}_{t-1}) = \frac{\partial^n \boldsymbol{C}(F(y_{1t}|\mathcal{F}_{t-1}), \ldots, F(y_{nt}|\mathcal{F}_{t-1})|\mathcal{F}_{t-1})}{\partial y_{1t} \ldots \partial y_{nt}}, \qquad (2)$$

However, if the random variables Y_{kt} are discrete, $F(y_{kt}|\mathcal{F}_{t-1})$ are step functions and the copula function is uniquely defined not on $[0, 1]^n$, but on the Cartesian product of the ranges of the n marginal distribution functions, i.e. $\bigotimes_{k=1}^{n} \text{Range}(F_{kt})$ so that it is impossible to derive the multivariate density function using Eq. (2).

Two approaches have been proposed to overcome this problem. The first is the continuation method suggested by Stevens (1950) and Denuit and Lambert (2005), which is based upon generating artificially continued variables $Y_{1t}^*, \ldots, Y_{nt}^*$ by adding independent random variables $U_{1t}, \ldots, U_{nt}$ (each of them being uniformly distributed on the set $[-1, 0]$) to the discrete count variables $Y_{1t}, \ldots, Y_{nt}$ and which does not change the concordance measure between the variables (Heinen and Rengifo (2003)).

The second method, on which we rely, has been proposed by Meester and MacKay (1994) and Cameron et al. (2004) and is based on finite difference approximations of the derivatives of the copula function, thus

$$f(y_{1t}, \ldots, y_{nt}|\mathcal{F}_{t-1}) = \Delta_n \ldots \Delta_1 \boldsymbol{C}(F(y_{1t}|\mathcal{F}_{t-1}), \ldots, F(y_{nt}|\mathcal{F}_{t-1})|\mathcal{F}_{t-1}), \qquad (3)$$

where Δ_k, for $k \in \{1, \ldots, n\}$, denotes the kth component first order differencing operator being defined through

$$\begin{aligned} &\Delta_k \boldsymbol{C}(F(y_{1t}|\mathcal{F}_{t-1}), \ldots, F(y_{kt}|\mathcal{F}_{t-1}), \ldots, F(y_{nt}|\mathcal{F}_{t-1})|\mathcal{F}_{t-1}) \\ &\quad = \boldsymbol{C}(F(y_{1t}|\mathcal{F}_{t-1}), \ldots, F(y_{kt}|\mathcal{F}_{t-1}), \ldots, F(y_{nt}|\mathcal{F}_{t-1})|\mathcal{F}_{t-1}) \\ &\qquad - \boldsymbol{C}(F(y_{1t}|\mathcal{F}_{t-1}), \ldots, F(y_{kt} - 1|\mathcal{F}_{t-1}), \ldots, F(y_{nt}|\mathcal{F}_{t-1})|\mathcal{F}_{t-1}). \end{aligned}$$

The conditional density of Y_t can therefore be derived by specifying the cumulative distribution functions $F(y_{1t}|\mathcal{F}_{t-1}), \ldots, F(y_{nt}|\mathcal{F}_{t-1})$ in Eq. (3).

2.2 Marginal Processes

The integer count hurdle (ICH) model that we propose for the modelling of the marginal processes is based on the decomposition of the process of the discrete integer valued variable into two components, i.e. a process indicating whether the integer variable is negative, equal to zero or positive (the direction process) and a process for the absolute value of the discrete variable irrespective of its sign (the size process). We present here the simplest form of the ICH model and we refer to Liesenfeld et al. (2006), reprinted in this volume, for a more elaborate presentation.

Let π^k_{jt}, $j \in \{-1, 0, 1\}$ denote the conditional probability of a negative $P(Y_{kt} < 0|\mathcal{F}_{t-1})$, a zero $P(Y_{kt} = 0|\mathcal{F}_{t-1})$ or a positive $P(Y_{kt} > 0|\mathcal{F}_{t-1})$ value of the integer variable Y_{kt}, $k = 1, \ldots, n$, at time t. The conditional density of Y_{kt} is then specified as

$$
\begin{aligned}
&f(y_{kt}|\mathcal{F}_{t-1}) \\
&= {\pi^k_{-1t}}^{\mathbb{1}_{\{Y_{kt}<0\}}} \cdot {\pi^k_{0t}}^{\mathbb{1}_{\{Y_{kt}=0\}}} \cdot {\pi^k_{1t}}^{\mathbb{1}_{\{Y_{kt}>0\}}} \cdot f_s(|y_{kt}|\,|Y_{kt} \neq 0, \mathcal{F}_{t-1})^{(1-\mathbb{1}_{\{Y_{kt}=0\}})},
\end{aligned}
$$

where $f_s(|y_{kt}|\,|Y_{kt} \neq 0, \mathcal{F}_{t-1})$ denotes the conditional density of the size process, i.e. conditional density of an absolute change of Y_{kt}, with support $\mathbb{N} \setminus \{0\}$. To get a parsimoniously specified model, we adopt the simplification of Liesenfeld et al. (2006) that the conditional density of an absolute value of a variable stems from the same distribution irrespective of whether the variable is positive or negative.

The conditional probabilities of the direction process are modelled with the autoregressive conditional multinomial model (ACM) of Russell and Engle (2002) using a logistic link function given by

$$
\pi^k_{jt} = \frac{\exp(\Lambda^k_{jt})}{\sum_{j=-1}^{1} \exp(\Lambda^k_{jt})} \tag{4}
$$

where $\Lambda^k_{0t} = 0$, $\forall t$ is the normalizing constraint. The resulting vector of log-odds ratios $\Lambda^k_t \equiv (\Lambda^k_{-1t}, \Lambda^k_{1t})' = (\ln[\pi^k_{-1t}/\pi^k_{0t}], \ln[\pi^k_{1t}/\pi^k_{0t}])'$ is specified as a multivariate ARMA(1,1) model:

$$
\Lambda^k_t = \mu + B_1 \Lambda^k_{t-1} + A_1 \xi^k_{t-1}. \tag{5}
$$

μ denotes the vector of constants, and B_1 and A_1 denote 2×2 coefficient matrices. In the empirical application, we put the following symmetry restrictions $\mu_1 = \mu_2$, as well as $b^{(1)}_{11} = b^{(1)}_{22}$ and $b^{(1)}_{12} = b^{(1)}_{21}$ on the B_1 matrix to obtain a parsimonious model specification. The innovation vector of the ARMA model is specified as a martingale difference sequence in the following way:

$$
\xi^k_t \equiv (\xi^k_{-1t}, \xi^k_{1t})', \quad \text{where} \quad \xi^k_{jt} \equiv \frac{x^k_{jt} - \pi^k_{jt}}{\sqrt{\pi^k_{jt}(1 - \pi^k_{jt})}}, \quad j \in \{-1, 1\}, \tag{6}
$$

and

$$x_t^k \equiv (x_{-1t}^k, x_{1t}^k)' = \begin{cases} (1,0)' & \text{if } Y_{kt} < 0 \\ (0,0)' & \text{if } Y_{kt} = 0 \\ (0,1)' & \text{if } Y_{kt} > 0, \end{cases} \tag{7}$$

denotes the state vector, whether Y_{kt} decreases, stays equal or increases at time t. Thus, ξ_t^k represents the standardized state vector x_t^k.

The conditional density of the size process is modelled with an at-zero-truncated Negative Binomial (NegBin) distribution:

$$\begin{aligned} &f_s(|y_{kt}|\,|Y_{kt} \neq 0, \mathcal{F}_{t-1}) \\ &\equiv \frac{\Gamma(\kappa^k + |y_{kt}|)}{\Gamma(\kappa^k)\Gamma(|y_{kt}|+1)} \left(\left[\frac{\kappa^k + \omega_t^k}{\kappa^k} \right]^{\kappa^k} - 1 \right)^{-1} \left(\frac{\omega_t^k}{\omega_t^k + \kappa^k} \right)^{|y_{kt}|}, \end{aligned} \tag{8}$$

where $|y_{kt}| \in \mathbb{N} \setminus \{0\}$, $\kappa^k > 0$ denotes the dispersion parameter and scaling parameter ω_t^k is parameterized using the exponential link function with a generalized autoregressive moving average model (GLARMA(p,q)) of Shephard (1995) in the following way:

$$\ln \omega_t^k = \delta \tilde{D}_t + \tilde{\lambda}_t^k \quad \text{with} \quad \tilde{\lambda}_t^k = \tilde{\mu} + S^k(\nu, \tau, K) + \beta_1 \tilde{\lambda}_{t-1}^k + \alpha_1 \tilde{\xi}_{t-1}^k.$$

where $\tilde{D}_t \in \{-1, 1\}$ indicates a negative or positive value of Y_{kt} at time t with the corresponding coefficient denoted by δ. $\tilde{\mu}$ denotes the constant term. β_1 as well as α_1 denote coefficients and $\tilde{\xi}_t^k$ being constructed as

$$\tilde{\xi}_t^k \equiv \frac{|Y_{kt}| - \mathrm{E}(|Y_{kt}|\,|Y_{kt} \neq 0, \mathcal{F}_{t-1})}{\mathrm{V}(|Y_{kt}|\,|Y_{kt} \neq 0, \mathcal{F}_{t-1})^{1/2}},$$

is the innovation term that drives the GLARMA model in λ_t^k. The conditional moments of the at-zero-truncated NegBin distribution are given by

$$\mathrm{E}(|Y_{kt}|\,|Y_{kt} \neq 0, \mathfrak{F}_{t-1}) = \frac{\omega_t^k}{1 - \vartheta_t^k},$$

$$\mathrm{V}(|Y_{kt}|\,|Y_{kt} \neq 0, \mathfrak{F}_{t-1}) = \frac{\omega_t^k}{1 - \vartheta_t^k} - \left(\frac{\omega_t^k}{(1 - \vartheta_t^k)} \right)^2 \left(\vartheta_t^k - \frac{1 - \vartheta_t^k}{\kappa^k} \right),$$

where ϑ_t^k is given by $\vartheta_t^k = [\kappa^k / (\kappa^k + \omega_t^k)]^{\kappa}$.

$$S^k(\nu, \tau, K) \equiv \nu_0 \tau + \sum_{l=1}^{K} \nu_{2l-1} \sin(2\pi(2l-1)\tau) + \nu_{2l} \cos(2\pi(2l)\tau) \tag{9}$$

is a Fourier flexible form used to capture diurnal seasonality, where τ is the intraday time standardized to $[0, 1]$ and ν is a $2K+1$ dimensional parameter vector.

3 Bivariate modelling of exchange rate changes

Data description

We apply our model to 1 min mid quote changes of the EUR/GBP and the EUR/USD exchange rates. The data has been provided by Olsen Financial Technologies and contains quotes from the electronic foreign exchange interbank market. The period under study spreads between October 6 (Monday), 2003, 0:01 EST, and October 10 (Friday), 2003, 17:00 EST, resulting in 6,780 observations for both time series. The sampling frequency of 1 min is, on the one side, sufficiently high to maintain the discrete nature of the data, whereas on the other side, it is low enough to preserve a significant correlation between the two marginal processes.

The histograms of the two marginal processes are presented in the Fig. 2. Both distributions reveal a fairly large support between −20 and 20 ticks for the EUR/GBP and between −30 and 30 ticks for the more volatile EUR/USD exchange rate. Thus, the discreteness of the quote changes combined with a high number of zero quote movements (about 13% for the EUR/GBP and about 7.5% for the EUR/USD) justifies the ICH-model approach of Liesenfeld et al. (2006).

We associate Y_{1t} and Y_{2t} with the changes of the EUR/GBP and the EUR/USD currency pairs, respectively, and present in Figs. 3 and 4 the dynamic features of these variables in the form of the multivariate autocorrelograms of the vectors x_t^1 and x_t^2, which indicate the change of the direction of the EUR/GBP and the EUR/USD exchange rates, as defined in Eq. (7).

We observe that there is a certain dynamic pattern, which should be explained by the ACM part of the ICH model. As indicated by the negative first-order autocorrelation and the positive first-order cross correlation coefficients, the probability of an upward movement of each exchange rate following a downward movement is significantly more probable than two subsequent negative or positive movements. In Fig. 5 the autocorrelograms for the absolute value of the non-zero exchange rate changes are presented. The high degree of persistence characterizing the processes should be explained by the GLARMA part of the ICH model.

The interdependence between the two marginal processes can be seen from Fig. 6, where we plotted the multivariate autocorrelogram of Y_{1t} and Y_{2t}. The two marginal processes are positively correlated, with the correlation coefficient of about 0.35.

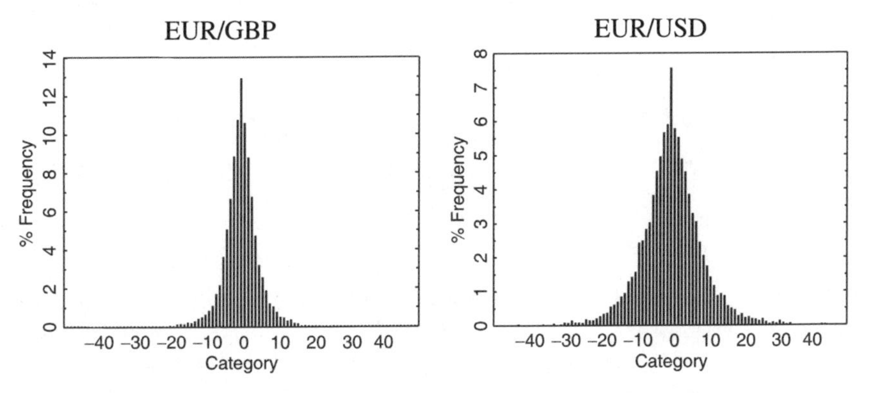

Fig. 2 Histograms of the tick changes of the EUR/GBP and the EUR/USD exchange rates.

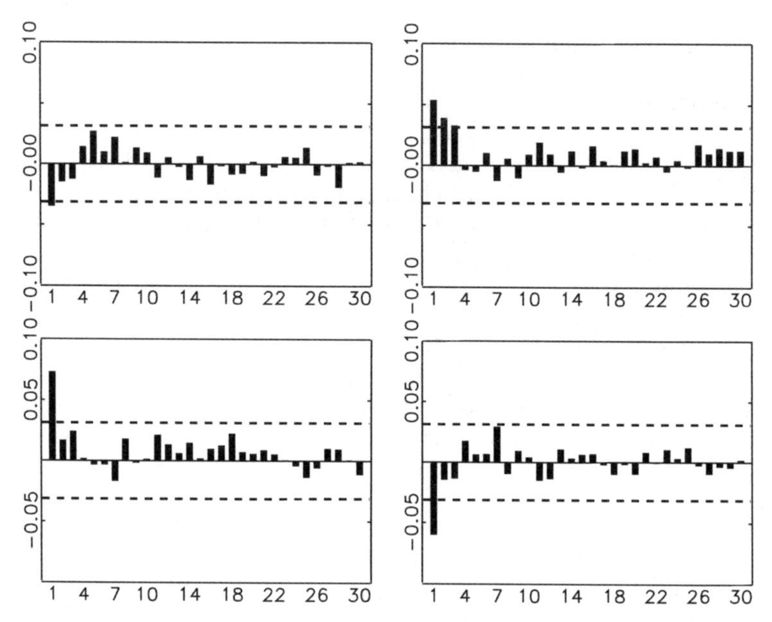

Fig. 3 Multivariate autocorrelation function for the EUR/GBP mid quote direction. *Upper left panel*: $\text{corr}(x^1_{-1,t}, x^1_{-1,t-l})$; *upper right panel*: $\text{corr}(x^1_{-1,t}, x^1_{1,t-l})$; *lower left panel*: $\text{corr}(x^1_{-1,t-l}, x^1_{1,t})$ and *lower right panel*: $\text{corr}(x^1_{1,t}, x^1_{1,t-l})$. The *dashed lines* mark the approximate 99% confidence interval $\pm 2.58/\sqrt{T}$.

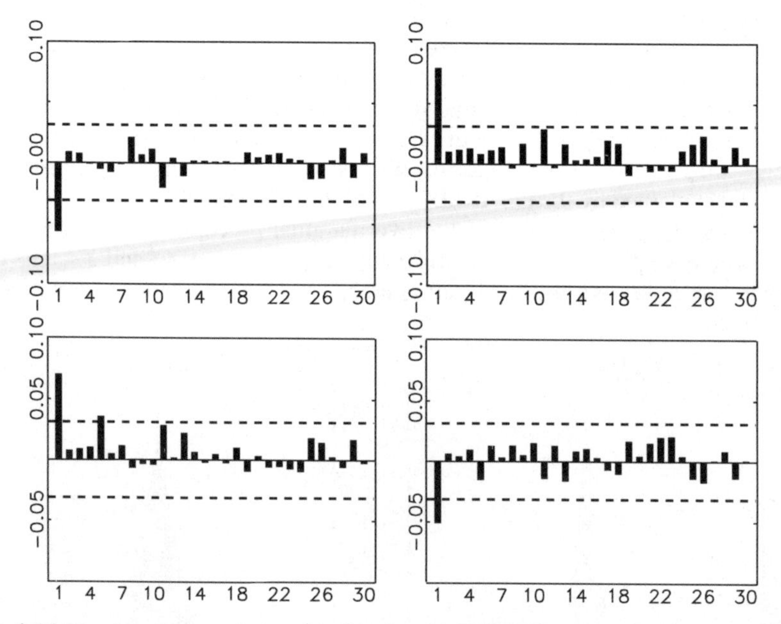

Fig. 4 Multivariate autocorrelation function for the EUR/USD mid quote direction. *Upper left panel*: $\text{corr}(x^2_{-1,t}, x^2_{-1,t-l})$; *upper right panel*: $\text{corr}(x^2_{-1,t}, x^2_{1,t-l})$; *lower left panel*: $\text{corr}(x^2_{-1,t-l}, x^2_{1,t})$ and *lower right panel*: $\text{corr}(x^2_{1,t}, x^2_{1,t-l})$. The *dashed lines* mark the approximate 99% confidence interval $\pm 2.58/\sqrt{T}$.

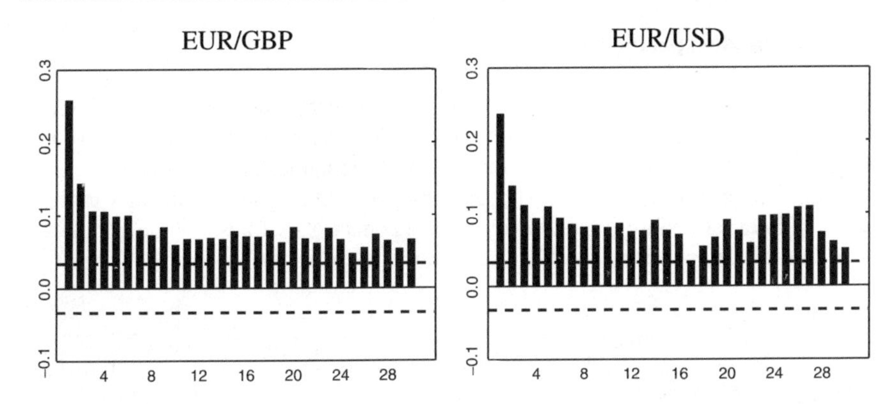

Fig. 5 Autocorrelation function of the non-zero absolute EUR/GBP and EUR/USD mid quote changes. The *dashed line* marks the approximate 99% confidence interval $\pm 2.58/\sqrt{\tilde{T}}$, where $\tilde{T}$ is the number of non-zero quote changes.

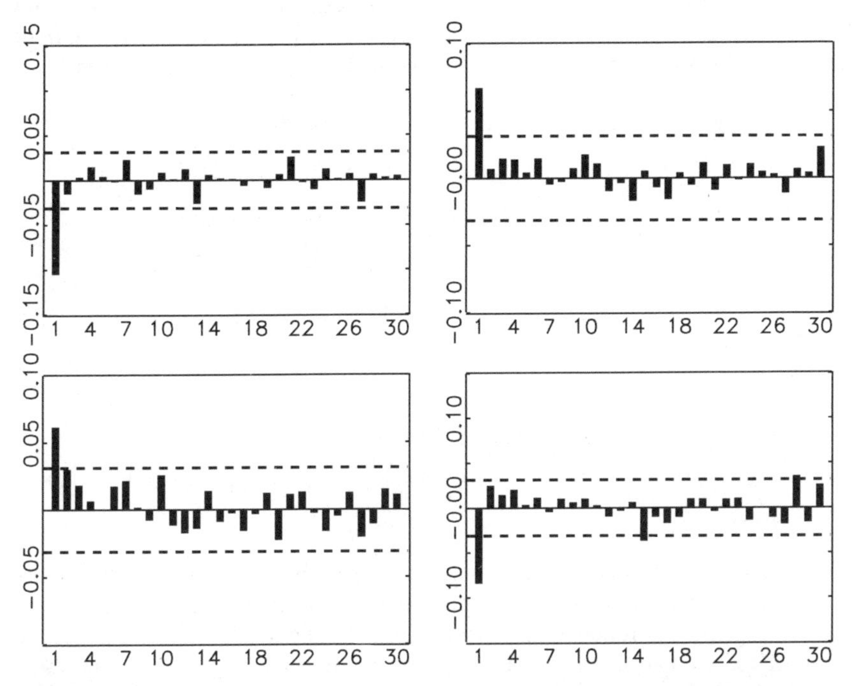

Fig. 6 Multivariate autocorrelation function for the EUR/GBP and EUR/USD mid quote changes. *Upper left panel*: $\mathrm{corr}(Y_{1t}, Y_{1t-l})$; *upper right panel*: $\mathrm{corr}(Y_{1t}, Y_{2t-l})$; *lower left panel*: $\mathrm{corr}(Y_{1t-l}, Y_{2t})$ and *lower right panel*: $\mathrm{corr}(Y_{2t}, Y_{2t-l})$. The *dashed lines* mark the approximate 99% confidence interval $\pm 2.58/\sqrt{T}$.

Bivariate model specification

The copula concept allows one to model the bivariate density without forcing the direction of the dependence upon the data generating process. We choose the standard Gaussian copula function since its single dependency parameter can easily be estimated and it allows for a straightforward sampling algorithm of variables from the bivariate conditional density, which is necessary to assess the goodness-of-fit of our specification. The Gaussian copula is given by

$$\mathbf{C}(u_t, v_t; \rho) = \int_{-\infty}^{\Phi^{-1}(u_t)} \int_{-\infty}^{\Phi^{-1}(v_t)} \frac{1}{2\pi\sqrt{1-\rho^2}} \exp\left(\frac{2\rho uv - u^2 - v^2}{2(1-\rho^2)}\right) du dv, \tag{10}$$

where $u_t = F(y_{1t}|\mathcal{F}_{t-1})$, $v_t = F(y_{2t}|\mathcal{F}_{t-1})$ and ρ denotes the time-invariant parameter of the Gaussian copula, which is the correlation between $\Phi^{-1}(u_t)$ and $\Phi^{-1}(v_t)$. Since ρ is chosen to be fixed over time $\boldsymbol{C}(F(y_{1t}|\mathcal{F}_{t-1}), F(y_{2t}|\mathcal{F}_{t-1})|\mathcal{F}_{t-1}) = \boldsymbol{C}(F(y_{1t}|\mathcal{F}_{t-1}), F(y_{2t}|\mathcal{F}_{t-1}))$ and the conditional bivariate density of Y_{1t} and Y_{2t} can be inferred from Eq. (3) as

$$\begin{aligned} f(y_{1t}, y_{2t}|\mathcal{F}_{t-1}) = {} & \boldsymbol{C}(F(y_{1t}|\mathcal{F}_{t-1}), F(y_{2t}|\mathcal{F}_{t-1})) \\ & - \boldsymbol{C}(F(y_{1t}-1|\mathcal{F}_{t-1}), F(y_{2}|\mathcal{F}_{t-1})) \\ & - \boldsymbol{C}(F(y_{1t}|\mathcal{F}_{t-1}), F(y_{2t}-1|\mathcal{F}_{t-1})) \\ & + \boldsymbol{C}(F(y_{1t}-1|\mathcal{F}_{t-1}), F(y_{2t}-1|\mathcal{F}_{t-1})). \end{aligned} \tag{11}$$

The cumulative distribution function $F(y_{1t}|\mathcal{F}_{t-1})$ (and analogously $F(y_{2t}|\mathcal{F}_{t-1})$) can be written as

$$\begin{aligned} & F(y_{1t}|\mathcal{F}_{t-1}) \\ & = \sum_{k=-50}^{y_{1t}} \pi_{-1t}^{1\ \mathbb{1}_{\{k<0\}}} \cdot \pi_{0t}^{1\ \mathbb{1}_{\{k=0\}}} \cdot \pi_{1t}^{1\ \mathbb{1}_{\{k>0\}}} \cdot f_s(|k|\,|k \neq 0, \mathcal{F}_{t-1})^{(1-\mathbb{1}_{\{k=0\}})} \end{aligned}$$

where we set the lower bound of the summation to -50 and where the probabilities of the downward, zero and upward movement of the exchange rate are specified with the logistic link function, as shown in Eq. (4), and the density for the absolute value of the change is specified along the conditional NegBin distribution, as presented in Eq. (8).

Estimation and simulation results

Our estimates are obtained by a one-step maximum likelihood estimation, whereas the log-likelihood function is taken as a logarithm of the bivariate density presented in the Eq. (11). Estimation results for the ACM part of ICH model are presented in Table 1 and for the GLARMA part of the ICH model in Table 2. The estimate of the dependency parameter $\hat{\rho}$ for the Gaussian copula equals to 0.3588 with standard

Table 1 ML estimates of the ACM-ARMA part of ICH model. Multivariate Ljung-Box statistic for L lags, $Q(L)$, is computed as $Q(L) = n \sum_{\ell=1}^{L} \text{tr}\left[\Gamma(\ell)'\Gamma(0)^{-1}\Gamma(\ell)\Gamma(0)^{-1}\right]$, where $\Gamma(\ell) = \sum_{i=\ell+1}^{n} v_t v_{t-\ell}'/(n-\ell-1)$. Under the null hypothesis, $Q(L)$ is asymptotically χ^2-distributed with degrees of freedom equal to the difference between 4 times L and the number of parameters to be estimated

	EUR/GBP		EUR/USD	
Parameter	Estimate	Standard deviation	Estimate	Standard deviation
μ	0.0639	0.0177	0.0837	0.0222
$b_{(11)}^{(1)}$	0.6583	0.0856	0.4518	0.0426
$b_{(12)}^{(1)}$	0.2910	0.0540	0.5054	0.0635
$a_{11}^{(1)}$	0.1269	0.0324	0.2535	0.0465
$a_{12}^{(1)}$	0.2059	0.0323	0.3739	0.0472
$a_{21}^{(1)}$	0.2009	0.0271	0.3350	0.0466
$a_{22}^{(1)}$	0.0921	0.0312	0.2586	0.0477
Resid. mean	$(-0.003, 0.002)$		$(0.003, 0.009)$	
Resid. variance	$\begin{pmatrix} 0.655 & 0.803 \\ 0.803 & 2.631 \end{pmatrix}$		$\begin{pmatrix} 1.413 & 2.306 \\ 2.306 & 4.721 \end{pmatrix}$	
Resid. $Q(20)$	72.359 (0.532)		89.054 (0.111)	
Resid. $Q(30)$	102.246 (0.777)		122.068 (0.285)	
Log-lik.	−6.2125			

Table 2 ML estimates of the GLARMA part of ICH model

	EUR/GBP		EUR/USD	
Parameter	Estimate	Standard deviation	Estimate	Standard deviation
$\kappa^{0.5}$	0.7862	0.0192	0.7952	0.0130
$\tilde{\mu}$	0.3363	0.0438	0.7179	0.0814
β_1	0.6567	0.0335	0.6085	0.0428
α_1	0.1675	0.0100	0.1455	0.0097
ν_0	0.0981	0.0633	−0.0396	0.0510
ν_1	−0.0712	0.0117	−0.0430	0.0100
ν_2	−0.0060	0.0091	0.0388	0.0099
ν_3	−0.0501	0.0105	−0.0258	0.0093
ν_4	0.0852	0.0238	0.0954	0.0215
ν_5	−0.0100	0.0129	−0.0234	0.0120
ν_6	0.0449	0.0115	0.0297	0.0108
Resid. mean	0.013		0.007	
Resid. variance	1.001		1.025	
Resid. $Q(20)$	26.408 (0.067)		42.332 (0.001)	
Resid. $Q(30)$	64.997 (0.000)		87.641 (0.000)	
Log-lik.	−6.2125			

deviation 0.0099, representing a strong positive correlation between the modelled marginal processes.

Regarding the estimates for the ACM submodel, we observe a significant persistency pattern ($\hat{B}_1$ matrix) of the direction processes and we can conclude, that if the probability of an exchange rate change has been high in the previous period, it is also supposed to be considerably high in the next period. Moreover, the obtained relations $a_{11}^{(1)} < a_{12}^{(1)}$ and $a_{21}^{(1)} > a_{22}^{(1)}$ between the innovation coefficients suggest the existence of some bounce or mean-reverting pattern in the evolution of the exchange rate process. The parsimonious dynamic specification seems to describe the dynamic structure very well, as the multivariate Ljung-Box statistics for the standardized residuals of the ACM model do not differ significantly from zero.

Regarding the estimation results for the GLARMA part of the ICH model, we observe that the values of the dispersion parameters $\kappa^{k-0.5}$ are significantly different from zero, allowing the rejection of the null hypothesis of at-zero-truncated Poisson distributions in favor of at-zero-truncated NegBin ones. The diagnostics statistics of the GLARMA submodel are quite satisfying. Although some Ljung-Box statistics (Q) for the standardized residuals still remain significantly different from zero, a large part of the autocorrelation structure of the size processes has been explained by the simple GLARMA(1,1) specification.

Jointly significant coefficients of the seasonal components $S(\nu, \tau, K)$ for $K = 3$ indicate that there exist diurnal seasonality patterns, which are plotted in Fig. 7, in the absolute changes of the exchange rates. We observe for every minute of the day that the mean of the non-zero absolute tick changes of the USD against the EUR is considerably higher than the mean for the GBP against the EUR. It confirms the results of the descriptive study presented previously, as the support of the EUR/USD distribution is more dispersed and the exchange rate is more volatile. The shapes of the diurnal seasonality functions for both exchange rates are quite similar. They evidence the existence of at least two very active trading periods, about 3.00 EST and 10.00 EST, which corresponds to the main trading periods of the European and the American Foreign Exchange market, respectively.

In order to verify the goodness-of-fit of our model in a more elaborate way, we simulate the conditional density of the bivariate process at every

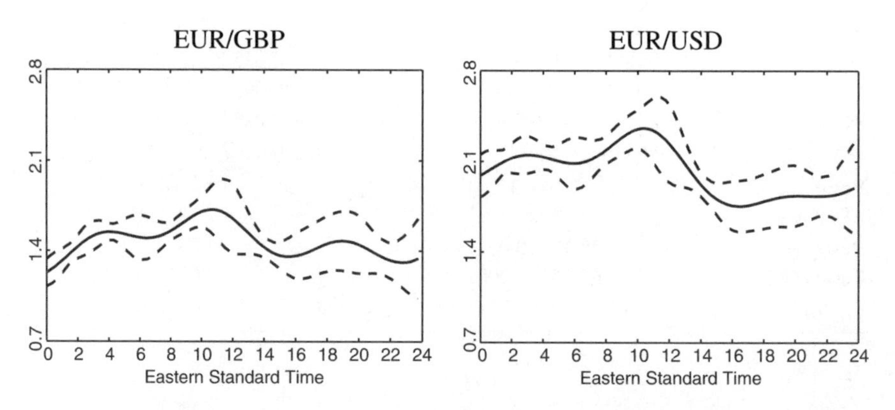

Fig. 7 Estimated diurnal seasonality function of the non-zero absolute EUR/GBP and EUR/USD tick changes. The *dashed line* mark the approximate 99% confidence interval.

point t, $t = 1, \ldots, T$. Such an approach enables us to verify whether the proposed density specification is able to explain the whole conditional joint density of the underlying data generating process. Relying on the simulated distributions at every time point available, we can easily address this point applying the modified version of the Diebold et al. (1998) density forecasting test for discrete data of Liesenfeld et al. (2006).
Moreover, we are able to compare the residuals of both marginal processes.
We use here the standard sampling method proposed for Gaussian copula functions, which can be summarized as:

For every t:

- Compute the Cholesky decomposition $\hat{A}$ (2×2) of estimated correlation matrix $\hat{R}$, where $\hat{R} = \begin{pmatrix} 1 & \hat{\rho} \\ \hat{\rho} & 1 \end{pmatrix}$
- Simulate $x_t = (x_{1t}, x_{2t})'$ from a bivariate standard normal distribution

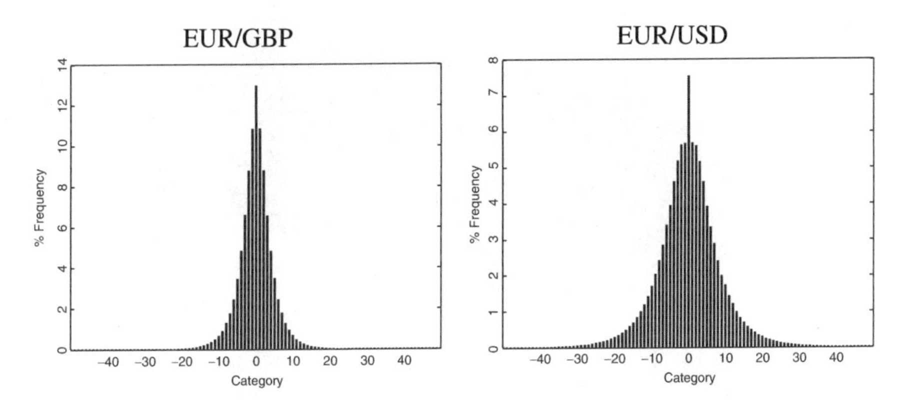

Fig. 8 Histogram of simulated tick changes of exchange rates.

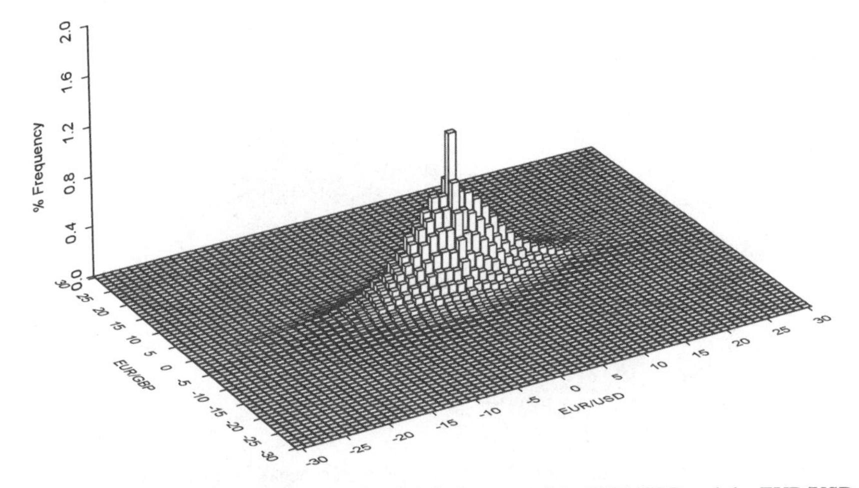

Fig. 9 Bivariate histogram of the simulated tick changes of the EUR/GBP and the EUR/USD exchange rates.

- Set $\hat{z}_t = \hat{A}x_t$
- Set $\hat{u}_{1t} = \Phi(\hat{z}_{1t})$ and $\hat{u}_{2t} = \Phi(\hat{z}_{2t})$ where Φ denotes the univariate standard normal distribution function
- Set $\hat{Y}_{1t} = \hat{F}_1^{-1}(\hat{u}_1|\mathcal{F}_{t-1}))$ and $\hat{Y}_{2t} = \hat{F}_2^{-1}(\hat{u}_2|\mathcal{F}_{t-1}))$ where $\hat{F}_1$ and $\hat{F}_2$ denote the estimated marginal cumulative distribution functions of the EUR/GBP and the EUR/USD changes, respectively

Figure 8 contains the plots of the unconditional histograms of the simulated marginal processes. Their shapes seem to agree with those of the raw data series already presented in Fig. 2.

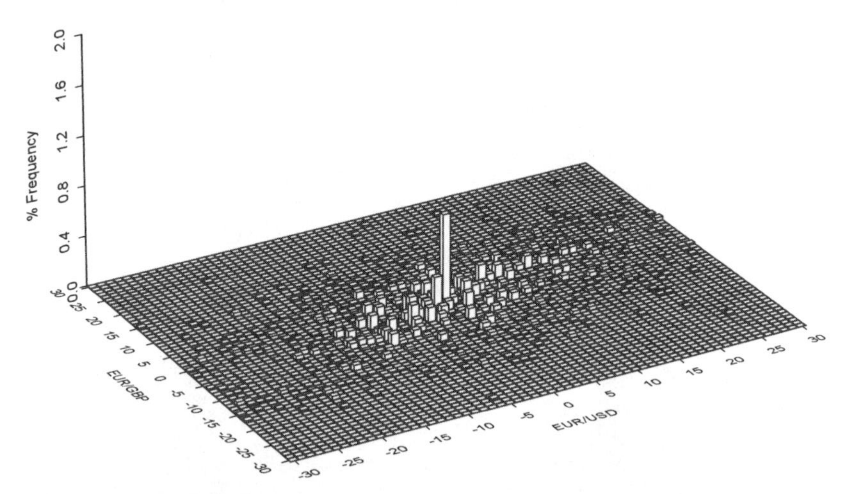

Fig. 10 Bivariate histogram of the positive differences between the empirical and the simulated bivariate histogram of the EUR/GBP and the EUR/USD exchange rate changes.

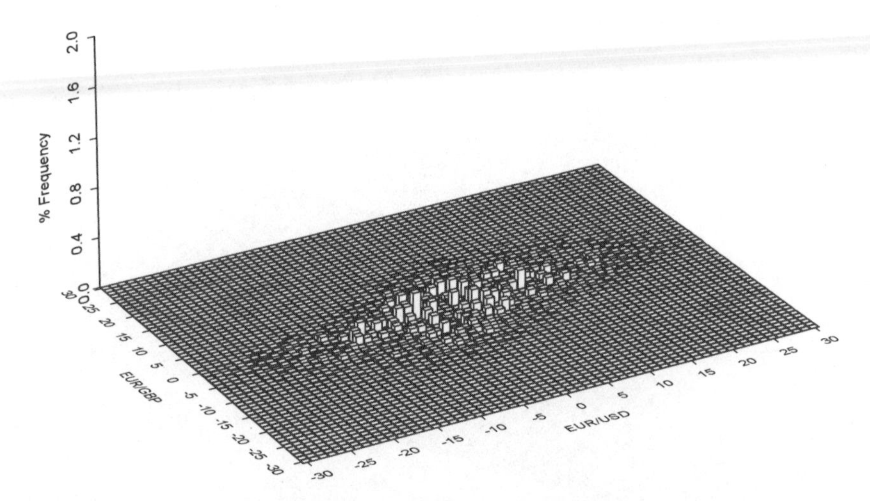

Fig. 11 Bivariate histogram of the absolute values of the negative differences between the empirical and the simulated bivariate histogram of the the EUR/GBP and the EUR/USD exchange rate changes.

The unconditional bivariate histogram of the simulated time series is presented in Fig. 9. Although the positive dependence between the marginal processes is reflected, the shape of the histogram does not correspond to the empirical one in full (see Fig. 1). In particular the frequency of the outcome $(0, 0)$ has been considerably underestimated. We compute the differences between the histograms of the empirical and the simulated data to infer in which points (i, j) the observed and the estimated probabilities disagree. To assess these differences graphically, we plotted in Fig. 10 only positive differences and in Fig. 11 only absolute negative differences. Besides the outcome probability of $(0, 0)$, the probabilities for points (i, j) concentrated around (0,0) are a little bit underestimated (positive differences in Fig. 10) as well, and the probabilities for points (i, j) which are a little further away from (0,0) are a little overestimated (negative differences in Fig. 11). Thus, we conclude that we underestimate the kurtosis of the empirical distribution. The real data is much more concentrated in the outcome (0,0), as well as evidencing much fatter tails. There is a clear signal for a tail dependency in the data generating process, as the extreme positive or negative movements of the exchange rates take place much more often than could be explained by a standard Gaussian copula function (see Fig. 10).

Additionally, we can address the goodness-of-fit of the conditional bivariate density by considering the bivariate autocorrelation function of the residual series $\hat{\varepsilon}_t = (\hat{\varepsilon}_t^1, \hat{\varepsilon}_t^2)'$ depicted in Fig. 14 and the quantile–quantile (QQ) plots of the modified density forecast test variables for the implied marginal processes in Fig. 12. We have mapped these modified density forecast test variables into a standard normal distribution, so that under the correct model specification, these normalized variables should be i.i.d. standard normally distributed. Figure 13 plots the autocorrelation functions of these normalized density forecast variables. Both plots indicate that the processes are almost uncorrelated. The deviation from normality, especially for the EUR/USD exchange rate changes and in the upper tail of the normalized density forecast variables indicated by the QQ-plots, reveals that our specification has difficulties to characterize extreme exchange rate changes appropriately.

The bivariate autocorrelation function of the residual series (Fig. 14) shows significant cross-correlations at lag 1. Although, we manage to explain a large

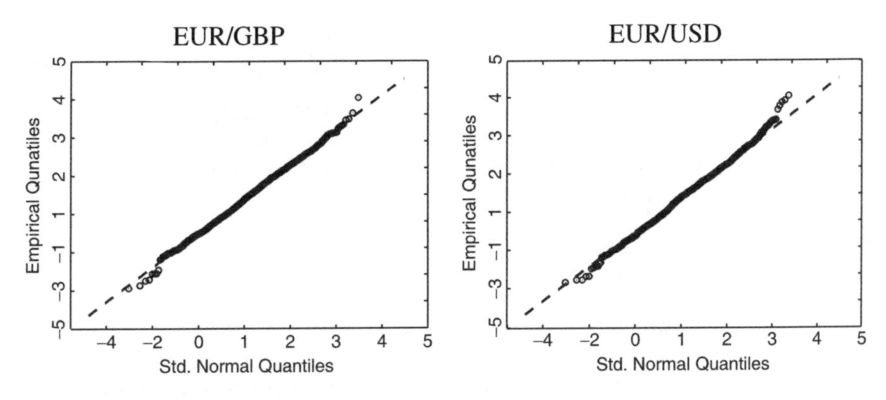

Fig. 12 QQ plot of the normalized density forecast variables.

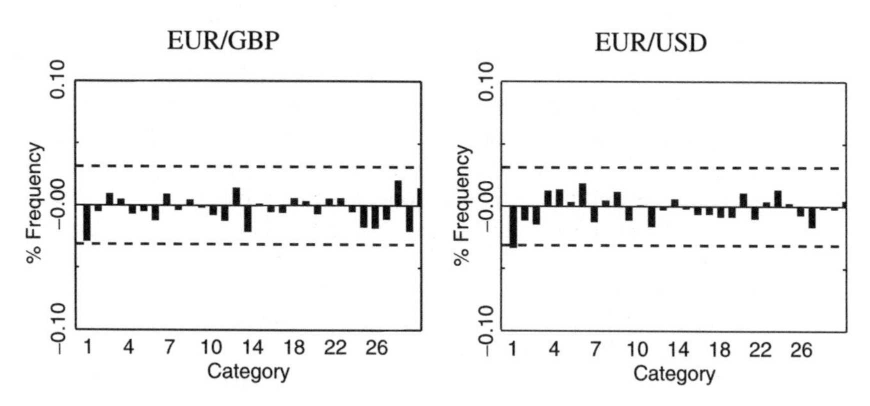

Fig. 13 Autocorrelation function of the normalized density forecast variables.

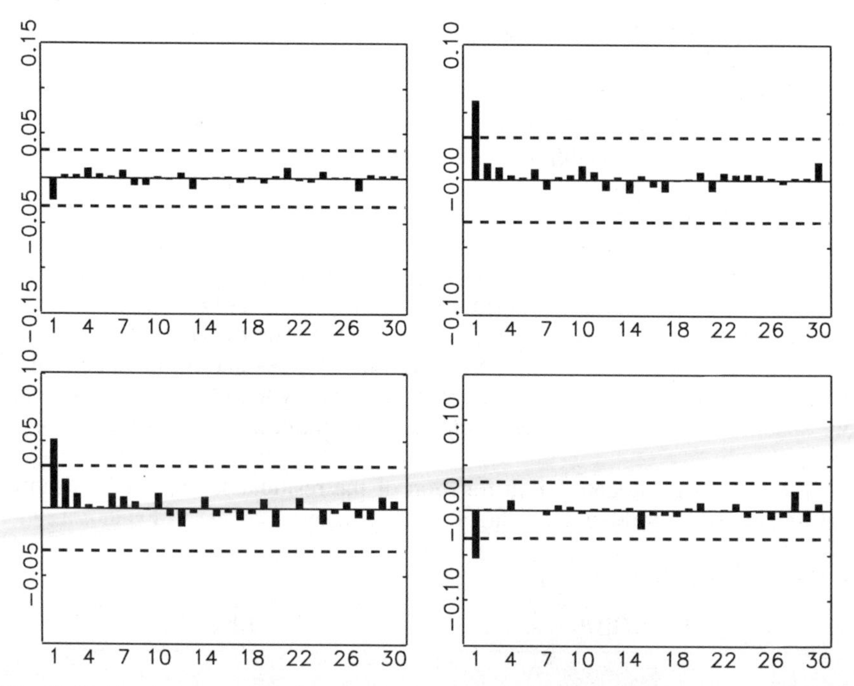

Fig. 14 Multivariate autocorrelation function for residuals of the EUR/GBP and the EUR/USD tick changes. *Upper left panel*: $\text{corr}(\varepsilon_t^1, \varepsilon_{t-l}^1)$; *upper right panel*: $\text{corr}(\varepsilon_t^1, \varepsilon_{t-l}^2)$; *lower left panel*: $\text{corr}(\varepsilon_{t-l}^2, \varepsilon_t^1)$ and *lower right panel*: $\text{corr}(\varepsilon_t^2, \varepsilon_{t-l}^2)$. The *dashed lines* mark the approximate 99% confidence interval $\pm 2.58/\sqrt{T}$.

part of the correlation structure of the processes for the exchange rate changes (compare Fig. 6), there is some room to improve the model specification. These results are also emphasized by the multivariate Ljung-Box statistics for the bid and ask change process and its residuals in Table 3.

Table 3 Multivariate Ljung-Box statistic for the residuals of the simulated bivariate process

	Real exchange rate changes	Residuals
$Q(20)$	473.6279	159.8679
$Q(30)$	521.8745	207.2940
$Q(50)$	588.0909	296.7522

4 Conclusion

In this paper we propose an approach that is capable of modelling complex multivariate processes for discrete random variables. Combining the approach by Cameron et al. (2004) for copulas of discrete random variables with the ICH model by Liesenfeld et al. (2006), we model the joint process for two integer count variables.

As an illustration of the explanatory power of our approach we estimate the joint distribution of the EUR/GBP and the EUR/USD exchange rate changes at the 1 min level. Even without detailed specification search, our model describes the exchange rate dynamics fairly well. Moreover, the marginal distributions which are characterized by inflated outcomes are also estimated satisfactorily.

In order to pick up the obvious excess kurtosis in the joint empirical distribution, we have tried out more flexible parametric alternatives to the Gaussian copula, such as the t-student copula, which allows for symmetric lower and upper tail dependency and an excessive concentration in (0,0) and the symmetrized Joe–Clayton copula, which has a quite parsimonious functional form and allows for asymmetric tail dependence. Although both specifications improve the goodness-of-fit of our model in some aspects the application of the t-student copula or the symmetrized Joe–Clayton copula has been by no means clearly superior to the simple Gaussian copula, so that we conclude that simply applying more flexible copula functions is not the proper remedy to capture the large excess kurtosis. An obvious alternative path of future research is to keep the Gaussian copula and to inflate the outcome (0,0) along the lines of zero inflated count data models.

Last but not least, the potential merits of the approach should be checked in the light of real world applications such as the measurement of multivariate conditional volatilities and the quantification of liquidity or value-at-risk applications. Obviously, our approach can easily be extended to the most general case of mixed multivariate distributions for continuous and discrete random variables. For financial market research at the high frequency level, such an extension is attractive for the joint analysis of several marks of the trading process (volumes, price and volume durations, discrete quote changes, etc.).

Acknowledgements The work is supported by the European Community's Human Potential Program under contract HPRN-CT-2002-00232, Microstructure of Financial Markets in Europe; and by the Fritz Thyssen Foundation through the project 'Dealer-Behavior and Price-Dynamics on the Foreign Exchange Market'. We thank Richard Olsen and Olsen Financial Technologies for providing us with the data.

References

Amilon H (2003) GARCH estimation and discrete stock prices: an application to low-priced Australian stocks. Econ Lett 81(2):215–222

Andersen TG, Bollerslev T, Diebold FX, Labys P (1999) (Understanding, optimizing, using and forecasting) realized volatility and correlation, New York University, Leonard N. Stern School Finance Department Working Paper, No. 99–061

Antoniou A, Vorlow CE (2005) Price clustering and discreteness: is there chaos behind the noise? Physica A 348:389–403

Ball C (1988) Estimation bias induced by discrete security prices. J Finance 43:841–865

Brock WA, Dechert WD, Scheinkman JA, LeBaron B (1996) A test for independence based on the correlation dimension. Econ Rev 15(3):197–235

Cameron C, Li T, Trivedi P, Zimmer D (2004) Modelling the differences in counted outcomes using bivariate copula models with application to mismesured counts. Econ J 7:566–584

Crack TF, Ledoit O (1996) Robust structure without predictability: the "compass rose" pattern of the stock market. J Finance 51(2):751–762

Denuit M, Lambert P (2005) Constraints on concordance measures in bivariate discrete data. J Multivariate Anal 93:40–57

Diebold FX, Gunther TA, Tay AS (1998) Evaluating density forecasts, with applications to financial risk management. Int Econ Rev 39:863–883

Fang Y (2002) The compass rose and random walk tests. Comput Stat Data Anal 39:299–310

Gleason KC, Lee CI, Mathur I (2000) An explanation for the compass rose pattern. Econ Lett 68(2):127–133

Hansen PR, Lunde A (2006) Realized variance and market microstructure noise. J Bus Econ Stat 24:127–218

Heinen A, Rengifo E (2003) Multivariate autoregressive modelling of time series count data using copulas, Center for Operations Research and Econometrics, Catholique University of Luvain

Huang RD, Stoll HR (1994) Market microstructure and stock return predictions. Rev Financ Stud 7(1):179–213

Johnson N, Kotz S, Balakrishnan N (1997) Discrete multivariate distributions. Wiley, New York

Kocherlakota S, Kocherlakota K (1992) Bivariate discrete distributions. Dekker, New York

Krämer W, Runde R (1997) Chaos and the compass rose. Econ Lett 54(2):113–118

Lee CI, Gleason KC, Mathur I (1999) A comprehensive examination of the compass rose pattern in futures markets. J Futures Mark 19(5):541–564

Liesenfeld R, Nolte I, Pohlmeier W (2006) Modelling financial transaction price movements: a dynamic integer count data model. Empir Econ 30:795–825

Meester S, MacKay J (1994) A parametric model for cluster correlated categorical data. Biometrics 50:954–963

Oomen RCA (2005) Properties of bias-corrected realized variance under alternative sampling schemes. J Financ Econ 3:555–577

Patton A (2001) Modelling time-varying exchange rate dependence using the conditional copula. Discussion Paper, UCSD Department of Economics

Russell JR, Engle RF (2002) Econometric analysis of discrete-valued irregularly-spaced financial transactions data. University of California, San Diego, Revised Version of Discussion Paper, No. 98–10

Shephard N (1995) Generalized linear autoregressions. Working Paper, Nuffield College, Oxford

Sklar A (1959) Fonctions de répartition à n dimensions et leurs marges. Public Institute of Statistics at the University of Paris 8:229–231

Stevens W (1950) Fiducial limits of the parameter of a discontinuous distribution. Biometrika 37:117–129

Szpiro GG (1998) Tick size, the compass rose and market nanostructure. J Bank Finance 22(12):1559–1569

Vorlow CE (2004) Stock price clustering and discreteness: the "compass rose" and predictability. Working Paper, University of Durham

Alvaro Escribano · Roberto Pascual

Asymmetries in bid and ask responses to innovations in the trading process

Abstract This paper proposes a new approach to jointly model the trading process and the revisions of market quotes. This method accommodates asymmetries in the dynamics of ask and bid quotes after trade-related shocks. The empirical specification is a vector error correction (VEC) model for ask and bid quotes, with the spread as the co-integrating vector, and with an endogenous trading process. This model extends the vector autoregressive (VAR) model introduced by Hasbrouck (Hasbrouck J (1991) Measuring the information content of stock trades. J Finance 46:179–207). We provide evidence against several symmetry assumptions, very familiar among microstructure models. We report asymmetric adjustments of ask and bid prices to trade-related shocks, and asymmetric impacts of buyer and seller-initiated trades. In general, buys are more informative than sells. The likelihood of symmetric quote responses increases with volatility. We show that our findings are robust across different model specifications, time frequencies, and trading periods. Moreover, we find similar asymmetries in markets with different microstructures.

Keywords Market microstructure · Bid and ask time series · VEC models · Adverse-selection costs · Asymmetric dynamics

JEL Classification G1

This paper has benefited from the support of the Spanish DGICYT project #PB98-0030 and the European Project on VPM-Improving Human Research Potential, HPRN-CT-2002-00232. The authors are grateful for the comments received from an anonymous referee and from Mikel Tapia, Ignacio Peña, Winfried Pohlmeier and the attendants to the Econometrics Research Seminar at C.O.R.E., Université Catholique de Louvain, Belgium. We also appreciate the suggestions of participants at the CAF Market Microstructure and High Frequency Data in Finance Workshop, August 2001, Sønderborg (Denmark), and the European Financial Association Meeting, August 2001, Barcelona (Spain)

A. Escribano (✉)
Departament of Economics, Universidad Carlos III de Madrid, C/Madrid 126, Getafe, 28903 Madrid, Spain
E-mail: alvaroe@eco.uc3m.es

R. Pascual
Departamento de Economía de la Empresa, Universidad de las Islas Baleares, Madrid, Spain
E-mail: rpascual@uib.es

1 Introduction

In this paper, we propose a new econometric approach to jointly model the time series dynamics of the trading process and the revisions of ask and bid prices. We use this model to test the validity of certain symmetry assumptions very common among microstructure models. Namely, we test whether ask and bid quotes respond symmetrically to trade-related shocks, and whether buyer-initiated trades and seller-initiated trades are equally informative. In essence, the procedure we propose generalizes Hasbrouck's (1991) vector autoregressive model for signed trades and changes in the quote midpoint by relaxing the implicit symmetry assumptions in his model.

The properties of the empirical model are derived from a structural dynamic model for ask and bid prices. In this model, ask and bid prices share a common lung-run component, the efficient price. The long-term value of the stock varies due to buyer-initiated shocks, seller-initiated shocks, and trade-unrelated shocks. The transitory components of ask and bid prices are characterized by two correlated and trade-dependent stochastic processes, whose dynamics are allowed to differ. The trading process is endogenous. Buyer and seller-initiated trades are generated by two idiosyncratic but mutually dependent stochastic processes. The generating processes of quotes and trades both depend on several exogenous variables that feature the trades and the market conditions.

We demonstrate that the empirical counterpart of this theoretical model is an extended vector error correction (VEC) model with four dependent variables: changes in the ask price, changes in the bid price, buyer-initiated trades, and seller-initiated trades. The bid–ask spread is the error correction term. Our VEC model reverts to the Hasbrouck's (1991) bivariate VAR model when: (a) ask and bid responses to trade-related shocks perfectly match; (b) the generating processes of buyer and seller-initiated trades are equivalent, and (c) the trade sign only matters as far as the direction of the quote adjustments is concerned.

For robustness purposes, we implement the model using three different subsamples: the 11 most frequently traded NYSE-listed stocks in 1996 and 2000, and the 11 most active stocks at the Spanish Stock Exchange (SSE) in 2000. With the two NYSE subsamples, we show that our main findings are not period-specific. We also show that our findings are unaltered by the dramatic increase in trading activity and the progressive decrease in the minimum price variation experienced by the NYSE from 1996 to 2000. With the Spanish data, we show that our findings are not limited to the particular microstructure of the NYSE. We perform additional robustness test considering alternative specifications of the empirical model.

We find two main patterns characterizing the dynamics of market quotes. On the one hand, ask and bid quotes do not respond symmetrically after trade-related shocks. They tend to be revised in the same direction, but not by the same amount. We show, however, that the likelihood of observing a symmetric response increases with volatility. On the other hand, ask and bid prices error-correct after a trade, which causes the spread to revert towards the minimum. The speed of reversion is significantly non-linear. The wider the bid–ask spread, the quicker the response of ask and bid quotes. These patterns result in two simultaneous but opposite effects on the price dynamics: information-induced positive cross-serial correlation and liquidity-induced negative cross-serial correlation.

For the NYSE samples, we report that buyer-initiated trades are more informative than seller-initiated trades. Namely, we find that the average long-term impact of a buyer-initiated trade on the ask quote is larger than the average long-term impact of a similar seller-initiated trade on the bid quote. In the SSE, however, no statistical difference is found between the impact of buyer and seller-initiated trades.

In general, our findings evidence that the dynamics between quotes and trades are more complex than suggested by classical microstructure models of quote formation. We also show that asymmetries are not exclusive of the NYSE, since they are also found in an electronic order-driven market without market makers, the SSE. In addition, our findings demonstrate that there is an important loss of information in averaging the dynamics of ask and bid quotes through the quote midpoint instead of jointly modeling them.

The remainder of the paper proceeds as follows. In Section 2, we review the literature and motivate the model. In Section 3, we present the theoretical dynamic model and its empirical counterpart. In Section 4, we describe the data. In Section 5, we analyze in detail a representative stock: IBM. In Section 6, we perform several robustness tests. Finally, in Section 7, we conclude.

2 Motivation

A large part of market microstructure research builds on the notion that trades convey new information that updates the market's expectation about the long-run value of the stock. This trade-related information causes simultaneous revisions of market quotes (e.g., Hasbrouck 1996). Ask and bid quotes are usually modeled as the result of adding a premium and subtracting a discount to the efficient price (e.g., Glosten 1987). The magnitude of these perturbations depends on certain market frictions, such as price discreteness, and market making costs.[1]

For simplicity purposes, many classic theoretical models of price formation impose, in some degree, what we will call in this paper the "symmetry assumption". First, the symmetry assumption implies that offer and demand quotes are posted symmetrically about the efficient price. Thus, in some models the transitory components of ask and bid prices are constant and equal-sized (e.g., Roll 1984; Madhavan et al. 1997; Huang and Stoll 1997); in some other cases, their dynamics are characterized by the same stochastic process (e.g., Glosten and Harris 1988; Lin et al. 1995; Hasbrouck 1999b). Second, the symmetry assumption implies that ask and bid quotes respond identically after a trade-related shock. Thus, a common premise in theoretical models is that ask and bid prices are simultaneously revised upward or downward, usually by the same amount, after a trade-related shock (e.g., Glosten and Milgrom 1985; Stoll 1989). Finally, the symmetry assumption implies that whether a trade is buyer or seller-initiated matters to determine the direction, but not the magnitude, of ask and bid updates (e.g., Easley and O'Hara 1992). That is, buys and sells are assumed to be equally informative.[2]

---

[1] See O'Hara (1995) for a review of the basics of this literature.

[2] The symmetry assumption is sometimes relaxed. For example, Easley and O'Hara (1987) allow different sequences of trades to have different price impacts.

Jang and Venkatesh (1991) reports that, in the NYSE, one-step-ahead revisions in ask and bid quotes right after trades do not generally support the theoretical prediction of symmetry. Moreover, quote-revision patterns strongly depend on the level of the outstanding spread, with more symmetric adjustments as the spread augments. They argue the symmetry assumption may be violated in practice because certain theoretical premises are simply unrealistic. Thus, it is usually assumed that quotes can be adjusted in a continuous fashion, which is not possible because of the minimum price variations. In addition, it is usually supposed that posted quotes are always for the specialist own account. However, in most stock exchanges, as is the case of the NYSE, quotes reflect the interest of several traders that may be selectively offering one-sided liquidity (see Madhavan and Sofianos 1998; Kavajecz 1999; Chung et al. 1999). Since different agents may be subject to different trading costs, the offer and demand components of the spread may vary asymmetrically about the efficient price.

Hasbrouck (1999a) points out that the usual theoretical premise of equal market making costs at the offer and demand sides of the market is reasonable if the same quote-setter is active on both sides. However, even if the specialist would take all trades, it is not clear that she would adjust quotes simultaneously and by the same amount after a trade. Thus, if the trade signals the presence of informed traders, a natural response of the specialist would be to post a wider spread. Therefore, she would update ask and bid quotes asymmetrically. Moreover, the costs of ask and bid exposure might not necessarily be balanced. Thus, a specialist offering liquidity in times of an upward price pressure would suffer from higher exposure costs on the ask side than on the bid side of the market.

Biais et al. (1995) shows that asymmetries between ask and bid quotes are not an exclusivity of the NYSE. Using data on a pure order-driven market, the Paris Bourse, they show that ask and bid one-step-ahead adjustments after trades are also asymmetric. Their findings hint that the ask (bid) quote may lead the adjustment of the bid (ask) price after an order to buy (sell). These authors conclude that "there is additional information in analyzing the dynamics of ask and bid prices jointly rather than averaging them through the quote midpoint" (pg. 1679).

There is also evidence suggesting that the impact of a buyer-initiated trade may not be merely the reverse of the price impact of seller-initiated trade. Empirical work on block trading (e.g., Holthausen et al. 1987; Griffiths et al. 2000; Koski and Michaely 2000) shows that orders to sell and orders to buy may have different permanent and transitory impacts on prices. Chan and Lakonishok (1993, 1995) use a broad range of trade sizes to evidence differences in the behavior of prices after institutional purchases and sales. Keim and Madhavan (1995) conclude that large buys take longer to execute than equivalent sells because traders perceive that price impacts of buys are greater than sells. Huang and Stoll (1996) report different realized spreads for buyer and seller-initiated trades. Similarly, Lakonishok and Lee (2001) observe that the information content of insider's activities come from purchases while insider selling appears to have no predictive ability.[3]

[3] These asymmetries might not be exclusive of large-sized trades. Hasbrouck (1988, 1991), Barclay and Warner (1993), Kempf and Korn (1999), among others, evidence the relationship between trade size and price impact is increasing, but concave. Therefore, informed traders may concentrate their trades in medium sizes.

In all these papers, buyer-initiated trades are usually found to be more informative than seller-initiated trades. Price responses to buyer-initiated versus seller-initiated trades may be asymmetric for a variety of reasons. Firstly, short selling restrictions may prevent insiders from exploiting negative information (Kempf and Korn 1999). Secondly, since an investor typically does not hold the market portfolio, the choice of a particular stock to sell does not necessarily convey negative information. On the contrary, the choice of a particular stock to buy, out of the numerous possibilities on the market, is likely to convey favorable firm-specific news (Chan and Lakonishok 1993). Finally, a sell order representing a small fraction of the initiator's known position may be considered as more liquidity-motivated than a similar buy order from an investor without current holdings in the security (Keim and Madhavan 1995).[4]

Alternative empirical approaches have been proposed to jointly model the generating processes of quotes and trades. The most influential is probably due to Hasbrouck (1991).[5] He suggests the following vector autoregressive (VAR) specification,

$$\begin{aligned} \Delta q_t &= \sum_{i=1}^{\infty} a_i \Delta q_{t-1} + \sum_{i=0}^{\infty} b_i x_{t-i} + v_{1,t} \\ x_t &= \sum_{i=1}^{\infty} c_i \Delta q_{t-1} + \sum_{i=1}^{\infty} d_i x_{t-i} + v_{2,t} \end{aligned} , \qquad (2.1)$$

where $\Delta q_t=(q_t-q_{t-1})$ represents the revision in the quote midpoint (q_t) after a trade at t and x_t is a trade indicator that equals 1 for buyer-initiated trades and -1 for seller-initiated trades. The terms $v_{1,t}$ and $v_{2,t}$ are mutually and serially uncorrelated white noises that represent trade-unrelated and trade-related shocks respectively. This econometric approach covers the dynamics of many structural microstructure models as special cases (see Hasbrouck 1991, 1996).

Hasbrouck builds on a "weak symmetry assumption:" the quote midpoint must revert to the efficient price as the end of trading approaches. Therefore, ask and bid prices may not be symmetrically posted around the efficient price. However, since the quote dynamics are averaged through the quote midpoint (q_t), the VAR model is not a valid framework to accommodate and evaluate possible asymmetries in the dynamics of ask and bid prices. Similarly, the trade dynamics are averaged through

[4] Other aspects of the trading process that may produce a lack of balance between the impact of buys and sells are transitory market conditions and the trade durations. Thus, a larger market pressure to sell than to buy might increase the expected impact of a seller-initiated market order (e.g., Goldstein and Kavajecz 2004) versus a similar buyer-initiated order. Easley et al. (1997) and Dufour and Engle (2000) show that trade durations (time between consecutive trades) partly explain the long-term impact of trades, even when trade size is accounted for. If a market overreacts to bad news, it might produce shorter time durations after seller-initiated trades than after buyer-initiated trades, and therefore asymmetries in the responses of quotes between buys and sells.

[5] Other econometric approaches, parametric and semi-parametric respectively, to model the relationship between the trading process and the price changes are Hausman et al. (1992), that used ordered probit models, and Kempf and Korn (1999) that employed a neural networks type model.

the trade indicator (x_t). Therefore, the expected impact of a buyer-initiated shock is exactly the reverse of a seller-initiated shock.

So as to allow for asymmetric dynamics, next section modifies the structural model used by Hasbrouck (1991) to motivate his VAR model. The empirical counterpart of this most flexible structural model will be a VEC model for ask and bid revisions in response to buyer and seller-initiated shocks, which generalizes Eq. (2.1). Given that ask and bid quotes have a common non-stationary long-run component, the efficient stock price, they must be co-integrated time series (see Hasbrouck 1995). The VEC model is the most common efficient parameterization of vector autoregressive models with co-integrated variables (e.g., Engle and Granger 1987). In this case, the co-integration relationship is known a priori, which lets setting a very general parameterization of the model. We will also allow quote revisions after trade-related shocks to follow non-linear patterns due to trade features, such as size and durations, and market conditions, such as volatility and liquidity.

Our approach connects with other econometric applications in microstructure research. Hasbrouck (1995) uses a common-trend representation to simultaneously model the quotes of both the NYSE and the regional markets. The model has an associated VEC representation. Hasbrouck is aimed to measure relative contributions to price discovery. Pascual et al. (2005) build on the model introduced in the next sections to incorporate the markets' trading processes into Hasbrouck's (1995) methodology. In this way, they are able to isolate, for each market, the trade-related contribution from the trade-unrelated contribution.

Hasbrouck (1999b) estimates an unobserved-components model for the best market quotes. In this model, ask and bid quotes have a common random walk component, and the respective transitory terms are modeled as two unobserved and identical first-order autoregressive processes. This model features discreteness, clustering, and stochastic volatility effects. However, it does not incorporate the trading process, and the transitory components of ask and bid quotes are assumed mutually independent. This analysis does not deal with asymmetric dynamics either. In a recent paper, Zhang et al. (2005) use Hasbrouck's (1999b) methodology to decompose the bid–ask spread of a single stock (General Electric) into its ask and bid exposure costs constituents. They show that ask and bid components of the spread change asymmetrically about the efficient price. Our findings, based on a less sophisticated but more widespread econometric approach, are totally consistent.

Even closer to the purpose of this paper is the independent study by Engle and Patton (2004). These authors (henceforth, EP) also estimate an error correction model for ask and bid quotes using data on a large set of NYSE-listed stocks. There are, however, remarkable differences between both empirical specifications. In EP's model, the trading process is exogenous. Thus, they model the dynamics of ask and bid quotes, but not the feedback from quotes to trades. As shown by Hasbrouck (1991), to accurately measure the informativeness of trades, we need to model not only how quotes evolve after the trade, but also how trading responds to the progressive adjustment in quotes. Since we are aimed to compare the information content of buyer and seller-initiated trades, we propose a model that accommodates a broader set of dynamic interactions between quotes and trades than EP's model. Moreover, EP's model is quote-driven, meaning that there is a new observation each time there is a change in quotes. Since we are interested in the price impact of trades, it makes sense to define our model in trade time. In this manner, we filter those quote changes that are not directly linked to the trading process.[6] These technical disparities may explain the discrepant findings we will report later on regarding the information content of buys versus sells. Nevertheless, in our opinion, EP's paper and the present paper are complementary, since both evidence the relevance of modeling ask and bid prices jointly rather than averaging them through the quote midpoint.

3 The model

3.1 A structural dynamic model of quote formation

In this subsection, we build on Hasbrouck (1991) to develop a dynamic model for ask and bid quotes. The model allows for asymmetric adjustment paths going after trades. Two main features differentiate our structural model from those previously found in the literature. First, ask and bid prices share a common long-run component, the efficient price, which is updated due to trade-related and trade-unrelated informative shocks. The empirical evidence previously revised, however, suggests that buyer and seller-initiated trades (henceforth, "buys" and "sells") may not be equally informative. To accommodate this empirical observation, we will distinguish between trade-related shocks to buy and trade-related shocks to sell. Second, quotes result from adding or subtracting a transitory component (w_t), due to market frictions and market-making costs, to the efficient price (m_t). Since the evidence at hand points to asymmetric short-term adjustments of the ask quote versus the bid quote, in our specification we will allow the short-term components of ask and bid quotes to differ ($\Delta w_t^{\mathrm{a}} \neq \Delta w_t^{\mathrm{b}}$).

We use the same notation as in Hasbrouck (1991). The model is defined in trade time. Thus, the subscript t denotes the t-th trade in the chronological sequence of trades. Hereafter, the superscript a means "ask quote," b means "bid quote," B refers to buys, and S refers to sells. m_t is the efficient price after the t-th trade, which could be either a buy (x_t^B) or a sell (x_t^S). Similarly, a_t and b_t are the quotes posted right after the t-th trade. The adjustment in the posted quotes after the t-th trade are $\Delta a_t = a_t - a_{t-1}$ and $\Delta b_t = b_t - b_{t-1}$.

The efficient price follows the random walk process in Eq. (3.1). Three types of stochastic shocks update m_t: trade-unrelated shocks ($v_{1,t}$), and trade-related shocks due to buyer-initiated trades ($v_{2,t}^B$) or seller-initiated trades ($v_{2,t}^S$). We let $v_{1,t}$ be mutually and serially uncorrelated with $v_{2,t}^B$ and $v_{2,t}^S$, while $v_{2,t}^B$ and $v_{2,t}^S$ are serially uncorrelated but, perhaps, mutually correlated. The parameters λ^B and λ^S measure the average amount of private information (adverse selection costs) conveyed by buys and sells, respectively. If $\lambda^B = \lambda^S$ buy shocks would have the same information content than sell shocks.

$$m_t = m_{t-1} + \lambda^B v_{2,t}^B + \lambda^S v_{2,t}^S + v_{1,t}. \tag{3.1}$$

[6] Another minor difference is that, unlike EP's model, the properties of our empirical specification are derived from a theoretical framework presented in the next section. The empirical model we derive from the structural model is an "extended" VECM, that is, it incorporates lagged values of the error correction term. The EP's model is a standard VEC.

The generating processes of market quotes are given by Eqs. (3.2) and (3.3),

$$\begin{aligned} a_t &= m_t + w_t^a = \\ &= m_t + \alpha_m^a (a_{t-1} - m_{t-1}) + A_{x,t}(L)' x_t + \alpha_a^{EC}(a_{t-1} - b_{t-1}) + \varepsilon_t^a \end{aligned} \quad (3.2)$$

$$\begin{aligned} b_t &= m_t - w_t^b = \\ &= m_t + \alpha_m^b (m_{t-1} - b_{t-1}) + B_{x,t}(L)' x_t + \alpha_b^{EC}(a_{t-1} - b_{t-1}) + \varepsilon_t^b. \end{aligned} \quad (3.3)$$

In Eqs. (3.2)–(3.3), a_t and b_t are the result of adding a time-varying stationary premium (w_t^a) and subtracting a time-varying stationary discount (w_t^b), respectively, to the efficient price. These transitory components are time-varying because we assume they are determined by the recent history of trades and quotes. We allow the magnitudes of these two components to differ. Therefore, a_t and b_t may not be symmetrically posted about m_t.[7] Moreover, we impose $0<\alpha_m^a<1$ and $0<\alpha_m^b<1$, implying that, in the absence of trading, a_t and b_t revert to the efficient price. The noise terms ε_t^a and ε_t^b are idiosyncratic errors reflecting market frictions and model misspecifications. Finally, the vectors $A_{x,t}(L)'=(A_{x,t}^B(L), A_{x,t}^S(L))$ and $B_{x,t}(L)'=(B_{x,t}^B(L), B_{x,t}^S(L))$ are finite order polynomials in the lag operator $L(L^k y_t=y_{t-k})$ with time-varying components. These polynomials would capture the transitory effect of trades on quotes. The dynamic structure denotes that a_t and b_t adjustments to trade-related shocks are progressive.[8]

The vector $x_t'=(x_t^B, x_t^S)$ in Eqs. (3.2)–(3.3) includes the time series of buys and sells. The trading process is endogenous. The idiosyncratic, but mutually dependent, stochastic processes in Eqs. (3.4) and (3.5) generate buys and sells,

$$x_t^B = \mu^B (a_{t-1} - m_{t-1}) + \pi^B (a_{t-1} - b_{t-1}) + v_{2,t}^B \quad (3.4)$$

$$x_t^S = \mu^S (m_{t-1} - b_{t-1}) + \pi^S (a_{t-1} - b_{t-1}) + v_{2,t}^S. \quad (3.5)$$

In Eqs. (3.4)–(3.5), the likelihood of observing a new trade decreases with its specific exposure costs ($\mu^B<0$, $\mu^S<0$) and the costs of executing a round-trip ($\pi^B<0$, $\pi^S<0$), defining downward sloping demand schedules. The terms $v_{2,t}^B$ and $v_{2,t}^S$ are the mutually correlated unexpected components of buys and sells, respectively.

The third element on the RHS of Eqs. (3.2)–(3.3) is decomposed in terms of buys and sells as follows,

$$A_{x,t}(L)' x_t = A_x^B(L) f_a^B(MC_t, D_t) x_t^B + A_x^S(L) f_a^S(MC_t, D_t) x_t^S$$

$$B_{x,t}(L)' x_t = B_x^B(L) f_b^B(MC_t, D_t) x_t^B + B_x^S(L) f_b^S(MC_t, D_t) x_t^S,$$

where $A_x^B(L)$, $A_x^S(L)$, $B_x^B(L)$, and $B_x^S(L)$ are finite time-invariant order polynomials in the lag operator L, having all roots outside the unit circle.

[7] Hasbrouck (1999b) models the exposure costs for bid and ask quotes as two independent stochastic processes. In our case, the transitory components could be mutually correlated because of the common components.

[8] Stabilizing NYSE rules (see Hasbrouck et al. 1993) and heterogeneous priors among traders (e.g., Harris and Raviv 1993) may explain the lagged effects of trades.

The terns $f_i^B(MC_t, D_t)$ and $f_i^S(MC_t, D_t)$, $i \in \{a,b\}$, are functional forms of two vectors of variables. The first vector (MC_t) includes exogenous variables that characterize the trade and the market environment. The second vector (D_t) control for trading-time regularities. The particular functional form considered is given in Eq. (3.6). We impose linearity for simplicity reasons. The price impact of a given trade is conditioned on these set of exogenous and deterministic variables, that we will specify latter on.

$$f_i^j(MC_t, D_t) = 1 + \sum_{k=1}^{n} \lambda_k^{i,j} MC_t^k + \sum_{h=1}^{n'} \gamma_h^{i,j} D_t^h, \; i \in \{a, b\}, j \in \{B, S\} \tag{3.6}$$

From Eqs. (3.1) to (3.3), a_t and b_t are nonstationary, integrated of order one, processes. Nonstationarity comes from the common long-run component (m_t), implying that the time series a_t and b_t must be co-integrated.[9] Our application has the unusual advantage that the co-integration relationship has a known co-integration vector (1,−1). The co-integration relationship is, therefore, $a_t - b_t$, the bid–ask spread (henceforth, s_t).

An increase in s_t represents a departure from the long-run equilibrium relationship between a_t and b_t. The error correction mechanism produces simultaneous revisions in both ask and bid quotes that correct such deviations. For this reason, we incorporate s_t into Eqs. (3.2)–(3.3) as a determinant of the transitory components of a_t and b_t. The coefficients α_a^{EC} and α_b^{EC} show how quickly do a_t and b_t revert to their common long-run equilibrium value.

3.2 The empirical model

The most common efficient parameterization of a vector autoregressive (VAR) model with co-integrated variables is, from Granger's representation theorem in Engle and Granger (1987), a vector error correction (VEC) model. In the Appendix I, we give an explicit derivation of the VEC model in Eq. (3.7) from the structural model in the previous subsection,

$$\begin{pmatrix} 1 & 0 & A^*_{aB,t} & A^*_{aB,t} \\ 0 & 1 & A^*_{bB,t} & A^*_{bS,t} \\ 0 & 0 & 1 & 0 \\ 0 & 0 & 0 & 1 \end{pmatrix} \begin{pmatrix} \Delta a_t \\ \Delta b_t \\ x_t^B \\ x_t^S \end{pmatrix} = \begin{pmatrix} \gamma_a^{EC}(L) \\ \gamma_b^{EC}(L) \\ \gamma_B(L) \\ \gamma_S(L) \end{pmatrix} s_{t-1} + A_t(L) \begin{pmatrix} \Delta a_{t-1} \\ \Delta b_{t-1} \\ x_{t-1}^B \\ x_{t-1}^S \end{pmatrix} + \begin{pmatrix} u_t^a \\ u_t^b \\ u_t^B \\ u_t^S \end{pmatrix}, \tag{3.7}$$

[9] Engle and Granger (1987), Stock and Watson (1988), Johansen (1991), and Escribano and Peña (1994), among others, provide formal derivations of this result.

with

$$A_t(L) = \begin{pmatrix} A_{aa}(L)A_{ab}(L)A_{aB,t}(L)A_{aS,t}(L) \\ A_{ba}(L)A_{bb}(L)A_{bB,t}(L)A_{bS,t}(L) \\ A_{Ba}(L)A_{Bb}(L)A_{BB,t}(L)A_{BS,t}(L) \\ A_{Sa}(L)A_{Sb}(L)A_{SB,t}(L)A_{SS,t}(L) \end{pmatrix}.$$

This model echoes the main features of the structural model in the previous subsection. First, the bid–ask spread $s_t = a - b_t$ is the error correction term. Second, the matrix on the left-hand side of Eq. (3.7) reflects that the theoretical model is trade-driven. Thus, trades have a contemporaneous effect on ask and bid quotes. The reverse, however, is not true. Third, the matrix of autoregressive polynomials $A_t(L)$ depicts the dynamical structure of the theoretical model. Moreover, $A_{ij}(L)$, for all i, $j \in \{a,b,B,S\}$, has its roots outside the unit circle. Thus, the influence of past quotes and trades decays with time. Finally, the polynomials $A_{ij,t}(L)$ are time-varying because they depend on a set of exogenous variables (MC_t) and trading-time dummies (D_t). The following expression makes explicit the type of dependence,

$$A_{ij,t}(L)x_{t-1} = A_{ij}^B(L)f_{ij}^B(MC_{t-1}, D_{t-1})x_{t-1}^B + A_{ij}^S(L)f_{ij}^S(MC_{t-1}, D_{t-1})x_{t-1}^S.$$

The polynomials $A_{ij,t}^B(L)$ and $A_{ij,t}^S(L)$ have all the roots outside the unit circle. Finally, $A_{ij,t}^* = -A_{ij,t}(0)$.

A salient feature of the VEC model Eq. (3.7) is the extra lags in the error correction term. This type of specification is called an extended vector error correction (EVEC) model. Arranz and Escribano (2000) show that extended error correction models are robust to the presence of structural breaks under partial co-breaking. Co-breaks represent those situations characterized by having breaks (level shifts, changes in trend etc.) occurring simultaneously in some variables, so that certain linear combinations of those variables have no breaks. The common lung-run trend jointly with their discrete type of moves makes a_t and b_t the perfect example of co-integrated time series that are partially co-breaking. Thus, this property of the model is consistent with the properties of the time series of ask and bid prices. The error correction terms $\gamma_a^{EC}(L)s_{t-1}$ and $\gamma_b^{EC}(L)s_{t-1}$ should be such that $\gamma_a^{EC}(1) - \gamma_b^{EC}(1) < 0$, in order to impose the error correction characterization on the spread. Extended error correction parameterizations of VAR models with co-integrated variables could be formally justified using the Smith-MacMillan decomposition introduced by Engle and Yoo (1991).

The individual error terms u_t^i in Eq. (3.7) $i = \{a,b,B,S\}$ are assumed to be serially uncorrelated random variables with zero mean and constant variance. We show in Appendix I that they cannot be treated as mutually uncorrelated since they have common components. Hence, the system of Eq. (3.7) is an example of seemingly unrelated regression equations, which can be efficiently estimated by SURE (see Zellner 1962). Estimating a system by SURE is equivalent to estimating it equation by equation by OLS when all equations have the same number of variables. In other case, all equations should be simultaneously estimated by SURE to get efficiency. Notice also that, under the restrictions imposed on the structural model in the previous section, Eq. (3.7) is exactly identified.

Next, we proceed with the estimation of the VEC model Eq. (3.7). In a preliminary step, however, we consider a base-line version of Eq. (3.7) where

$f_i^j(MC_t, D_t)=1$ for all i and j. In this case, the matrix of autoregressive polynomials is time invariant, $A_t(L)=A(L)$, and the impact of trades on quotes is perfectly linear. We will show that this model suffices to illustrate the essentials of the dynamic relationship between trades and quotes. However, some other aspects of this relationship can only be captured by considering the more general case.

Following Hasbrouck (1991), we characterize the trading processes using indicator variables. Namely, x_t^B equals one for buys and zero otherwise, and x_t^S equals one for sells and zero otherwise. The discreteness of these variables, however, may introduce some problems in the estimation process.[10] To control for these potential problems, we also estimate Eq. (3.7) using the trade size $(\widetilde{x}_t^i)$ to characterize each transaction. In particular, we define $\widetilde{x}_t^i = x_t^i \log(V_t)$, where V_t is the size of the t-th trade in shares.

4 Data

The database comprises high frequency data on trades and quotes from two markets with remarkably different microstructures: the NYSE and the SSE. The NYSE is a peculiar mixture of microstructure types. It combines an electronic limit order book, only partially transparent, with monopolist market makers, and an intensive trading activity at the floor market. The SSE, on the contrary, is a representative example of an electronic order-driven venue. Liquidity provision depends exclusively on a fully transparent open limit order book. Twenty levels of the book are nowadays visible in real time for all market participants. There are no market makers, no floor trading, price improvements are not possible, and all the orders are submitted through vendor feeds, and stored or matched electronically.

We use data on two different markets to show that asymmetric dynamics between ask and bid quotes in response to trades are not exclusive of the NYSE. In addition, trades in the SSE always involve a market order (or equivalent), the initiating side, and one or more limit orders stored on the book. Therefore, trades are straightforwardly classified as either buyer or seller-initiated by simply identifying the side of the book the market order hits. Thus, with the SSE data we do not bear the ambiguity and misclassification problems that appear when traditional trade-direction algorithms, such as Lee and Ready (1991), are applied to NYSE data (see Ellis et al. 2000, and Odders-White 2000). Finally, using Spanish data we do not have reporting delays neither in trades nor in quotes since the book and trade files are updated simultaneously and in real time. Therefore, we avoid the use ad hoc rules to match trades and quotes, like the classical "five-second rule" applied to NYSE data.[11]

NYSE data is obtained from the TAQ database. We consider two different sample periods, January to March 1996 and 2000. Several details in the mi-

[10] Our model is nonlinear and well behaved around the mean (nonlinear). Like standard linear probability models (LPM), only in the extremes it can give predictions out of the zero and one interval. The corresponding estimation theory for dynamic models with weakly dependent variables is covered in White (1994) and Wooldridge (1994). Park and Phillips (2000) extend it to cover nonlinear cointegration cases with limited dependent variables.

[11] Blume and Goldstein (1997) shows that the "five-second rule" could not be generalized to all sample periods and markets. However, Odders-White (2000) shows that this rule does not seem to explain much of the bias induced by the Lee and Ready's (1991) algorithm.

crostructure of the NYSE changed from 1996 to 2000. Particularly interesting for the purposes of this paper is the progressive decrease of the minimum price variation or tick, from US$1/8 in 1996 to 1 cent in 2000. Jang and Venkatesh (1991) remarks that symmetric responses of ask and bid quotes are impaired by the discreteness of quote changes. Therefore, a small tick should decrease the probability of observing asymmetric adjustments of ask versus bid quotes after trade-related shocks. Trading activity in the NYSE has sharply increased since 1996. For example, from January to March 1996 IBM transacted 130,620 times; during the same interval in 2000 the number of trades was 234,766. For GE, the number of trades increased from 106,347 in 1996 to 350,795 in 2000. By considering these two NYSE subsamples, we have the opportunity to check whether microstructure and trading activity changes have influenced the dynamical relationship between trades and quotes.

The NYSE sample includes the 11 most frequently traded stocks in 1996 and 2000, respectively, excluding stocks that experienced splits. The complete set of stocks is listed in Appendix II. We consider trades from both the primary market (NYSE) and regional markets. However, we only keep NYSE quotes because the evidence suggests that regional quotes only follow with some delay those of the primary market (Blume and Goldstein 1997).[12] Trades not codified as "regular trades", such as trades out of sequence or reported with error, have been discarded. Trades from the same market, with the same price, and with the same time stamp are treated as just one trade. All quote and trade registers prior to the opening and after the close are dropped. The overnight changes in quotes are treated as missing values. Quotes with bid–ask spreads lower than or equal to zero or quoted depth equal to zero have also been eliminated. After these adjustments, around 3% of all trades have been eliminated. Finally, we follow Blume and Goldstein (1997) in deleting quoted spreads that exceed 20% of the quote midpoint, and quote updates that exceed 50% of the prior quote. Prices and quotes are coupled using the "five-second rule" (Lee and Ready 1991). This rule assigns to each trade the first quote stamped at least fives before the trade itself.

A trade is classified as buyer (seller) initiated when the transaction price is closer to the ask (bid) price than to the bid (ask) price. Trades with price equal to the quote midpoint are not classified. The trade indictor x_t^B (x_t^S) equals one for buys (sells) and zero otherwise; for a midpoint trade, both indicators equal zero. A change in quotes, either Δa_t or Δb_t, is the difference between the quote prevailing right before the t-th trade takes place and the quote prevailing right before the next trade in time.

The SSE database contains the 11 most frequently traded stocks in 2000, listed in Appendix II. Spanish data is supplied by the SSE Interconnection System (SIBE). We retrieve trades and quotes from July to September 2000 because data from January to March is not available. We apply the same filters as for the US

[12] Hasbrouck (1995) concludes that the contribution of the regional markets to the price discovery process of NYSE-listed stocks is negligible. Harris et al. (1995), however, observe that both the NYSE and the regional markets error correct to deviations from each other, therefore suggesting that the regional quotes do are informative. Tse (2000) compares the methodologies used in these papers. He concludes that the discrepant findings are only due to the choice of quotes (Hasbrouck) versus trade prices (Harris et al.). Tse suggests that trades in the regional markets could contribute to price discovery even when quotes were non-informative. This conclusion fundaments our choice of discarding regional quotes while keeping regional trades.

data. Quote changes and trade indicators are computed analogously to the NYSE case. Since price improvements are impossible in the SSE, there are no transaction prices inside the bid–ask spread. Therefore, we do not require trade-direction algorithms to classify the trades as either buys or sells.

We rely on the theoretical and empirical research in market microstructure to determine the exogenous variables to be included in the vector MC_t. Easley and O'Hara (1987) formally show that large-sized trades are more informative. Empirically, Hasbrouck (1991) and Barclay and Warner (1993), among others, show that this relationship is increasing but concave. In Easley and O'Hara's (1992) model, higher trading intensity signals new information. Consistently, Easley et al. (1997) and Dufour and Engle (2000) find that shorter trade durations are associated with larger price impacts. Subrahmanyam (1997) finds that trades in the regional markets are less informative than NYSE trades, arguably because they attract liquidity-motivated traders (see Bessembinder and Kaufman 1997). Price instability means uncertainty about the true value of the stock (e.g., Bollerslev and Melvin 1994). Finally, a positive (negative) order imbalance between limit orders to sell and limit orders to buy may signal an overvalued (undervalued) stock (e.g., Huang and Stoll 1994).

The following variables are defined so as to capture the relationships detailed above. The trade size (V_t) is measured in shares. Trade durations (T_t) are computed as the time in seconds between two consecutive trades. A dummy variable (M_t) identifies regional trades. Order imbalance (OI_t) is computed as the difference between ask depth and bid depth. Finally, short-term volatility (R_t) is computed as the sum of the square changes of the quote midpoint $\sum_{k=1}^{z} (\Delta q_k)^2$ in a 5-min interval before each trade.[13]

Finally, we construct eight trading-time dummies for the NYSE session: one for trades during the first half-hour of trading, five for each trading hour between 10:00 A.M. and 3:00 P.M. and, finally, two for the last trading hour, divided in two half-hour intervals. Similarly, for the SSE session (9:00 A.M. to 5:30 P.M.), we construct nine dummy variables: one for the first half-hour of trading, another one for the second half-hour, six for each trading hour between 10:00 and 5:00 P.M. and, finally, one for the last half-hour.

5 Estimation of the baseline model for IBM in 1996

In this section, we present the details of estimating a restricted version of model Eq. (3.7) described in Section 3 where $f_i^j(MC_t,D_t)=1$ for all i and j. We use data on a representative NYSE stock, IBM, in 1996. In the next section, we check whether the dynamic patterns about to be reported for IBM can be generalized to other stocks, other markets, other time periods, and other model specifications, including the unrestricted model Eq. (3.7).

We consider two alternative specifications of the model. The first one uses the trade-sign indicators x_t^B and x_t^S to represent the trading process. The second one uses the trade-size indicators $\tilde{x}_t^B = x_t^B \log(V_t)$ and $\tilde{x}_t^S = x_t^S \log(V_t)$. The polynomials in the autoregressive matrix $A(L)$ are all truncated at lag five, as in

[13] This variable is not defined for trades performed during the first 5 min of trading. In these cases, we treat volatility as missing.

Hasbrouck (1991). The system is estimated by SURE, using the Feasible Generalized Least Squares (FGLS) algorithm, described, for example, in Green (1997, pp. 674–688).

Preliminary tests indicate that the following null,

$$A(L) = \begin{pmatrix} A_{aa}(L) & 0 & A_{aB}(L) & A_{aS}(L) \\ 0 & A_{bb}(L) & A_{bB}(L) & A_{bS}(L) \\ A_{Ba}(L) & 0 & A_{BB}(L) & A_{BS}(L) \\ 0 & A_{Sb}(L) & A_{SB}(L) & A_{SS}(L) \end{pmatrix} \tag{5.1}$$

cannot be rejected. This null means that Δa_t (Δb_t) depends of its own lags but not on Δb_t (Δa_t) lags. This restriction prevents for multicolinearity problems. Additionally, buys (sells) do not depend on lagged values of Δb_t (Δa_t).

Table 1 summarizes the estimation of the baseline version of Eq. (3.7) with the restrictions in Eq. (5.1). Panel A (B) reports the estimated coefficients and the residual correlation matrix for the model with trade-sign (trade-size) indicators. The noise terms $\left(\widetilde{u}_t^a, \widetilde{u}_t^b\right)$ in Eq. (3.7) are positively correlated (0.4362 in Panel A and 0.4324 in Panel B). This shows that the trade-unrelated shocks tend to move ask and bid quotes in the same direction. The noise terms $\left(\widetilde{u}_t^S, \widetilde{u}_t^B\right)$ are negatively correlated (−0.6804 in Panel A and −0.6038 in Panel B). This shows that unexpected increases in the buy pressure are usually coupled with unexpected decreases in the sell pressure. All these correlations are statistically significant at the 1% level. The remaining cross-equation correlation coefficients are statistically equal to zero. Therefore, as Appendix I suggested, the coefficients of Eqs. (3.7)–(5.1) cannot be efficiently estimated equation by equation.

The dynamics of ask and bid quotes after a trade are characterized by two simultaneous effects. First, ask and bid quotes error correct after a trade. The coefficients of the lagged bid–ask spread, s in Table 1, reveal that deviations between quotes induce simultaneous corrections in ask and bid prices. This dynamic effect causes the current spread to mean-revert as the ask decreases and the bid increases. Therefore, the model shows that changes in the spread are transient. This dynamic effect indicates that liquidity suppliers, either market markers or limit order traders, provide liquidity when it is valuable (see Biais et al. 1995).[14]

Second, the estimated coefficients for the trading process, x^B and x^S in Table 1, evidence that at least on dimension of the symmetry assumption is not satisfied at all. Ask and bid prices do not move symmetrically through time. This finding generalizes the one-step-ahead evidence in Jang and Venkatesh (1991). Table 1 – Panel A reports that after a unitary buy shock, both the ask quote and the bid quote tend to increase. However, on average, the ask price is raised an accumulated US $0.0247 five trade-time periods later. The bid price is raised a remarkably lower US $0.0038. Similarly, after a sell both quotes tend to be revised downwards. Nonetheless, the accumulated decrease in the ask price after five trade-time intervals is −US$0.0010 while the bid price decreases a far larger −US$0.0190.

[14] The adjustment path that leads to the long-run equilibrium between the ask price and the bid price is not necessarily linear. Following Escribano and Granger (1998), we have replaced the linear error correction term in Eq. (3.7) by a non-linear one, a cubic polynomial on the contemporaneous spread, $\eta_{1,1}^{j} s_{t-1} + \eta_{1,2}^{j} s_{t-1}^{2} + \eta_{1,3}^{j} s_{t-1}^{3}$. We find that all the coefficients are significant, indicating that the quote adjustment is faster the wider the quoted spread. However, we do not get too much improvement in terms of model adjustment.

Table 1 The base-line VEC model for IBM

Equation	Variable	Panel A: trade-sign indicators	Panel B: trade-size indicators
Δa_t	s	−0.0537	−0.0503
	Δa	0.0633	0.0285
	x^B	0.0247	0.0040
	x^S	−0.0010	−0.0004
Δb_t	s	0.0490	0.0541
	Δb	0.0940	0.0668
	x^B	0.0038	0.0007
	x^S	−0.0190	−0.0035
$x^B{}_t$	s	0.0934	0.6925
	Δa	−2.7348	−17.0895
	x^B	0.7042	0.6968
	x^S	0.1300	0.1265
$x^S{}_t$	s	0.1575	1.1918
	Δb	2.8268	18.0433
	x^B	0.1544	0.1267
	x^S	0.7422	0.7339
	Obs.	130,620	130,620
	R^2		
	Δa_t	0.0726	0.0826
	Δb_t	0.0604	0.0722
	$x^B{}_t$	0.4586	0.4375
	$x^S{}_t$	0.5123	0.4905
Residual correlation matrix			
$Cov(u_t^a,u_t^b)$		0.4362	0.4324
$Cov(u_t^B,u_t^S)$		−0.6804	−0.6038
$Cov(u_t^a,u_t^S)$		0.0000	0.0001
$Cov(u_t^b,u_t^B)$		0.0003	0.0002

This table summarizes the estimation of the VEC model,

$$\begin{pmatrix} 1 & 0 & A^*_{aB} & A^*_{aB} \\ 0 & 1 & A^*_{bB} & A^*_{bS} \\ 0 & 0 & 1 & 0 \\ 0 & 0 & 0 & 1 \end{pmatrix} \begin{pmatrix} \Delta a_t \\ \Delta b_t \\ x_t^B \\ x_t^S \end{pmatrix} = \begin{pmatrix} \gamma_a^{EC}(L) \\ \gamma_b^{EC}(L) \\ \gamma_B(L) \\ \gamma_S(L) \end{pmatrix} s_{t-1} + A(L) \begin{pmatrix} \Delta a_{t-1} \\ \Delta b_{t-1} \\ x_{t-1}^B \\ x_{t-1}^S \end{pmatrix} + \begin{pmatrix} u_t^a \\ u_t^b \\ u_t^B \\ u_t^S \end{pmatrix}$$

with the restrictions,

$$A(L) = \begin{pmatrix} A_{aa}(L) & 0 & A_{aB}(L) & A_{aS}(L) \\ 0 & A_{bb}(L) & A_{bB}(L) & A_{bS}(L) \\ A_{Ba}(L) & 0 & A_{BB}(L) & A_{BS}(L) \\ 0 & A_{Sb}(L) & A_{SB}(L) & A_{SS}(L) \end{pmatrix}$$

The model is defined in trade time and truncated at 5 lags. We use data for IBM from January to March 1996. The model is estimated by SURE. We report, for each variable, the sum of all lags whenever the coefficients are statistically significant at the 1% level. We also provide the R^2 for each equation in the system and information about the residual correlation matrix. Panel A uses trade-sign indicators to characterize the trading process. The trade-sign indictor x_t^B (x_t^S) equals 1 for buys (sells) and zero otherwise. For midpoint trades both variables equal zero. The error-correction term is the bid-ask spread; Δa_t (Δb_t) is the change in the ask (bid) quote between two consecutive trades. Panel B replaces the trade-sign indicators by the following trade-size indicators, and , where V_t is the trade size in number of shares

Figure 1 represents the impulse–response function (IRF) of the model in Panel A derived by dynamic simulation. These curves represent the responses of ask and bid prices after both a unitary buyer-initiated shock (increasing curves) and a unitary seller-initiated shock (decreasing curves). These trade-related shocks occur after a steady state period characterized by constant quotes, no trades, and a null bid–ask spread. The IRF measure the long-run impact of a particular trade-related shock on both quotes when the whole dynamic structure of the model is taken into account. Always on average terms, buys have a larger impact on the ask quote and sells have a larger impact on the bid quote. Statistical tests performed over the estimated VEC model corroborate that these differences are statistically significant. Briefly, quotes tend to be revised in the same direction but not by the same amount after a trade. This result suggests that ask (bid) quotes may lead the adjustment of the quoted prices after a buy (sell) shock.

Table 1 evidences two opposite and simultaneous effects associated with trade-related shocks on the time series dynamics ask and bid quotes. First, trade-related

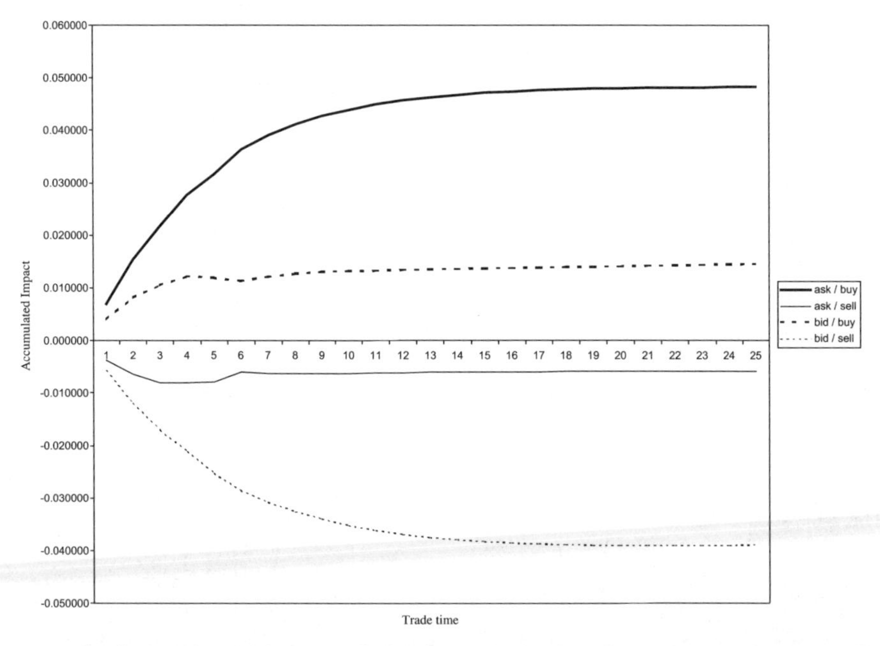

Fig. 1 Baseline VEC model for IBM: Impulse–Response Function after an unexpected unitary trade related shock. This figure display s the Impulse–Response Function (IRF) of ask and bid quotes to an unexpected unitary buyer-initiated shock (increasing paths) and seller-initiated shock (decreasing paths) according to the following VEC model, estimated using IBM data from January to March 1996,

$$\begin{pmatrix} 1 & 0 & A_{aB}* & A_{aB}* \\ 0 & 1 & A_{bB}* & A_{bS}* \\ 0 & 0 & 1 & 0 \\ 0 & 0 & 0 & 1 \end{pmatrix} \begin{pmatrix} \Delta a_t \\ \Delta b_t \\ x_t^B \\ x_t^S \end{pmatrix} = \begin{pmatrix} \gamma_a^{EC}(L) \\ \gamma_b^{EC}(L) \\ \gamma_B(L) \\ \gamma_S(L) \end{pmatrix} s_{t-1} + A(L) \begin{pmatrix} \Delta a_{t-1} \\ \Delta b_{t-1} \\ x_{t-1}^B \\ x_{t-1}^S \end{pmatrix} + \begin{pmatrix} u_t^a \\ u_t^b \\ u_t^B \\ u_t^S \end{pmatrix}$$ with the restrictions, $A(L) = \begin{pmatrix} A_{aa}(L) & 0 & A_{aB}(L) & A_{aS}(L) \\ 0 & A_{bb}(L) & A_{bB}(L) & A_{bS}(L) \\ A_{Ba}(L) & 0 & A_{BB}(L) & A_{BS}(L) \\ 0 & A_{Sb}(L) & A_{SB}(L) & A_{SS}(L) \end{pmatrix}$ The model is defined in trade time and truncated at 5 lags. We use data for IBM from January to March 1996. The initial shock is simulated after a steady state characterized by no trades, no changes in quotes, and a zero bid-ask spread

shocks induce positive cross-serial correlation as both quotes tend to be adjusted in the same direction. Second, a trade-related shock tends to increase the bid–ask spread as ask and bid quotes adjust asymmetrically. This effect sets in motion the error correction mechanism, which causes negative cross-serial correlation between quotes. The first one is an information-motivated effect. The second one is a liquidity-motivated effect. Our VEC model is able to identify and separate these two effects. A model that would summarize the quote dynamics through the quote midpoint, however, would confound them. Our model therefore supports Biais et al. (1995) intuition that there is additional information in analyzing the dynamics of ask and bid prices jointly rather than averaging them through the quote midpoint.

Finally, we use classical Wald tests (e.g., Davidson and MacKinnon 1993) on the coefficients of the two estimated VEC models for IBM to look for asymmetries in the average price impact of buys and sells. Namely, we compare the accumulated coefficients of x_t^B in the Δa_t equation with the accumulated coefficients of x_t^S in the Δb_t equation. In absolute terms, the impact of a buy shock on the ask price is larger than an equivalent impact of a sell shock on the bid price. In Panel A, $\widetilde{A}_{aB}(1) = 24.72$ vs. $\widetilde{A}_{bS}(1) = -18.95$ – estimated coefficients multiplied by 10^3. Similarly, the average response of the bid price to a buy shock, $\widetilde{A}_{bB}(1) = 3.822$, is statistically larger than the response of the ask price to a sell shock, $\widetilde{A}_{aS}(1) = -0.551$. The results for the model in Panel B are similar. Based on this test, we should conclude that on average, buyer-initiated shocks for IBM were more informative that similar seller-initiated shocks. Nonetheless, this test does not take into account the complete dynamics between trades and quotes captured in the VEC model (Eq. 3.7). In next sections, we will perform a more precise test based on IRFs to compare the information content of buys and sells.

The dynamics of the trading process are not the focus of this paper, but they show the patterns previously reported in other studies (e.g., Hasbrouck 1991). Particularly relevant is the strong positive autocorrelation in signed trades. Purchases are more likely followed by new purchases and sales are more likely followed by additional sales. Clusters of signed trades may be explained by traders successively reacting to new information, informed traders strategically splitting orders so as to ameliorate the price impact, imitative behavior among different traders, etc. Unfortunately, our model does not help in discerning the appropriate explanation.

Previous findings are unaltered when we control for intra-daily regularities by letting $f_i^j(MC_t, D_t)=1+\sum_{h\neq 4}\gamma_h^{i,j}\mathrm{D}_t^h$.

6 Robustness

In this section, we perform several robustness analyses to assess the degree of generality of the results obtained for IBM in the previous section. In the first subsection, we summarize the estimation of the baseline model for the remaining stocks in Appendix II. In the second subsection, we summarize the estimation of the unrestricted model Eq. (3.7). In the third subsection, we simulate the dynamics of the unrestricted model Eq. (3.7) to get a more precise understanding of the asymmetries evidenced in preceding subsections. Finally, in the fourth subsection we consider the effect of time aggregation.

6.1 Estimation of the baseline model for the complete sample

Table 2 summarizes the estimation of the baseline model for the three sets of stocks described in Appendix II. To be concise, we only report the results for the model with trade-size indicators. Table 2 contains the average coefficients across the 11 stocks in each subsample. In addition, it includes the number of stocks for which the aggregated coefficients for a given variable in the model are significant and positive/negative.[15]

In general, the results are highly consistent with those previously reported for IBM'96: (a) Ask and bid quotes error-correct, and (b) buys (sells) have a larger impact on the ask (bid) quote than sells (buys); thus, ask and bid quotes react asymmetrically to a trade-related shock. Table 2 corroborates the existence of the simultaneous information-related effects and liquidity-related effects associated trade-related shocks that were reported for IBM'96 in Section 5. Notice that these two effects coexist because of the asymmetric adjustments of bid and ask quotes after trades.

As with IBM'96, we have simulated unitary buyer and seller initiated shocks on the estimated VEC models to obtain the IRF of ask and bid quotes. Table 3 reports the main findings. Namely, we test the null of equality of the absolute IRFs of (a) ask and bid quotes after a unitary buyer-initiated shock; (b) ask and bid quotes after a unitary seller-initiated shock, and (c) ask quote after a unitary buyer-initiated shock and bid quote after a unitary seller-initiated shock. The two first nulls are strongly rejected. As previously shown in Table 2, buys (sells) have a larger impact on the ask (bid) quotes. The third hypothesis, however, is only accepted for the NYSE data, at the 5% for the 1996 sample and at the 10% level for the 2000 sample. Therefore, Table 3 shows that buys are more informative than sells in the NYSE, but the same is not true for the SSE.[16]

6.2 Estimation of the unrestricted VEC model

We proceed now with the estimation of the unrestricted VEC model Eq. (3.7). We test whether the asymmetries evidenced with the base-line model persist once we add additional structure to the model. We consider the model with trade-size indicators. Thus, the function $f_i^j(MC_t, D_t)$ includes trade durations (T_t), the dummy for regional market trades (M_t), the order imbalance (OI_t), and the short-term volatility (R_t). These exogenous variables interact with the trade-size indicators, introducing non-linear patterns in the impact of trades on ask and bid quotes. We maintain the restrictions in Eq. (5.1), now applied to the time-variant autoregressive matrix $A_t(L)$.

[15] More detailed results are available from the authors upon request.

[16] In order to gauge the importance of modelling the trading processes, we have performed the same exercise but considering only the dynamics captured by the quote equations. We find that, on average, for the NYSE'00 sample, the impact of a unitary buyer-initiated shock is underestimated by a 65%. Similarly, the impact of a seller-initiated shock is underestimated by a 59%. Similar percentages are found for the other subsamples. This evidences the relevance of considering the complete set of dynamical interactions and fee-backs between trades and quotes.

Estimated coefficients are not reported because of space limitations, but they are consistent with theoretical predictions. They are also regular across markets.[17] We have already shown in previous sections that a larger trade size increases the price impact of trades. In addition, a buy (sell) of any size executed in a high volatile period, as measured by R_t, has a larger impact on the ask (bid) quote. The trading activity in the regional markets is less informative than in the NYSE; both buys and sells have a lower impact on quotes when they are worked trough the regional venues. A positive order imbalance on the book, that is, more volume on the offer side than on the demand side, decreases (increases) the impact of an incoming buy (sell) on quotes. Finally, shorter durations increase the impact of buys (sells) on the ask (bid) quote, though this relationship is the weakest.

Next, we show that the asymmetries between ask and bid responses to trade related-shocks evidenced with the baseline model persist with this more complex specification. As in the previous subsection, we use the estimated coefficients of the unrestricted VEC model to simulate the impact of unitary trade-related shocks on ask and bid quotes. Also in this case, shocks occur after a steady state characterized by no trades, no changes in quotes, and a zero bid–ask spread. In this analysis, we are interested in the linear effect of a trade in quotes; hence, the exogenous variables are set equal to zero. We will investigate the consequences of altering the level of the exogenous variables in the next subsection.

Table 4 summarizes our findings. Compared with Tables 3, 4 not only corroborates the asymmetries observed with the baseline model, but it reinforces them since the statistical tests provide stronger support to the alternative hypothesis that NYSE buys are more informative than sells. This hypothesis is this time reject at the 1% level for the NYSE'96 subsample and at the 5% level for the NYSE'00 subsample. For the SSE, however, the null of equal informativeness of buys and sells still cannot be rejected.

6.3 A closer look to the asymmetry assumption

In this subsection, we obtain the responses of ask and bid quotes to trade-related shocks using model Eq. (3.7) when we let the level of the variables in MC_t to vary. The goal is to obtain additional insights on the asymmetries evidenced in previous subsections. We consider the model with the trade-size indicators $\widetilde{x}_t^B$ and $\widetilde{x}_t^S$.

We proceed as follows. As is previous simulation exercises, an unexpected trade happens after a steady state period with no prior trades, stable quotes, and zero spreads. For each exogenous variable, we compute the 25, 75, and 95% percentiles of its stock-specific empirical distribution. These values define three different levels of the variable: small (S), medium (M), and large (L) respectively. We assume that each variable in MC_t follows a general probabilistic process, exogenous to the VEC model Eq. (3.7), that we approximate by an AR(p) model.[18] This model is estimated

[17] These results are available upon request from the authors.

[18] For the regional dummy, we simply compare the impact of a regional trade with the impact of a NYSE trade. The auto-regressive order p is determined using likelihood-ratio tests, starting with p=7.

Table 2 The base-line VEC model for the entire sample

Equation	Variable	NYSE Sample 1996				NYSE Sample 2000				SSE Sample 2000			
		Coefficients average	Std.	(+)	(−)	Coefficients average	Std.	(+)	(−)	Coefficients average	Std.	(+)	(−)
Δa_t	s	−0.0627	0.0234	0	11	−0.0401	0.0201	0	11	−0.0795	0.0298	0	11
	Δa	−0.0639	0.0568	2	9	0.0298	0.0475	9	2	−0.1935	0.1255	0	10
	x^B	0.0038	0.0015	11	0	0.0034	0.0022	11	0	0.0011	0.0010	11	0
	x^S	−0.0004	0.0006	3	7	−0.0013	0.0012	0	11	−0.0004	0.0005	1	10
Δb_t	s	0.0541	0.0226	11	0	0.0438	0.0187	11	0	0.1016	0.0140	11	0
	Δb	−0.0558	0.0629	2	9	0.0183	0.0412	8	3	−0.1834	0.0669	0	10
	x^B	0.0002	0.0005	6	4	0.0006	0.0006	10	0	0.0004	0.0004	11	0
	x^S	−0.0040	0.0019	0	11	−0.0040	0.0027	0	11	−0.0012	0.0012	0	11
$x^B{}_t$	s	0.3611	1.2205	8	2	2.0670	1.1449	11	0	−2.6797	9.3107	4	7
	Δa	−34.9566	15.6377	0	11	−18.4364	13.8204	0	11	−79.9208	63.2079	0	11
	x^B	0.7402	0.0457	11	0	0.6917	0.0625	11	0	0.7300	0.0796	11	0
	x^S	0.1702	0.0297	11	0	0.2086	0.0410	11	0	0.2509	0.0574	11	0
$x^S{}_t$	s	1.2202	0.8637	10	1	1.8800	1.2852	11	0	15.5386	17.0832	10	1
	Δb	34.6731	16.7873	11	0	16.9249	14.1931	11	0	87.8538	70.3804	11	0
	x^B	0.1351	0.0429	11	0	0.1671	0.0360	11	0	0.2124	0.0511	11	0
	x^S	0.6935	0.0653	11	0	0.6325	0.0454	11	0	0.7058	0.0575	11	0

Table 2 (continued)

Equation	Variable	NYSE Sample 1996				NYSE Sample 2000				SSE Sample 2000			
		Coefficients average	Std.	(+)	(−)	Coefficients average	Std.	(+)	(−)	Coefficients average	Std.	(+)	(−)
R^2													
	Δa_t	0.0826	0.0294			0.0537	0.0260			0.1444	0.0489		
	Δb_t	0.0802	0.0299			0.0555	0.0263			0.1478	0.0493		
	x^B_t	0.5157	0.0658			0.4860	0.0547			0.4845	0.1212		
	x^S_t	0.4621	0.0798			0.3977	0.0298			0.5241	0.1128		

This table summarizes the estimation of the VEC model, $\begin{pmatrix} 1 & 0 & A_{aB}* & A_{aB}* \\ 0 & 1 & A_{bB}* & A_{bS}* \\ 0 & 0 & 1 & 0 \\ 0 & 0 & 0 & 1 \end{pmatrix} \begin{pmatrix} \Delta a_t \\ \Delta b_t \\ \widetilde{x}_t^B \\ \widetilde{x}_t^S \end{pmatrix} = \begin{pmatrix} \gamma_a^{EC}(L) \\ \gamma_b^{EC}(L) \\ \gamma_B(L) \\ \gamma_S(L) \end{pmatrix} s_{t-1} + A(L) \begin{pmatrix} \Delta a_{t-1} \\ \Delta b_{t-1} \\ \widetilde{x}_{t-1}^B \\ \widetilde{x}_{t-1}^S \end{pmatrix} + \begin{pmatrix} u_t^a \\ u_t^b \\ u_t^B \\ u_t^S \end{pmatrix}$ with the restrictions, $A(L) = \begin{pmatrix} A_{aa}(L) & 0 & A_{aB}(L) & A_{aS}(L) \\ 0 & A_{bb}(L) & A_{bB}(L) & A_{bS}(L) \\ A_{Ba}(L) & 0 & A_{BB}(L) & A_{BS}(L) \\ 0 & A_{Sb}(L) & A_{SB}(L) & A_{SS}(L) \end{pmatrix}$

The model is defined in trade time and truncated at 5 lags. We use data on 11 NYSE-listed stocks from January to March 1996, 11 NYSE-listed stocks from January to March 2000, and 11 SSE stocks from July to September 2000. The model is estimated by SURE. We report, for each sample and variable, the cross-sectional average of the sum of all lags whenever the coefficients are statistically significant at the 1% level. We also provide the cross-sectional average R^2 for each equation in the system. Finally, we include the number of stocks for which the coefficient of the corresponding variable is statistically positive/negative at the 1% level. The error-correction term is the bid-ask spread; Δa_t (Δb_t) is the change in the ask (bid) quote between two consecutive trades. The trade-size indicators are $\widetilde{x}_t^B = x_t^B \log(V_t)$ and $\widetilde{x}_t^S = x_t^S \log(V_t)$, where V_t is the trade size in number of shares, and x_t^B (x_t^S) equals 1 for buys (sells) and zero otherwise. For midpoint trades both variables equal zero

Table 3 Simulation of the base-line VEC model for the entire sample

Absolute IRF×100				
Buy/ask vs. buy/bid		Ask	Bid	Ask-bid t-test
NYSE 1996	Mean	0.9749	0.1042	0.8707*
	Std.	(0.3889)	(0.0964)	(0.3241)
NYSE 2000	Mean	1.0570	0.2129	0.8441*
	Std.	(0.8160)	(0.2195)	(0.6900)
SSE 2000	Mean	0.3643	0.1239	0.2404*
	Std.	(0.4103)	(0.1644)	(0.2518)
Total	Mean	0.7987	0.1470	0.6517*
	Std.	(0.6375)	(0.1695)	(0.5374)
Sell/ask vs. sell/bid		Bid	Ask	Bid-ask
NYSE 1996	Mean	0.8844	0.1024	0.7821*
	Std.	(0.3028)	(0.1102)	(0.2147)
NYSE 2000	Mean	1.0043	0.3447	0.6596*
	Std.	(0.7174)	(0.3187)	(0.4584)
SSE 2000	Mean	0.3021	0.1026	0.1994*
	Std.	(0.2695)	(0.1002)	(0.1705)
Total	Mean	0.7303	0.1832	0.5470*
	Std.	(0.5560)	(0.2283)	(0.3925)
Buy/ask vs. sell/bid		Buy/ask	Sell/bid	Difference
NYSE 1996	Mean	0.9749	0.8844	0.0905**
	Std.	(0.3889)	(0.3028)	(0.1585)
NYSE 2000	Mean	1.0570	1.0043	0.0528***
	Std.	(0.8160)	(0.7174)	(0.1193)
SSE 2000	Mean	0.3643	0.3021	0.0622
	Std.	(0.4103)	(0.2695)	(0.1719)
Total	Mean	0.8445	0.7935	0.0510**
	Std.	(0.6277)	(0.5513)	(0.1200)

This table reports statistical tests on the average impulse–response functions (IRFs) of the VEC model

$$\begin{pmatrix} 1 & 0 & A_{aB}* & A_{aB}* \\ 0 & 1 & A_{bB}* & A_{bS}* \\ 0 & 0 & 1 & 0 \\ 0 & 0 & 0 & 1 \end{pmatrix} \begin{pmatrix} \Delta a_t \\ \Delta b_t \\ \widetilde{x}_t^B \\ \widetilde{x}_t^S \end{pmatrix} = \begin{pmatrix} \gamma_a^{EC}(L) \\ \gamma_b^{EC}(L) \\ \gamma_B(L) \\ \gamma_S(L) \end{pmatrix} s_{t-1} + A(L) \begin{pmatrix} \Delta a_{t-1} \\ \Delta b_{t-1} \\ \widetilde{x}_{t-1}^B \\ \widetilde{x}_{t-1}^S \end{pmatrix} + \begin{pmatrix} u_t^a \\ u_t^b \\ u_t^B \\ u_t^S \end{pmatrix}$$

with the restrictions, $A(L) = \begin{pmatrix} A_{aa}(L) & 0 & A_{aB}(L) & A_{aS}(L) \\ 0 & A_{bb}(L) & A_{bB}(L) & A_{bS}(L) \\ A_{Ba}(L) & 0 & A_{BB}(L) & A_{BS}(L) \\ 0 & A_{Sb}(L) & A_{SB}(L) & A_{SS}(L) \end{pmatrix}$

We compare (a) the impact of a unitary buyer-initiated shock on the ask and bid quotes; (b) the impact of unitary seller-initiated shock on the ask and bid quotes, and (c) the impact of a unitary buyer-initiated shock on the ask quote with the impact of a unitary seller-initiated shock on the bid quote. We use data on 11 NYSE-listed stocks from January to March 1996, 11 NYSE-listed stocks from January to March 2000, and 11 SSE stocks from July to September 2000. We report statistical tests of the null of equality of the absolute IRFs against the alternative of a positive difference. The error-correction term is the bid-ask spread. Δa_t (Δb_t) is the change in the ask (bid) quote between two consecutive trades. The trade-size indicators are $\widetilde{x}_t^B = x_t^B \log(V_t)$ and $\widetilde{x}_t^S = x_t^S \log(V_t)$, where V_t is the trade size in number of shares, and x_t^B (x_t^S) equals 1 for buys (sells) and zero otherwise. For midpoint trades both variables equal zero

*Statistically greater than zero at the 1% level

**Statistically greater than zero at the 5% level

***Statistically greater than zero at the 10% level

by Generalized Least Squares (GLS), controlling for deterministic intraday patterns. The AR(p) models are used in the simulation exercise to generate the future values of each exogenous variable. Then, we compute the response of ask and bid quotes after a unitary trade-related shock (either buyer or seller-initiated) conditional on the level of one of the exogenous variables in MC_t while the others are kept equal to zero.

We compare the IRF after a unitary shock when all the exogenous variables are zero (Table 4) with the same IRF when the level of a given exogenous variable increases from zero to S, from zero to M, and from zero to L. Table 5 reports the relative change of the absolute value of the IRF 500 periods after the shock. We also provide the number of stocks in each subsample for which the impact is positive/negative and significantly different from zero.

Table 5 shows that asymmetries after regional trades are less important than after NYSE trades. As previously indicated, the positive (negative) impact of a buy (sell) shock on the ask (bid) quote is weaker when it comes from the regional markets. Moreover, bid (ask) quotes are not usually altered after an unexpected regional buy (sell). Therefore, after a regional trade ask and bid quotes do not move symmetrically either, but the asymmetry is less remarkable than after a NYSE trade.

Table 5 also evidences that the probability of observing asymmetric adjustments of ask and bid quotes decreases as volatility increases. For the NYSE'96 sample, the impact of a buyer-initiated shock on the ask quote is 3.41% larger when volatility is high (L). The impact on the bid quote, however, is 50.84% larger. Similarly, the impact of a seller-initiated shock on the bid quote is 5.38% larger in the more volatile scenario, but the impact on the ask quote is 42.44% larger. A similar finding is found for the SSE'00 sample. Therefore, the adjustments of ask and bid quotes after a trade-related shock are more balanced in periods of high volatility. This finding would suggest that trades executed during volatile periods transmit more unambiguous signals, since they cause ask and bid quotes to be adjusted symmetrically more often than usual.

The results for trade durations and order imbalances are not conclusive.

6.4 Time aggregation

So far, we have shown that asymmetries exist when ask and bid quotes measured in trade-time respond to trade-related shocks. In this section, we study whether these asymmetries persist when we consider different time scales. In particular, we aggregate our time series of quotes and trades into 1-min and 5-min intervals. Thus, a change in quotes in now given by the difference between the final and the initial quote in each time interval. Similarly, the buyer (seller) initiated volume is the sum of the size of all buys (sells) executed during each time interval. Finally, the bid–ask spread is given by the posted quotes at the end of each time interval. We estimate the baseline model Eq. (3.7) with the restrictions in Eq. (5.1) with t meaning either a 1-min or a 5-min interval. The model is truncated at three lags and estimated by SURE.

Table 6 provides our findings for the NYSE’00 sample. For the NYSE’96 and SSE’00 samples, results are similar and available upon request. Table 6 shows that the dynamics observed in trade-time remain in these alternative scales. Ask and bid quotes error-correct to deviations between them, causing the bid–ask spread to revert towards narrow levels. Ask (bid) quotes are also more sensible than bid quote to buys (sells) with aggregated data. In addition, buys are more informative than sells at the 5% level in the 1-min frequency and at the 1% level at the 5-min periodicity. This suggests that the asymmetry in ask and bid responses to trades is not just a high frequency phenomena.

7 Conclusions

This paper has introduced a new econometric approach to jointly model the time series dynamics of the trading process and the revisions of ask and bid prices. This model represents a generalization of the VAR model introduced by Hasbrouck (1991). We use this approach to check a very common theoretical assumption among microstructure models: the symmetry assumption. The symmetry assumption asserts that ask and bid quotes respond symmetrically to trades, that ask and bid quotes are posted symmetrically about the efficient price, and that buys and sells are equally informative.

Our model accommodates (not imposes) asymmetric responses of ask and bid prices to trade-related shocks. It also captures asymmetric impacts of buyer and seller-initiated trades. This is possible because it incorporates the co-integration relationship between the ask price and the bid price, because buys and sells are generated by idiosyncratic but mutually dependent processes, and because these trading processes are endogenous. The properties of the empirical model are derived directly from a structural dynamic model for ask and bid prices. The model is estimated using data from two different markets, the NYSE and the SSE.

Table 4 Simulation of the unrestricted VEC model

Absolute IRF×100				
Buy/ask vs. buy/bid		Ask	Bid	Ask-bid t-test
NYSE 1996	Mean	1.0279	0.1423	0.8855*
	Std.	(0.3546)	(0.1247)	(0.2854)
NYSE 2000	Mean	1.1401	0.4990	0.6411*
	Std.	(0.8348)	(0.3706)	(0.5025)
SSE 2000	Mean	0.2843	0.0549	0.2294*
	Std.	(0.2840)	(0.0832)	(0.2350)
Total	Mean	0.8174	0.2321	0.5853*
	Std.	(0.6565)	(0.2966)	(0.4441)
Sell/ask vs. sell/bid		Bid	Ask	Bid-ask
NYSE 1996	Mean	0.9170	0.1741	1.0911*
	Std.	(0.3180)	(0.1302)	(0.4404)
NYSE 2000	Mean	0.9945	0.5595	1.5540*
	Std.	(0.6248)	(0.4194)	(1.0304)
SSE 2000	Mean	0.2412	0.0518	0.2930*
	Std.	(0.2087)	(0.0573)	(0.2575)
Total	Mean	0.7176	0.2618	0.9794*
	Std.	(0.5341)	(0.3310)	(0.8324)

Table 4 (continued)

Absolute IRF×100				
Buy/Ask vs. Sell/Bid		Buy/Ask	Sell/Bid	Difference
NYSE 1996	Mean	1.0279	0.9170	0.1108*
	Std.	(0.3546)	(0.3180)	(0.1703)
NYSE 2000	Mean	1.1401	0.9945	0.1456**
	Std.	(0.8348)	(0.6248)	(0.2357)
SSE 2000	Mean	0.2843	0.2412	0.0431
	Std.	(0.2840)	(0.2087)	(0.1126)
Total	Mean	0.8174	0.7176	0.0998*
	Std.	(0.6565)	(0.5341)	(0.1796)

This table reports statistical tests on the impulse–response functions (IRFs) of the VEC model,

$$\begin{pmatrix} 1 & 0 & A_{aB,t*} & A_{aB,t*} \\ 0 & 1 & A_{bB,t*} & A_{bS,t*} \\ 0 & 0 & 1 & 0 \\ 0 & 0 & 0 & 1 \end{pmatrix} \begin{pmatrix} \Delta a_t \\ \Delta b_t \\ \widetilde{x}_t^B \\ \widetilde{x}_t^S \end{pmatrix} = \begin{pmatrix} \gamma_a^{EC}(L) \\ \gamma_b^{EC}(L) \\ \gamma_B(L) \\ \gamma_S(L) \end{pmatrix} s_{t-1} + A_t(L) \begin{pmatrix} \Delta a_{t-1} \\ \Delta b_{t-1} \\ \widetilde{x}_{t-1}^B \\ \widetilde{x}_{t-1}^S \end{pmatrix} + \begin{pmatrix} u_t^a \\ u_t^b \\ u_t^B \\ u_t^S \end{pmatrix} \text{ with}$$

the restrictions, $A_t(L) = \begin{pmatrix} A_{aa}(L) & 0 & A_{aB,t}(L) & A_{aS,t}(L) \\ 0 & A_{bb}(L) & A_{bB,t}(L) & A_{bS,t}(L) \\ A_{Ba}(L) & 0 & A_{BB,t}(L) & A_{BS,t}(L) \\ 0 & A_{Sb}(L) & A_{SB,t}(L) & A_{SS,t}(L) \end{pmatrix}$

We compare (a) the impact of a unitary buyer-initiated shock on the ask and bid quotes; (b) the impact of unitary seller-initiated shock on the ask and bid quotes, and (c) the impact of a unitary buyer-initiated shock on the ask quote with the impact of a unitary seller-initiated shock on the bid quote. We use data on 11 NYSE-listed stocks from January to March 1996, 11 NYSE-listed stocks from January to March 2000, and 11 SSE stocks from July to September 2000. We report statistical tests of the null of equality of the absolute IRFs against the alternative of a positive difference. We keep all the polynomials in the autoregressive matrix constant during the simulation, even when they are time-variant due to exogenous variables. The error-correction term is the bid-ask spread; Δa_t (Δb_t) is the change in the ask (bid) quote between two consecutive trades; $\widetilde{x}_t^B = x_t^B \log(V_t)$, and $\widetilde{x}_t^S = x_t^S \log(V_t)$, where V_t is the trade size in number of shares, and x_t^B (x_t^S) equals 1 for buys (sells) and zero otherwise. For midpoint trades both variables equal zero

*Statistically greater than zero at the 1% level
**Statistically greater than zero at the 5% level
***Statistically greater than zero at the 10% level

The dynamics of ask and bid prices are characterized by two findings, robust across markets, trading periods, and model specifications. First, we show that these quotes do not follow symmetric patterns after trades. Ask and bid prices tend to be revised in the same direction but not by the same amount. Ask (bid) quotes are more sensible to buyer (seller) initiated shocks than bid (ask) quotes. We evidence, however, that the likelihood of symmetric responses increases with volatility. In addition, we show that asymmetries are less frequent in the NYSE after regional trades. In addition to the former information-motivated trade-related effect, we also observe a liquidity-motivated trade-related effect. Ask and bid quotes error correct to mutual deviations, which causes a strong mean reversion in the bid–ask spread. These two findings produce simultaneous but opposite effects in the dynamics of ask and bid prices: information-induced positive cross-serial correlation and liquidity-induced negative cross-serial correlation.

Table 5 Impulse–response functions conditional on the exogenous variables

Variable	Ask										Bid									
	Buy shock					Sell shock					Buy shock					Sell shock				
	Δabs(IRF) (%)			Signif. (1%)		Δabs(IRF) (%)			Signif. (1%)		Δabs(IRF) (%)			Signif. (1%)		Δabs(IRF) (%)			Signif. (1%)	
NYSE 1996	S	M	L	I	D	S	M	L	I	D	S	M	L	I	D	S	M	L	I	D
T_t	−0.39	−1.71	−4.36	2	9	0.43	2.00	5.74	2	9	0.61	3.17	8.37	8	2	−0.29	−1.40	−3.61	1	10
R_t	0.18	1.35	3.41	10	1	3.98	21.88	50.84	8	2	2.53	16.98	42.44	9	1	0.29	1.97	5.38	9	2
OI_t	1.40	−2.53	−10.06	0	11	1.99	0.95	8.05	7	3	2.19	−0.80	−3.35	3	7	−1.29	2.31	9.95	11	0
M_t		−52.75		0	11		−93.26		0	3		−96.95		0	2		−49.66		0	11
NYSE 2000																				
T_t	−0.75	−2.10	−4.49	2	8	−0.47	−1.28	−2.96	8	2	−1.14	−3.18	−7.21	1	9	0.25	1.05	2.51	8	3
R_t	0.61	1.83	3.96	10	0	0.42	1.21	2.64	6	5	0.77	2.14	5.19	6	4	0.63	1.97	4.36	9	2
OI_t	0.47	−1.22	−5.06	1	10	−0.68	1.81	7.61	10	1	0.82	−2.19	−8.55	0	11	0.94	3.94	4.36	9	2
M_t		−47.77		0	11		−92.11		0	2		−96.01		0	1		−56.47		0	11
SSE 2000																				
T_t	−0.17	−1.16	−3.10	3	7	−3.75	−16.54	−5.76	1	8	0.45	2.67	7.98	7	3	−0.08	0.93	12.95	4	5
R_t	0.60	4.42	14.88	11	0	1.01	7.51	32.75	8	2	2.23	16.11	53.31	10	0	0.47	3.60	13.31	11	0
OI_t	0.65	−0.78	−3.49	0	11	−0.86	1.05	4.49	9	1	3.23	−3.42	−13.66	0	10	−0.28	0.33	1.41	9	2
M_t		–					–					–					–			

Table 5 (continued)

Variable	Ask										Bid									
	Buy shock					Sell shock					Buy shock					Sell shock				
	Δabs(IRF) (%)			Signif. (1%)		Δabs(IRF) (%)			Signif. (1%)		Δabs(IRF) (%)			Signif. (1%)		Δabs(IRF) (%)			Signif. (1%)	
NYSE 1996	S	M	L	I	D	S	M	L	I	D	S	M	L	I	D	S	M	L	I	D
All stocks																				
T_t	−0.44	−1.66	−4.00	7	24	−1.15	−4.76	−0.90	11	19	−0.03	0.89	3.05	16	14	−0.04	0.15	3.37	13	18
R_t	0.46	2.56	7.53	31	1	1.65	9.31	26.21	22	9	1.84	11.74	33.64	25	5	0.46	2.51	7.69	29	4
OI_t	0.84	−1.51	−6.20	1	32	0.12	1.29	6.75	26	5	3.80	−1.04	−4.68	3	28	−0.65	1.19	5.10	29	4
M_t		−49.80		0	22		−92.80		0	5		−96.64		0	3		−54.12		0	22

This table reports the result of simulating a unitary trade–related shock using the VEC model, $\begin{pmatrix} 1 & 0 & A_{aB,t}* & A_{aB,t}* \\ 0 & 1 & A_{bB,t}* & A_{bS,t}* \\ 0 & 0 & 1 & 0 \\ 0 & 0 & 0 & 1 \end{pmatrix} \begin{pmatrix} \Delta a_t \\ \Delta b_t \\ x_t^B \\ x_t^S \end{pmatrix} = \begin{pmatrix} \gamma_a^{EC}(L) \\ \gamma_b^{EC}(L) \\ \gamma_B(L) \\ \gamma_S(L) \end{pmatrix} s_{t-1} + A_t(L)$ $\begin{pmatrix} \Delta a_{t-1} \\ \Delta b_{t-1} \\ \widetilde{x}_{t-1}^B \\ \widetilde{x}_{t-1}^S \end{pmatrix} + \begin{pmatrix} u_t^a \\ u_t^b \\ u_t^B \\ u_t^S \end{pmatrix}$ with the restrictions, $A_t(L) = \begin{pmatrix} A_{aa}(L) & 0 & A_{aB,t}(L) & A_{aS,t}(L) \\ 0 & A_{bb}(L) & A_{bB,t}(L) & A_{bS,t}(L) \\ A_{Ba}(L) & 0 & A_{BB,t}(L) & A_{BS,t}(L) \\ 0 & A_{Sb}(L) & A_{SB,t}(L) & A_{SS,t}(L) \end{pmatrix}$ when we alter the level of the exogenous variables in the autoregressive matrix. The exogenous variables are: trade durations (T_t), short-term volatility (R_t), order imbalance (OI_t), and a dummy variable that identifies regional market trades (M_t). The error-correction term is the bid-ask spread. The endogenous variables are the change in the ask quote (Δa_t) and the bid quote (Δb_t), and trade-size indicators for buys $\left(\widetilde{x}_t^B\right)$ and for sells $\left(\widetilde{x}_t^S\right)$

We use data on 11 NYSE-listed stocks from January to March 1996, 11 NYSE-listed stocks from January to March 2000, and 11 SSE stocks from July to September 2000. For each stock, we compute the response of ask and bid prices to both a buyer-initiated shock and a seller-initiated shock, 500 periods ahead. The impact depends on the level of several exogenous variables that feature the trade and the market conditions. Three levels of each variable are considered: "small" (S), "medium" (M), and "large" (L), obtained from the 25%, 75%, and 95% percentiles of the empirical distribution. For the M_t dummy, we compare the impact of a NYSE trade with the impact of a regional trade. This table reports the average relative change (in %) across stocks in the total impact of a unitary trade-related shock when the exogenous variable increases from zero to S, from zero to M, and from zero to L. We also provide the number of stocks in each subsample for which the absolute IRF significantly increases/decreases as we increase the level of the corresponding exogenous variable

We also show that NYSE buyer-initiated trades are more informative that seller-initiated trades, both in 1996 and in 2000. This pattern persists even when we consider different model specifications, including different time periodicities. This finding, however, cannot be generalized to the SSE case.

This paper has also evidenced that market frictions like the minimum price variation are not enough to explain the violation of the symmetry assumption. We find similar asymmetric patterns in ask and bid responses in 1996, with a US$1/8 tick, and in 2000, with a US$0.01 tick. In addition, asymmetries are found in markets with very different microstructures, like the NYSE and the SSE. Since the specialist contribution is less essential in frequently traded NYSE stocks (Madhavan and Sofianos 1998), it sounds interesting to extend the analysis in this paper by considering a larger sample of NYSE and SSE stocks, stratified by trading frequency. This analysis would clarify the role that market makers play in explaining the asymmetric dynamics between ask and bid quotes. Moreover, our findings suggest that ask (bid) quotes may lead the price discovery process after buyer (seller) initiated trades. An interesting topic for future research would be evaluating the relative contribution of ask and bid quotes to price discovery, conditional on variables like recent market trends, accumulated net volume, order imbalances, and so on. Finally, the intriguing finding that buys are more informative

Table 6 The base-line VEC model for aggregated data

Equation	Variable	Panel A: 1-minute intervals				Panel B: 5-minute intervals			
		Coefficients average	Std.	(+)	(−)	Coefficients average	Std.	(+)	(−)
	Δb_t	−0.2231	0.0752	0	11	−0.3290	0.1177	0	11
	Δa	0.0633	0.0556	10	1	−0.0297	0.0473	10	1
	x^B	0.0172	0.0089	11	0	0.1316	0.0886	11	0
	x^S	−0.0133	0.0080	0	11	−0.1135	0.0746	0	11
	Δb_t	0.2608	0.0710	11	0	0.4421	0.0870	11	0
	Δb	0.0593	0.0696	9	2	−0.0326	0.0529	9	2
	x^B	0.0121	0.0074	11	0	0.1240	0.0846	11	0
	x^S	−0.0173	0.0089	0	11	−0.1241	0.0819	0	11
$x^B{}_t$	s	2.1055	2.3079	10	1	0.0361	0.4750	10	1
	Δa	−1.2400	1.2389	0	11	0.0155	0.3778	0	11
	x B	0.6748	0.1089	11	0	0.6472	0.1317	11	0
	x S	0.2963	0.0879	11	0	0.3579	0.1307	11	0
$x^S{}_t$	s	1.9812	2.0402	10	1	0.1100	0.6099	10	1
	Δb	0.4601	1.8280	4	7	0.0033	0.1560	4	7
	x^B	0.3985	0.1223	11	0	0.4275	0.1433	11	0
	x^S	0.5396	0.1476	11	0	0.5601	0.1496	11	0
	Obs.	23,484	186			4,688	62		
R_2									
	Δa_t	0.1621	0.0400			0.1913	0.0725		
	Δb_t	0.1847	0.0387			0.2030	0.0730		
	$x^B{}_t$	0.9532	0.0222			0.9945	0.0015		
	$x^S{}_t$	0.9328	0.0382			0.9938	0.0015		

Table 6 (continued)

Absolute IRF×100				
Buy/ask – buy/bid	Diference	p-value t-test	Diference	p-value t-test
Mean	1.7480	0.0017*	2.3780	0.0111**
Std.	(1.5230)		(2.9193)	
Sell/bid – sell/ask				
Mean	6.9870	0.0007*	49.1156	0.0024*
Std.	(2.7803)		(25.8798)	
Buy/ask – sell/bid				
Mean	1.9127	0.0105**	11.0851	0.0099*
Std.	(2.3180)		(13.2844)	

This table summarizes the estimation of the VEC model,
$$\begin{pmatrix} 1 & 0 & A^*_{aB} & A^*_{aB} \\ 0 & 1 & A^*_{bB} & A^*_{bS} \\ 0 & 0 & 1 & 0 \\ 0 & 0 & 0 & 1 \end{pmatrix} \begin{pmatrix} \Delta a_t \\ \Delta b_t \\ \tilde{x}^B_t \\ \tilde{x}^S_t \end{pmatrix} = \begin{pmatrix} \gamma^{EC}_a(L) \\ \gamma^{EC}_b(L) \\ \gamma_B(L) \\ \gamma_S(L) \end{pmatrix} s_{t-1} + A(L) \begin{pmatrix} \Delta a_{t-1} \\ \Delta b_{t-1} \\ \tilde{x}^B_{t-1} \\ \tilde{x}^S_{t-1} \end{pmatrix} + \begin{pmatrix} u^a_t \\ u^b_t \\ u^B_t \\ u^S_t \end{pmatrix}$$
with the restrictions, $A(L) = \begin{pmatrix} A_{aa}(L) & 0 & A_{aB}(L) & A_{aS}(L) \\ 0 & A_{bb}(L) & A_{bB}(L) & A_{bS}(L) \\ A_{Ba}(L) & 0 & A_{BB}(L) & A_{BS}(L) \\ 0 & A_{Sb}(L) & A_{SB}(L) & A_{SS}(L) \end{pmatrix}$
The model is defined in 1-minute (Panel A) and 5-minute intervals (Panel B), and truncated at 3 lags. We use data on 11 NYSE-listed stocks from January to March 2000. The model is estimated by SURE. We report for each variable the cross-sectional average of the sum of all lags whenever the coefficients are statistically significant at the 1% level. We also provide the cross-sectional average R^2 for each equation in the system. We include the number of stocks for which the coefficient of the corresponding variable is statistically positive/negative at the 1% level. Finally, we compare (a) the impact of a unitary buyer-initiated shock on the ask and bid quotes; (b) the impact of unitary seller-initiated shock on the ask and bid quotes, and (c) the impact of a unitary buyer-initiated shock on the ask quote with the impact of a unitary seller-initiated shock on the bid quote. We report the differences in the absolute impulse-response functions (IRF) and the result of a t-test on the null of equal IRFs against the alternative of a strictly positive difference. The error-correction term is the bid-ask spread. The endogenous variables are the change in the ask quote (Δa_t) and the bid quote (Δb_t) in each time interval, and trade-size indicators for buys $(\tilde{x}^B_t)$ and for sells $(\tilde{x}^S_t)$ are computed as the accumulated volume of buyer-initiated trades and seller-initiated trades, respectively, in each time interval

than sells in the NYSE but not in the SSE suggests that microstructure differences may be playing a role. This is a possibility that deserves a more exhaustive analysis.

1 Appendix I

1.1 Derivation of the VEC model (3.7)

From Eq. (3.2)

$$\left[1 - \alpha^a_m L\right](a_t - m_t) = A_{x,t}(L)' x_t + \alpha^{EC}_a (a_{t-1} - b_{t-1}) + \varepsilon^a_t.$$

As $0 < \alpha_m^a < 1, \alpha(L) = \left[1 - \alpha_m^a L\right]$ is a stationary polynomial in L. Then,

$$(a_t - m_t) = \alpha(L)^{-1} A_{x,t}(L)' x_t + \alpha(L)^{-1} \alpha_a^{EC} s_{t-1} + \alpha(L)^{-1} \varepsilon_t^a. \quad \text{(A.1)}$$

Let $\Delta=(1-L)$ be the first differencing operator. Pre-multiplying in (A.1) by Δ, and letting $\alpha(L)^{-1} A_{x,t}(L)\Delta = \widetilde{A}_{x,t}(L)$ and $\alpha(L)^{-1}\alpha_a^{EC}\Delta = \widetilde{\alpha}_a^{EC}(L)$ we obtain

$$\Delta a_t = \Delta m_t + \widetilde{A}_{x,t}(L)' x_t + \widetilde{\alpha}_a^{EC}(L) s_{t-1} + \theta(L)\varepsilon_t^a, \quad \text{(A.2)}$$

where $\theta(L)\varepsilon_t^a=(1-L)(1-\alpha_m^a L)\varepsilon_t^a$ which can be approximated by a moving average polynomial of finite order, say q, $\theta(L)\varepsilon_t^a \approx \widetilde{\theta}(L)\varepsilon_t^a = \left(1 - \widetilde{\theta}_a^1 L - \widetilde{\theta}_a^2 L^2 - \cdots - \widetilde{\theta}_a^q L^q\right)\varepsilon_t^a$. Similar expansions are made with $\widetilde{A}_{x,t}(L)$ and $\widetilde{\alpha}_a^{EC}(L)$. Substituting Eq. (3.1) in (A.2) we have

$$\Delta a_t = \widetilde{A}_{x,t}(L)' x_t + \widetilde{\alpha}_a^{EC}(L) s_{t-1} + \xi_t^a. \quad \text{(A.3)}$$

The error term $\xi_t^a = \widetilde{\theta}(L)\varepsilon_t^a + \lambda^B v_{2,t}^B + \lambda^S v_{2,t}^S + v_{1,t}$ has an invertible moving average (MA) representation. Inverting the MA or alternatively adding long-enough dynamics of the regressors of (A.3), Δa_t and also Δb_t (since they are highly correlated), the moving average structure disappears. Therefore, Eq. (A.3) could parsimoniously be approximated by

$$\Delta a_t = \widetilde{\alpha}_a^{EC}(L) s_{t-1} + A_{aa}(L)\Delta a_{t-1} + A_{ab}(L)\Delta b_{t-1} + \widetilde{A}_{x,t}(L)' x_t + u_t^a. \quad \text{(A.4)}$$

The errors are white noise, $E(u_t^a)=0$ and $E(u_t^a, u_{t-k}^a)=0\,\forall k\neq 0$, with the autoregressive polynomials $A_{ij}(L)$ having all roots outside the unit circle. Let,

$$\widetilde{A}_{x,t}(L)' x_t = A_{aB}^B(L) f_{aB}^B(MC_{t-1}, D_{t-1}) x_t^B + A_{aS}^S(L) f_{aS}^S(MC_{t-1}, D_{t-1}) x_t^S.$$

Equation (A.4) can now be written as

$$\Delta a_t = \widetilde{\alpha}_a^{EC}(L) s_{t-1} + A_{aa}(L)\Delta a_{t-1} + A_{ab}(L)\Delta b_{t-1} + A_{aB,t}(L) x_t^B + A_{aS,t}(L) x_t^S + u_t^a, \quad \text{(A.5)}$$

which is the first equation of the system Eq. (3.7).

The corresponding equation for Δb_t is similarly obtained by repeating the previous steps for Eq. (3.3) obtaining the equivalent expression of Eq. (A.3) for b_t

$$\Delta b_t = \widetilde{B}_{x,t}(L)' x_t + \widetilde{\alpha}_b^{EC}(L) s_{t-1} + \xi_t^b. \quad \text{(A.6)}$$

Notice that ξ_t^a and ξ_t^b have a component in common $(\lambda^B v_{2,t}^B + \lambda^S v_{2,t}^S + v_{1,t})$ and, therefore, they are mutually correlated. This correlation depends on the importance of the idiosyncratic components in each of the residuals. From the same arguments, we can obtain the equivalent model to (A.5) for Δb_t with white noise errors

$$\Delta b_t = \widetilde{\alpha}_b^{EC}(L) s_{t-1} + A_{ba}(L)\Delta a_{t-1} + A_{bb}(L)\Delta b_{t-1} + A_{bB,t}(L) x_t^B + A_{bS,t}(L) x_t^S + u_t^b. \quad \text{(A.7)}$$

As the errors u_t^a and u_t^b are mutually correlated and therefore efficient estimation requires at least a joint estimation of (A.5) and (A.7).

From Eq. (3.4) using (A.1)–(A.2) we obtain,

$$\begin{aligned} x_t^B &= \mu^B \widetilde{B}_{x,t}(L)' x_{t-1} + (\mu^B \widetilde{\alpha}_a^{EC}(L)L + \pi^B)s_{t-1} + \alpha(L)^{-1}\varepsilon_t^a + v_{2,t}^B = \\ &= \varphi_{x,t}^B(L)' x_{t-1} + \varphi_s^B(L)s_{t-1} + \xi_t^B \end{aligned} \tag{A.8}$$

where the error term $\xi_t^B = \alpha(L)^{-1}\varepsilon_t^a + v_{2,t}^B$ has an invertible moving average (MA) representation. As previously done, this moving average structure can be approximated by,

$$\begin{aligned} x_t^B &= \alpha_B^{EC}(L)s_t + A_{Ba}(L)\Delta a_{t-1} + A_{Bb}(L)\Delta b_{t-1} + A_{BB,t}(L)x_{t-1}^B \\ &+ A_{BS,t}(L)x_{t-1}^S + u_t^B \end{aligned} \tag{A.9}$$

where $E(u_t^B)=0$ and $E(u_t^B, u_{t-k}^B)=0 \forall k \neq 0$.

The corresponding equation for x_t^s is similarly obtained by repeating the previous steps with Eq. (3.5). We first obtain,

$$x_t^s = \varphi_{x,t}^S(L)' x_{t-1} + \varphi_s^S(L)s_{t-1} + \xi_t^S \tag{A.10}$$

where the error term $\xi_t^S = \alpha(L)^{-1}\varepsilon_t^b + v_{2,t}^S$. Following the argument stated right after Eq. (A.8), we get the last equation of the system (Eq. 3.7),

$$\begin{aligned} x_t^S &= \alpha_S^{EC}(L)s_{t-1} + A_{Sa}(L)\Delta a_{t-1} + A_{Sb}(L)\Delta b_{t-1} + A_{SB,t}(L)x_{t-1}^B \\ &+ A_{SS,t}(L)x_{t-1}^S + u_t^S \end{aligned} \tag{A.11}$$

Notice that Eqs. (A.9) and (A.11) have correlated errors if either $v_{2,t}^S$ and $v_{2,t}^B$ or ε_t^b and ε_t^a are correlated, which is a very likely event.

1 Appendix II

1.1 Sample

NYSE 1996 Stocks	Company	Observations (number of trades)
GE	General Electric Co	106,347
GT	Goodyear Tire Rubber Co	86,802
IBM	Int Business Machines Corp	130,620
JNJ	Jhonson & Johnson	64,607
KO	Coca-Cola Co	72,620
MO	Phillip Morris Companies Inc	91,938
MRK	Merck & Co Inc	96,425
PG	Procter & Gamble Co	52,326
T	ATT Corp	87,882
TX	Texaco Inc	76,912
WMT	Wal-Mart Stores Inc	102,660

NYSE 2000 Stocks		
AOL	America Online Inc	626,768
C	Citigroup Inc	260,149
EMC	EMC Corporation	282,196
GE	General Electric Co	350,795
IBM	Int Business Machines Corp	234,766
LU	Lucent Technologies Inc	705,948
MOT	Motorola Inc	195,067
NOK	Nokia Corp	221,005
NT	Nortel Networks Corp	239,094
PFE	Pfizer Inc	233,658
T	ATT Corp	236,662
SSE 2000 Stocks		
AMS	Amadeus Global Travel Distribution	49,824
BBVA	Banco Bilbao-Vizcaya Argentaria	123,687
ELE	Endesa	80,844
IBE	Iberdrola	35,811
REP	Repsol YPF	90,213
SCH	Banco Santander Central Hispano	196,880
TEF	Telefónica	424,327
TPI	Telefónica Publicidad e Información	52,253
TPZ	Telepizza	36,880
TRR	Terra Networks	146,285
ZEL	Zeltia	88,339

References

Arranz MA, Escribano A (2000) Cointegration testing under permanent breaks: a robust extended error correction model. Oxf Bull Econ Stat 62:23–52

Barclay MJ, Warner JB (1993) Stealth trading and volatility, which trades move prices? J Financ Econ 34:281–305

Bessembinder H, Kaufman HM (1997) A comparison of trade execution costs for NYSE and Nasdaq-listed stocks. J Financ Econ 46:293–319

Biais B, Hillion P, Spatt C (1995) An empirical analysis of the limit order book and the order flow in the Paris Bourse. J Finance 50:1655–1689

Blume ME, Goldstein MA (1997) Quotes, order flow, and price discovery. J Finance 52:221–244

Bollerslev T, Melvin M (1994) Bid–ask spreads and volatility in the foreign exchange market. An empirical analysis. J Int Econ 36:355–372

Chan LKC, Lakonishok J (1993) Institutional trades and intraday stock price behavior. J Financ Econ 33:173–199

Chan LKC, Lakonishok J (1995) The behavior of stock prices around institutional trades. J Finance 50(4):1147–1174

Chung KH, Van Ness BF, Van Ness RA (1999) Limit orders and the bid–ask spread. J Financ Econ 53:255–287

Davidson R, MacKinnon JG (1993) Estimation and Inference in Econometrics. Oxford University Press

Dufour A, Engle RF (2000) Time and the price impact of a trade. J Finance 55(6):2467–2498

Easley D, O'Hara M (1987) Price, trade size, and information in securities markets. J Financ Econ 19:69–90

Easley D, O'Hara M (1992) Time and the process of security price adjustment. J Finance 47 (2):577–605

Easley D, Kiefer NM, O'Hara M (1997) One day in the life of a very common stock. Rev Financ Stud 10:805–835

Ellis K, Michaely R, O'Hara M (2000) The accuracy of trade classification rules: evidence from Nasdaq. J Financ Quant Anal 35:529–551

Engle R, Granger C (1987) Co-integration and error correction: representation, estimation and testing. Econometrica 35:251–276

Engle RF, Patton AJ (2004) Impacts of trades in a error-correction model of quote prices. J Financ Mark 7:1–25

Engle RF, Yoo BS (1991) Cointegrated economic time series: a survey with new results. In: Granger CWJ, Engle RF (eds) Long-run economic relations. Readings in cointegration, Oxford University Press, pp 237–266

Escribano A, Granger CWJ (1998) Investigating the relationships between gold and silver prices. J Forecast (17):81–107

Escribano A, Peña D (1994) Cointegration and common factors. J Time Ser Anal 15:577–586

Glosten LR (1987) Components of the bid–ask spread and the statistical properties of transaction prices. J Finance 42:1293–1307

Glosten LR, Harris LE (1988) Estimating the components of the bid/ask spread. J Financ Econ 21:123–142

Glosten LR, Milgrom PR (1985) Bid, ask and transaction prices in specialist market with heterogeneously informed traders. J Financ Econ 14:71–100

Goldstein MA, Kavajecz KA (2004) Trading strategies during circuit breakers and extreme market movements. J Financ Mark 7:301–333

Green WH (1997) Econometric analysis. Prentice-Hall, Upper Saddle River, NJ

Griffiths MD, Smith BF, Alasdair D, Turnbull S, White RW (2000) The costs and determinants of order aggressiveness. J Financ Econ 56:65–88

Harris M, Raviv A (1993) Differences of opinion make a horse race. Rev Financ Stud 6:473–506

Harris FH, McInish TH, Shoesmith GL, Wood RA (1995) Co-integration, error correction, and price discovery on internationally linked security markets. J Financ Quant Anal 30:563–579

Hasbrouck J (1988) Trades, quotes and information. J Financ Econ 22:229–252

Hasbrouck J (1991) Measuring the information content of stock trades. J Finance 46:179–207

Hasbrouck J (1995) One security, many markets: determining the contributions to price discovery. J Finance 50(4):1175–1199

Hasbrouck J (1996) Modeling market microstructure time series. In: Maddala GS, Rao CR (eds) Handbook of statistics, vol 14. Statistical methods in finance. Elsevier, North-Holland, Amsterdam

Hasbrouck J (1999a) Security bid/ask dynamics with discreteness and clustering: simple strategies for modeling and estimation. J Financ Mark 2:1–28

Hasbrouck J (1999b) The dynamics of discrete bid and ask quotes. J Finance 54(6):2109–2142

Hasbrouck J, Sofianos G, Sosebee D (1993) New York Stock Exchange systems and trading procedures, NYSE Working Paper #93–01

Hausman JA, Lo AW, MacKinlay AC (1992) An ordered probit analysis of transaction costs prices. J Financ Econ 31:319–379

Holthausen RW, Leftwich RW, Mayers D (1987) The effect of large block transactions on security prices. J Financ Econ 19:237–267

Huang RD, Stoll HR (1994) Market microstructure and stock return predictions. Rev Financ Stud 7(1):179–213

Huang RD, Stoll HR (1996) Dealer versus auction markets: a paired comparison of execution costs on NASDAQ and the NYSE. J Financ Econ 41:313–357

Huang RD, Stoll HR (1997) The components of the bid–ask spread: a general approach. Rev Financ Stud 10:995–1034

Jang H, Venkatesh PC (1991) Consistency between predicted and actual bid–ask quote-revisions. J Finance 46:433–446

Johansen S (1991) Estimation and hypothesis testing of co-integration vectors in Gaussian vector autoregressive models. Econometrica 59:1551–1580

Kavajecz KA (1999) A specialist's quoted depth and the limit order book. J Finance 54:747–771

Keim DB, Madhavan A (1995) Anatomy of the trading process: empirical evidence on the behavior of institutional traders. J Financ Econ 37:371–398

Kempf A, Korn O (1999) Market depth and order size. J Financ Mark 2:29–48

Koski JL, Michaely R (2000) Prices, liquidity, and the information content of trades. Rev Financ Stud 13:659–696

Lakonishok J, Lee I (2001) Are insider trades informative? Rev Financ Stud 14(1):79–111

Lee CM, Ready MJ (1991) Inferring trade direction from intraday data. J Finance 46:733–746

Lin JC, Sanger GC, Booth GG (1995) Trade size and components of the bid–ask spread. Rev Financ Stud 8:1153–1183

Madhavan A, Sofianos G (1998) An empirical analysis of NYSE specialist trading. J Financ Econ 48:189–210

Madhavan A, Richardson M, Roomans M (1997) Why do security prices change? A transaction-level analysis of NYSE stocks. Rev Financ Stud 10:1035–1064
Odders-White ER (2000) On the occurrence and consequences of inaccurate trade classification. J Financ Mark 3:259–286
O'Hara M (1995) Market microstructure theory. Blackwell, Cambridge
Park JY, Phillips PCB (2000) Nonstationary binary choice. Econometrica 68:1249–1280
Pascual R, Pascual-Fuster B, Climent F (2005) Cross-listing, price discovery, and the informativeness of the trading process. J Financ Mark, forthcoming
Roll R (1984) A simple implicit measure of the effective bid–ask spread in a efficient market. J Finance 39:1127–1139
Stock J, Watson M (1988) Testing for common trends. J Am Stat Assoc 83:1097–1107
Stoll HR (1989) Inferring the components of the bid–ask spread: theory and empirical tests. J Finance 19:115–134
Subrahmanyam A (1997) Multi-market trading and the informativeness of stock trades: an empirical intraday analysis. J Econ Bus 49:515–531
Tse Y (2000) Further examination of price discovery on the NYSE and regional exchanges. J Financ Res 23:331–351
White H (1994) Estimation, inference and specification analysis. Cambridge University Press, Cambridge
Wooldridge JM (1994) Estimation and inference for dependent processes. In: Engle RF, McFadden DL (eds) Handbook of econometrics, vol IV. Elsevier, North Holland, Amsterdam
Zellner A (1962) An efficient method of estimating seemingly unrelated regressions and tests for aggregation bias. J Am Stat Assoc 57:348–368
Zhang MY, Russell JR, Tsay RS (2005) Determinants of bid and ask quotes and implications for the cost of trading, Working Paper, University of Chicago

Stefan Frey · Joachim Grammig

Liquidity supply and adverse selection in a pure limit order book market

Abstract This paper analyzes adverse selection costs and liquidity supply in a pure open limit order book market. We relax assumptions of the Glosten/Såndas modeling framework regarding marginal zero profit order book equilibrium and the parametric market order size distribution. We show that using average zero profit conditions considerably increases the empirical performance while a nonparametric specification for market order size combined with marginal zero profit conditions does not. A cross sectional analysis corroborates the finding that adverse selection costs are more severe for smaller capitalized stocks. We also find additional support for one of the central hypothesis put forth by the theory of limit order book markets, which states that liquidity supply and adverse selection costs are inversely related. Furthermore, adverse selection cost estimates based on our structural model and those obtained using popular model-free methods are strongly correlated. This indicates the robustness of the theory-based approach.

Keywords Limit order book market · Liquidity supply · Adverse selection

JEL Classification G10 · C32

Stefan Frey is also doctoral student at the Graduiertenkolleg "Unternehmensentwicklung, Marktprozesse und Regulierung in dynamischen Entscheidungsmodellen". Joachim Grammig is also research fellow at the Centre for Financial Research (CFR), Cologne. Earlier drafts of the paper were presented at the German Finance Association's 2004 annual meeting, and seminars at the Universities Cologne, Konstanz, Mannheim and Tübingen, and the 2005 meeting of the Verein fuer Socialpolitik Econometrics Section as well as the 2005 MicFin Workshop in Madrid.

S. Frey (✉) · J. Grammig
Faculty of Economics, Department of Statistics, Econometrics and Empirical Economics, University of Tübingen, Mohlstr. 36, 72074 Tübingen, Germany
E-mail: stefan.frey@uni-tuebingen.de
E-mail: joachim.grammig@uni-tuebingen.de

1 Introduction

Ten years after the question phrased in Glosten's (1994) celebrated paper: 'Is the electronic order book inevitable?' seems to be answered, given the triumphal procession of open order book systems in Continental Europe and recent developments in US stock markets.[1] A central feature of a pure limit order book market is the absence of dedicated market makers. Liquidity is supplied voluntarily by patient market participants who provide an inflow of limit buy and sell orders, the lifeblood of the trading process. The non-executed orders constitute the limit order book, the consolidated source of liquidity. As the viability and resiliency of such a market structure is in the interest of regulators, operators and individual investors it is not surprising that theoretical and empirical studies of limit order markets abound in the literature.[2] However, theoretical models explaining liquidity supply and demand in limit order book markets have not been very successful when confronted with real world order book data. Såndas (2001) extends the methodology proposed by DeJong et al. (1996) and estimates a version of Glosten's (1994) limit order book model allowing for real world features like discrete price ticks and time priority rules. The empirical results obtained using data from the Swedish stock exchange were not encouraging. Formal specification tests reject the model, transaction costs estimates are significantly negative, and book depth is systematically overestimated.

This paper shows how some potentially restrictive assumptions in the Glosten/ Såndas framework can be relaxed, while retaining suitable moment conditions for GMM estimation. We show that the revised econometric methodology considerably improves the empirical performance. The alternative approach is employed in a cross sectional analysis of adverse selection costs and liquidity supply in a limit order market.

Given the discontenting results reported in the previous literature, it is not surprising that many recent empirical papers analyzing limit order book market data have severed the close connection to the theoretical framework. Extending the approach of the early papers by Biais et al. (1995), Hall et al. (2003), Coppejans et al. (2003), Cao et al. (2004), Grammig et al. (2004), Pascual and Veredas (2004) and Ranaldo (2004) employ discrete choice and count data models to analyze the determinants of order submission activity and the interaction of liquidity supply and demand processes in limit order markets. Beltran et al. (2004) advocate a principal components approach to extract latent factors that explain the state of the order book. Gomber et al. (2004) and Degryse et al. (2003) conduct intra-day event studies to analyze the resiliency of limit order markets. These papers interpret the empirical results in the light of predictions of microstructure models. However, a structural interpretation of the parameter estimates cannot be delivered.

This paper returns to the theoretical basis for the empirical analysis of limit order book markets. We hypothesize that the discontenting empirical model

[1] In January 2002 the New York Stock Exchange (NYSE), known as a hybrid specialist market, adopted the key feature of electronic order book markets, namely the public display of all limit orders (NYSE open book program).

[2] Traditionally, market microstructure theory focussed on quote driven markets with one or more market makers (see O'Hara, 1995 for an overview). Recent papers by Parlour (1998), Seppi (1997), Foucault (1999) and Foucault et al. (2003) have changed the focus to the analysis of price and liquidity processes in order book markets.

performance is due to the following problems. First, the real world trading process might be organized in a way that deviates too much from the theoretical framework. Second, some of the underlying theoretical model's assumptions might be too restrictive. The Glosten/Såndas model imposes a zero expected profit condition for order book equilibrium which may not hold in a very active order market with discrete price ticks and time priority rules. Furthermore, the parametric distribution of market order sizes assumed by Såndas (2001), though leading to convenient closed form liquidity supply equations and GMM moment conditions, might be misspecified. Hasbrouck (2004) conjectures that the latter is responsible for the empirical failure of the model.

The original methodological contribution of this paper is to propose alternative estimation strategies which relax some allegedly restrictive assumptions in the Glosten/Såndas framework. First, we show that the parametric distributional assumption about market order sizes can be abandoned in favor of a straightforward nonparametric alternative that still delivers convenient closed form unconditional moment restrictions that can be used for GMM estimation. Second, we motivate a set of alternative moment conditions which replace the zero expected marginal profit conditions used by Såndas (2001). These moment conditions, referred to as average break even conditions, are derived from the assumption that the expected profit of the orders placed on a specific quote is zero.

We estimate the model using both the standard and the revised methodology based on reconstructed order book data from the Xetra electronic order book system which operates at various European exchanges. The data are tailor-made for the purpose of this paper since the trading protocol closely corresponds to the theoretical trading process from which the moment conditions used for the empirical methodology are derived.

We show that using average break even conditions instead of marginal break even conditions delivers a much better empirical performance. Encouraged by this result, we employ the methodology in a cross sectional analysis of adverse selection effects and liquidity in the Xetra limit order market. This is the original empirical contribution of the paper. The main results can be summarized as follows. First, we provide new evidence, from a limit order market, that adverse selection effects are more severe for smaller capitalized, less frequently traded stocks. This corroborates the results of previous papers dealing with different theoretical backgrounds, empirical methodologies, and market structures. Second, the empirical results support one of the main hypothesis of the theory of limit order markets, namely that book liquidity and adverse selection effects are inversely related. Finally, we compare the adverse selection components implied by the structural model estimates with popular ad hoc measures which are based on a comparison of effective and realized spreads. The latter approach is model-free, frequently used in practice and academia (see e.g. Boehmer (2004) and SEC (2001)) and requires publicly available trade and quote data only. The first approach is based on a structural model and permits an economic interpretation of the structural parameters, but the demand on the data is higher as reconstructed order books are needed. We show that both methodologies lead to quite similar conclusions. This result indicates the robustness of the structural model approach. It also provides a theoretical underpinning for using the ad-hoc method for the analysis of limit order data.

The remainder of the paper is organized as follows. Section 2 describes the market structure and data. Section 3 discusses the theoretical background and develops the empirical methodology. The empirical results are discussed in Section 4. Section 5 concludes with a summary and an outlook for further research.

2 Market structure and data

2.1 The Xetra open limit order book system

In our empirical analysis we use data from the automated auction system Xetra which operates at various European trading venues, like the Vienna Stock Exchange, the Irish Stock Exchange, the Frankfurt Stock Exchange (FSE) and the European Energy Exchange.[3] Xetra is a pure open order book system developed and maintained by the German Stock Exchange. It has operated since 1997 as the main trading platform for German blue chip stocks at the FSE. Since the Xetra/FSE trading protocol is the data generating process for this study we will briefly describe its important features.[4]

Between an opening and a closing call auction—and interrupted by another mid-day call auction—Xetra/FSE trading is based on a continuous double auction mechanism with automatic matching of orders based on the usual rules of price and time priority. During pre- and post-trading hours it is possible to enter, revise and cancel orders, but order executions are not conducted, even if possible. During the year 2004, the Xetra/FSE hours extended from 9 A.M. C.E.T to 5.30 P.M. C.E.T. For blue chip stocks there are no dedicated market makers like the Specialists at the New York Stock Exchange (NYSE) or the Tokyo Stock Exchange's Saitori. For some small capitalized stocks listed in Xetra there may exist so-called Designated Sponsors—typically large banks—who are required to provide a minimum liquidity level by simultaneously submitting competitive buy and sell limit orders. In addition to the traditional limit and market orders, traders can submit so-called iceberg (or hidden) orders. An iceberg order is similar to a limit order in that it has pre-specified limit price and volume. The difference is that a portion of the volume is kept hidden from the other traders and is not visible in the open book.

Market orders and marketable limit orders which exceed the volume at the best quote are allowed to 'walk up the book'.[5] In other words, market orders are guaranteed immediate full execution, at the cost of incurring a higher price impact on the trades. This is one of the key features of the stylized theoretical trading environment upon which the econometric modeling is based, but which may not necessarily be found in the real world trading process.[6]

---

[3] The Xetra technology was recently licensed to the Shanghai Stock Exchange, China's largest stock exchange.

[4] The Xetra trading system resembles in many features other important limit order book markets around the world like Euronext, the joint trading platform of the Amsterdam, Brussels, Lisbon and Paris stock exchanges, the Hong Kong stock exchange described in Ahn et al. (2001) and the Australian stock exchange, described in Cao et al. (2004).

[5] A marketable limit order is a limit order with a limit price that makes it immediately executable against the current book. In our study, 'real' market orders (i.e. orders submitted without an upper or lower price limit) and marketable limit orders are treated alike. Henceforth, both real market orders and marketable limit orders are referred to as market orders.

[6] For example, Bauwens and Giot (2001) describe how the Paris Bourse's trading protocol converted the volume of a market order in excess of the depth at the best quote into a limit order at that price which enters the opposite side of the order book.

Xetra/FSE faces some local, regional and international competition for order flow. The FSE maintains a parallel floor trading system, which bears some similarities with the NYSE, and, like in the US, some regional exchanges participate in the hunt for liquidity. Furthermore, eleven out of the thirty stocks we analyze in our empirical study are also cross listed at the NYSE, as an ADR or, in the case of Daimler Chrysler, as a globally registered share. However, the electronic trading platform clearly dominates the regional and international competitors in terms of market shares, at least for the blue chip stocks that we study in the present paper.

2.2 Data and descriptive analyses

The Frankfurt Stock Exchange granted access to a database containing complete information about Xetra open order book events (entries, cancelations, revisions, expirations, partial-fills and full-fills of market, limit and iceberg orders) which occurred during the first three months of 2004 (January, 2nd—March, 31st). The sample comprises the thirty German blue chip stocks constituting the DAX30 index. Based on the event histories we perform a real time reconstruction of the order book sequences. Starting from an initial state of the order book (supplied by the exchange), we track each change in the order book implied by entry, partial or full fill, cancelation and expiration of market, limit and iceberg orders in order to re-construct the order book at each point in time. Our reconstruction procedure permits distinguishing the visible and the hidden part of the order book. The latter consists of the hidden part of the non-executed iceberg orders. To implement the empirical methodology outlined below, we take snapshots of the visible order book entries whenever a market order triggers an execution against the book.

Table 1 reports descriptive statistics of the cross section of stocks. The activity indicators show an active market. Averaged across stocks, about 13,000 non-marketable limit orders per stock are submitted each day. Among those, almost 11,000 get canceled before execution. This indicates that the limit order traders closely monitor the book for profit opportunities which is in fact one of the core assumptions of the underlying theoretical model. The large trade sizes (on average over 40,000 euro per trade) indicate that Xetra/FSE is a trading venue for institutional traders and not a retail market. Averaged across stocks, 2,100 trades are executed per day. Table 1 also reports average effective and realized spreads. Following Huang and Stoll (1996) the average effective spread is computed by taking two times the absolute difference of the transaction price of a trade (computed as average price per share) and the prevailing midquote and averaging over all trades of a stock. Realized spreads are computed similarly, but instead of taking the prevailing midquote, the midquote five minutes after the trade is used.[7] Note that in an open order book market like Xetra, there is no possibility to trade inside the bid-ask spread. Orders are either executed at the best quote or they walk up the book until they are completely filled. Table 1 shows that on average 15% of the order volume walks up the book, i.e. part of the order is matched by standing limit orders beyond the best bid and ask. This implies that the effective spread is then, by definition, larger than or equal to the quoted spread. To ensure

[7] By choosing a five minutes lag we follow the previous literature, see e.g. SEC (2001).

Table 1 Sample descriptives

Company name	Ticker symbol	Turnover	Mkt. Cap.	$\overline{m}$	Percentage of aggr. trades (%)	Trades per day	LO sub.	LO canc.	$\overline{P}$	Eff. spread	Real. spread
TUI	TUI	26,281,175	2,025	24,723	17.6	1,063	6,767	5,714	18.7	0.125	0.015
CONTINENTAL	CONT	25,627,638	4,060	25,574	13.5	1,002	8,036	7,052	31.6	0.092	−0.011
MAN	MAN	27,685,031	2,434	26,189	13.0	1,057	7,214	6,235	27.7	0.096	0.003
METRO	MEO	38,874,669	5,018	31,480	15.7	1,235	7,975	6,702	35.0	0.089	0.000
LINDE	LIN	22,378,772	3,448	24,971	15.8	896	8,342	7,454	43.6	0.080	−0.009
LUFTHANSA	LHA	43,946,809	4,548	32,504	11.9	1,352	8,079	6,780	14.2	0.111	0.022
FRESENIUS	FME	12,850,947	1,944	20,680	16.7	621	5,764	5,195	54.0	0.098	0.010
THYSSEN-KRUPP	TKA	37,892,493	6,450	30,017	11.3	1,262	7,864	6,672	15.9	0.111	0.029
DEUTSCHE POST	DPW	43,836,617	6,806	33,330	11.0	1,315	6,861	5,666	18.2	0.097	0.018
HYPO-VEREINSB.	HVM	98,351,090	6,629	50,783	15.0	1,937	10,204	8,293	18.7	0.098	0.019
COMMERZBANK	CBK	53,171,668	7,569	36,659	12.6	1,450	11,922	10,476	15.4	0.100	0.023
ADIDAS-SALOMON	ADS	31,976,047	4,104	32,635	20.1	980	8,057	7,105	92.6	0.070	−0.002
DEUTSCHE BOERSE	DB1	35,696,903	4,847	36,359	18.4	982	6,598	5,698	46.9	0.075	0.003
HENKEL	HEN3	18,174,548	3,682	25,904	16.6	702	7,989	7,306	65.9	0.077	0.005
ALTANA	ALT	30,985,416	3,338	28,310	18.9	1,095	7,718	6,609	48.6	0.079	0.008
SCHERING	SCH	51,413,053	7,055	33,756	16.2	1,523	9,111	7,669	40.8	0.071	0.004
INFINEON	IFX	146,462,315	4,790	52,331	8.6	2,799	10,320	7,744	11.6	0.104	0.040
BAYER	BAY	88,776,121	15,911	36,994	12.4	2,400	15,258	12,988	23.1	0.076	0.012
RWE	RWE	97,655,566	12,653	42,203	13.0	2,314	14,438	12,355	33.8	0.062	0.002
BMW	BMW	87,854,358	12,211	41,639	14.4	2,110	14,736	12,764	34.7	0.060	0.003
VOLKSWAGEN	VOW	104,249,843	9,688	40,963	16.0	2,545	13,474	11,273	39.2	0.056	0.004
BASF	BAS	124,434,537	25,425	48,236	13.8	2,580	18,211	15,898	43.3	0.051	0.002
SAP	SAP	184,628,162	27,412	65,795	21.9	2,806	19,733	17,095	131.5	0.049	0.001
E.ON	EOA	160,625,983	33,753	55,950	13.6	2,871	18,899	16,468	52.5	0.048	0.003

Table 1 (continued)

Company name	Ticker symbol	Turnover	Mkt. Cap.	$\overline{m}$	Percentage of aggr. trades (%)	Trades per day	LO sub.	LO canc.	$\overline{P}$	Eff. spread	Real. spread
MUENCH.RUECK	MUV2	207,353,230	16,396	60,534	20.7	3,425	20,154	16,894	93.9	0.049	0.005
DAIMLERCHRYSLER	DCX	187,737,846	30,316	56,736	14.5	3,309	18,722	15,919	36.4	0.055	0.010
DEUTSCHE TELEKOM	DTE	350,627,866	34,858	78,884	5.0	4,445	14,498	11,009	15.7	0.072	0.031
DEUTSCHE BANK	DBK	309,282,831	38,228	78,083	19.3	3,961	23,169	19,772	67.2	0.044	0.004
ALLIANZ	ALV	289,980,556	33,805	64,114	21.4	4,523	29,791	25,882	100.1	0.049	0.010
SIEMENS	SIE	321,704,299	52,893	72,831	16.7	4,418	23,659	19,920	64.0	0.041	0.006
Average		108,683,880	14,076	42,972	15.2	2,099	12,785	10,887	44.5		

Mkt. cap. is the market capitalization in million euros at the end of December 2003, $\overline{m}$ is the average trade size (in euros). *Percentage of Aggr. trades (%)* gives the percentage of total trading volume that has not been executed at the best prices (that is, the order walked up the book). *Turnover* is the average trading volume in euros per trading day, *trades per day* is the average number of trades per day, *LO sub.* and *LO canc.*, respectively, denote the average number of non-marketable limit order submissions and cancelations per day. $\overline{P}$, *eff. spread* and *real. spread* refer to the sample averages of midquote, effective spread and realized spread, respectively. The average effective spread is computed by taking two times the absolute difference of the transaction price of a trade and the prevailing midquote and averaging over all trades of a stock. The average realized spread is computed similarly, but instead of taking the prevailing midquote, we use the midquote five minutes after the trade. To ensure comparability across stocks, we compute effective and realized spreads relative to the midquote prevailing at the time of the trade and multiply by 100 to obtain a % figure. The table is sorted in descending order by the difference of effective and realized spread. The sample ranges from January 2, 2004 to March 31, 2004

comparability across stocks, we compute effective and realized spreads relative to the midquote prevailing at the time of the trade. Analyzing effective and realized spreads is a straightforward way to assess and compare transaction costs and adverse selection effects across stocks or trading venues. The realized spread can be viewed as a transaction costs measure that is purged of informational effects while the difference of effective and realized spread (referred to as price impact) is a natural measure for the amount of informational content of the order flow.[8] Average effective spreads range from 0.04% to 0.13%. Realized spreads are considerably smaller. This implies that price impacts, computed as the difference between effective and realized spreads, are relatively large. In other words, a large fraction of the spread is due to informational order flow. This is not an unexpected result. In an open automated auction market there is no justification for inventory costs associated with market making or monopolistic power of a market maker, the other factors that may explain the spread. Furthermore, order submission fees, i.e. operational costs, are very small.

Table 1 shows that there is a considerable variation of price impacts, market capitalization and trading activity across stocks. The Spearman rank correlation between market capitalization and price impacts is −0.88 (p-value$<$0.001) and the correlation between price impacts and daily number of trades is −0.87 (p-value$<$0.001). Price impacts thus tend to be larger for smaller capitalized, less frequently traded stocks. We will come back to this result when discussing the empirical results based on the structural model.

3 Methodology

3.1 Såndas' basic framework

Såndas (2001) develops a variant of Glosten's (1994) limit order book model with discrete price ticks and time priority rules. The model delivers equations which predict that order book depth and adverse selection effects are inversely related. The associated empirical methodology is rooted in economic theory, and delivers structural parameter estimates of transaction costs and adverse selection effects in a limit order book market. Below we will briefly describe the assumptions of the basic model and the estimation strategy proposed in Såndas (2001). The fundamental asset value X_t is described by a random walk with innovations depending on an adverse selection parameter α, which gives the informational content of a signed market order of size m_t,

$$X_{t+1} = \mu + X_t + \alpha m_t + \eta_{X,t+1}. \tag{1}$$

Negative values of m_t denote sell orders, positive values buy orders. Furthermore, it is assumed that $E(X_t) = 0$. $\eta_{X,t+1}$ is an innovation orthogonal to X_t.

[8] Boehmer (2004) and SEC (2001) conduct exhaustive comparisons of transaction costs and adverse selection effects in US exchanges based on effective and realized spread analyses.

μ gives the expected change in the fundamental value. Market buy and sell orders are assumed to arrive with equal probability with a two-sided exponential density describing the distribution of order sizes m_t:[9]

$$f(m_t) = \begin{cases} \frac{1}{2\lambda} e^{\frac{-m_t}{\lambda}} & \text{if } m_t > 0 \text{ (market buy)} \\ \frac{1}{2\lambda} e^{\frac{m_t}{\lambda}} & \text{if } m_t < 0 \text{ (market sell).} \end{cases} \quad (2)$$

Risk neutral limit order traders face a order processing cost γ (per share) and have knowledge about the distribution of market order size and the adverse selection component α, but not about the true asset price. They choose limit order prices and quantities such that their expected profit is maximized. If the last unit at any discrete price tick exactly breaks even, i.e. has expected profit equal to zero, the order book is in equilibrium.

Denote the ordered discrete price ticks on the ask (bid) side by p_{+k} (p_{-k}) with k=1,2,... and the associated volumes at these prices by q_{+k} (q_{-k}). Given these assumptions and setting $q_{0,t} \equiv 0$, the equilibrium order book at time t can recursively be constructed as follows:

$$\begin{aligned} q_{+k,t} &= \frac{p_{+k,t} - X_t - \mu - \gamma}{\alpha} - Q_{+k-1,t} - \lambda \quad k = 1, 2, \ldots \text{ (ask side)} \\ q_{-k,t} &= \frac{X_t + \mu - p_{-k,t} - \gamma}{\alpha} - Q_{-k+1,t} - \lambda \quad k = 1, 2, \ldots \text{ (bid side),} \end{aligned} \quad (3)$$

where $Q_{+k,t} = \sum_{i=+1}^{+k} q_{i,t}$ and $Q_{-k,t} = \sum_{i=-1}^{-k} q_{i,t}$. Equation (3) contains the model's key message. Order book depth and informativeness of the order flow are inversely related. If the model provides a good description of the real world trading process, and if consistent estimates of the model parameters can be provided, one can use Eq. (3) to predict the evolution of the order book for a given stock and quantify adverse selection costs and their effect on order book depth.

Såndas (2001) proposes to employ GMM for parameter estimation and specification testing. Assuming mean zero random deviations from order book equilibrium at each price tick, and eliminating the unobserved fundamental asset value X_t by adding the resulting bid and ask side equations for quote $+k$ and $-k$, the following unconditional moment restrictions can be used for GMM estimation,

$$E(p_{+k,t} - p_{-k,t} - 2\gamma - \alpha(Q_{k,t} + 2\lambda + Q_{-k,t})) = 0 \; k = 1, 2, \ldots. \quad (4)$$

Since Eq. (4) follows from the assumption that the last (marginal) limit order at the respective quote has zero expected profit, it is referred to as 'marginal break even condition'. A second set of moment conditions results from eliminating X_t by

[9] In an alternative specification we allowed for additional flexibility by allowing the expected buy and sell market order sizes to be different. However, the parameter estimates and diagnostics changed only marginally. We therefore decided to stick to the specification in Eq. (2) which is more appealing both from a methodological and theoretical point of view.

subtracting the deviations from equilibrium depths at the kth quote at time t+1 and t and taking expectations which yields

$$\begin{aligned} E(\Delta p_{+k,t+1} - \alpha(Q_{+k,t+1} - Q_{+k,t}) - \mu - \alpha m_t) = 0 \quad k = 1, 2, \ldots \\ E(\Delta p_{-k,t+1} + \alpha(Q_{-k,t+1} - Q_{-k,t}) - \mu - \alpha m_t) = 0 \quad k = 1, 2, \ldots, \end{aligned} \tag{5}$$

where $\Delta p_{j,t+1}=p_{j,t+1}-p_{j,t}$. We refer to the equations in Eq. (5) as 'marginal update conditions'. They relate the expected changes in the order book to the market order flow. An obvious additional moment condition to identify the expected market order size is given by

$$E(|X_t| - \lambda) = 0. \tag{6}$$

Moment conditions Eqs. (4), (5) and (6) can conveniently be exploited for GMM estimation a la Hansen (1982).

Såndas (2001) derives the moment conditions from the basic model setup outlined by Glosten (1994). Both Glosten's framework and Såndas' empirical implementation entail a set of potentially restrictive assumptions that may be problematic when confronting the model with real world data. Maybe the most crucial assumption of the Glosten framework is that limit order traders are assumed to be uninformed and that private information is only revealed through the arrival of market orders. Recent literature, however, suggests that limit orders may also be information-motivated (Seppi (1997); Kaniel and Liu (2001); Cheung et al. (2003)). Bloomfield et al. (2005) observe in an experimental limit order market that informed traders use more limit orders than liquidity traders. Since both break even and update conditions are derived from the assumption of uninformed limit order traders, the rejection of the model when confronted with real world data might be a result from a violation of this fundamental assumption.[10] Another important consideration is the number of active liquidity providers. Glosten (1994) assumes perfect competition. Biais et al. (2000) propose solutions for oligopolistic competition.

The following section proposes a revised set of moment conditions which are derived from a relaxation of the expected marginal profit condition and the parametric assumption of the market order distribution. However, we leave the basic assumption of uninformed limit order traders intact. Its relaxation would entail a fundamental revision of the theoretical base model. This is left for further research.

3.2 Revised moment conditions

3.2.1 Alternatives to the distributional assumption on market order sizes

Reviewing the Såndas/Glosten framework Hasbrouck (2004) conjectures that the parametric specification for the market order size distribution Eq. (2) may be incorrect.[11] Indeed, the plot of the empirical market order distribution against the fitted exponential densities depicted in Figure 3 in Såndas (2001) sheds some doubt

[10] We are grateful to a referee for pointing this out.

[11] It should be noted that the exponential assumption in DeJong et al.'s (1996) implementation of the Glosten model did not seem to be a restrictive assumption.

on this distributional assumption. To provide a formal assessment, we have employed the nonparametric testing framework proposed by Fernandes and Grammig (2005) and found that the exponential distribution is rejected on any conventional level of significance for our sample of stocks. Hasbrouck (2004) argues that the misspecification of the exponential distribution could be responsible for the discontenting empirical results which have been reported when the model is confronted with real world data.

Of course, the exponential assumption is convenient both from a theoretical and an econometric perspective. It yields the closed form conditions for order book equilibrium (3) which, in turn, lend itself conveniently to GMM estimation. However, the parametric assumption can easily be dispensed with and a straightforward nonparametric approach can be pursued for GMM estimation. In the appendix we show that the zero expected profit condition for the marginal unit at ask price p_{+k} can be written as

$$p_{+k} - \gamma - \alpha E[m|m \geq Q_{+k}] - X - \mu = 0.^{12} \tag{7}$$

Assuming exponentially distributed market orders as in Eq. (2) we have $E[m \mid m \geq Q_{+k}] = Q_{+k} + \lambda$. Hence, Eq. (7) becomes

$$Q_{+k} = \frac{p_{+k} - X - \gamma - \mu}{\alpha}. \tag{8}$$

This is an alternative to Eq. (3) to describe order book equilibrium. Although the closed form expression implied by the parametric distributional assumption is convenient, it is not necessary for the econometric methodology to rely on it. Instead, we can rewrite Eq. (7) to obtain

$$E[m|m \geq Q_{+k}] = \frac{p_{+k} - X - \gamma - \mu}{\alpha}. \tag{9}$$

In order to utilize Eq. (9) for GMM estimation, one can simply replace $E[m \mid m \geq Q_{+k}]$ by the conditional sample means $\widehat{E}[m|m \geq Q_{+k}]$. Since the number of observations will be large for frequently traded stocks (which is the case in our application), conditional expectations can be precisely estimated by the conditional sample means. Nonparametric equivalents of the marginal break even and update conditions (4) and (5) can be derived in the same fashion as described in the previous section. GMM estimation is more computer intensive since evaluating the GMM objective function involves computation of the conditional sample means, but it is a straightforward exercise.

Empirical evidence suggests that market orders are timed in that market order traders closely monitor the state of the book when deciding on the size of the submitted market order (see e.g. Biais et al. (1995), Ranaldo (2004) and Gomber et al. (2004)). To account for state dependency, Såndas (2001) proposed using a set of instruments which scale the value of the λ parameter in Eq. (2). The nonparametric strategy developed here can be easily adapted to account for a

[12] For notational brevity we omit the subscripts. Market order size m and fundamental price X are observed at time t, and the equation holds for any price tick $p_{+k,t}$ with associated cumulative volume $Q_{+k,t}$, k=1,2....

market order distribution that changes with the state of the book. One only has to base the computation of the conditional upper tail expectation on a vector of state variables F, i.e. calculate $\widehat{E}[m|m \geq Q_{+k}, F]$. For the purpose of this study we focus on the unconditional market order distribution and leave modeling the conditional market order distribution as a topic for further research.

3.2.2 Average profit conditions

To justify the marginal zero expected profit assumption, one implicitly assumes a repetitive two phase trading process. In phase one, agents submit and cancel limit orders until the book is free of (expected) profit opportunities and no agent wants to submit, revise or cancel her order. Limit orders are sorted by price priority and, within the same price tick, by time priority. When the book is such an equilibrium the order book should display no 'holes', i.e. zero volumes in between two price ticks. In phase two, a single market order of a given size arrives and is executed against the equilibrium order book. After this event we go back to phase one, during which the book is replenished again until equilibrium is reached and another market order arrives and so forth. Can this be a reasonable description of a real world trading process? The descriptive statistics on the trading and order submission activity reported in Table 1 indicate a dynamic trading environment. For a large stock, like Daimler Chrysler, we have on average over 3,000 trade events per day, about 19,000 submissions of limit orders, of which over 80% are canceled before execution. One could argue that such an active limit order trader behavior indicates a thorough monitoring of the book which eliminates any profit opportunities. This is quite in line with the theoretical framework. However, with on average 10 seconds duration between trade events (for Daimler Chrysler) the time to reach the new equilibrium after a market order hits the book and before a new order arrives, seems a short span.

The marginal break even conditions can also be challenged by the following reasoning. The conditions imply nonzero expected profits for limit order units that do not occupy the last position of the respective price ticks. On the other hand, this implies that the whole book offers positive expected profits for traders acting as market makers. If market making provides nonzero expected profit opportunities, then this would attract new entrants and the competition between these would-be market makers ultimately eliminate any profit opportunities.

These considerations lead us to consider an alternative to the marginal profit conditions which does not rely on the assumption that limit order traders immediately cancel or adjust all their orders which show negative expected profit on a marginal unit, and that also acknowledges the effect of market maker competition on expected profits. For this purpose we retain most of the assumptions of the Glosten/Såndas framework. However, instead of evaluating the expected profit of the marginal profit for the last unit at each quote k, we assume that the expected profit of the whole block of limit orders at any quote is zero. The marginal zero profit condition is thus replaced by an 'average zero profit condition'. This assumption allows to differentiate between two types of costs associated with the submission of a limit order, a fixed cost component, like order submission and surveillance costs, and marginal costs (per share), like execution or clearing fees and opportunity costs of market making. In the appendix we show that the liquidity supply equations which are implied by the zero expected profit condition can be written as

$$q_{+k,t} = 2\left(\frac{p_{+k,t} - X_t - \gamma - \dfrac{\xi}{q_{+k,t}}}{\alpha} - \lambda - \mu\right) - Q_{+k-1,t} \quad k = 1, 2, \ldots \text{ (ask side)}$$

$$q_{-k,t} = 2\left(\frac{X_t - p_{-k,t} - \gamma - \dfrac{\xi}{q_{-k,t}}}{\alpha} - \lambda + \mu\right) - Q_{-k+1,t} \quad k = 1, 2, \ldots \text{ (bid side)}. \tag{10}$$

ξ denotes the fixed cost component which is assumed to be identical for each price tick in the order book. To derive the equations in Eq. (10), we have retained the parametric assumption about the distribution of trade sizes. Considering a nonparametric alternative along the lines described in the previous subsection is also feasible. Proceeding as above, i.e. by eliminating the unobserved fundamental asset value X_t by adding the bid and ask side equations for quote $+k$ and $-k$ yields the following unconditional moment restrictions which we refer to as average break even conditions,

$$E\left(\Delta p_{\pm k,t} - 2\gamma - \frac{\xi}{q_{+k,t}} - \frac{\xi}{q_{-k,t}} - \alpha\left(\frac{1}{2}Q_{+k,t} + 2\lambda + \frac{1}{2}Q_{-k,t}\right)\right) = 0 \quad k = 1, 2, \ldots, \tag{11}$$

where $\Delta p_{\pm k,t} = p_{+k,t} - p_{-k,t}$. Subtracting deviations from the implied depths at the kth quote at time $t+1$ and t and taking expectation yields the following equations which we refer to as average update conditions,

$$E\left(\Delta p_{+k,t+1} - \frac{\xi}{q_{+k,t}} + \frac{\xi}{q_{+k,t-1}} - \frac{\alpha}{2}(Q_{+k,t+1} - Q_{+k,t}) - \mu - \alpha m_t\right) = 0 \quad k = 1, 2, \ldots$$

$$E\left(\Delta p_{-k,t+1} + \frac{\xi}{q_{+k,t}} - \frac{\xi}{q_{+k,t-1}} + \frac{\alpha}{2}(Q_{-k,t+1} - Q_{-k,t}) - \mu - \alpha m_t\right) = 0 \quad k = 1, 2, \ldots, \tag{12}$$

where $\Delta p_{j,t+1} = p_{j,t+1} - p_{j,t}$. The average break even and update conditions replace the marginal break even and update conditions of Eqs. (4) and (5).

4 Empirical results

4.1 Performance comparisons

Using the DAX30 order book data we follow Såndas (2001) and estimate the model parameters exploiting the marginal break even conditions (4) and the marginal updating conditions (5) along with Eq. (6). To construct the moment conditions we use the respective first four best quotes, i.e. k=1,…,4 on the bid and the ask side of the visible order book. This yields thirteen moment conditions: four

Table 2 First stage GMM results baseline specification

Ticker	α	γ	λ	μ	$J(9)$	p-value
LIN	0.0228 (118.1)	−0.0185 (55.5)	0.5728 (134.6)	0.0003 (3.1)	0.5	1.000
DPW	0.0025 (149.8)	−0.0063 (47.0)	1.8362 (150.0)	−0.0001 (3.3)	21.0	0.013
HEN3	0.0415 (95.8)	−0.0178 (40.1)	0.3937 (114.0)	−0.0003 (1.9)	23.2	0.006
MEO	0.0132 (135.1)	−0.0163 (66.1)	0.9066 (166.7)	0.0000 (0.4)	25.3	0.003
LHA	0.0019 (159.7)	−0.0072 (63.6)	2.3210 (150.8)	0.0001 (4.4)	31.2	0.000
MAN	0.0104 (121.9)	−0.0151 (61.8)	0.9445 (134.5)	0.0004 (5.2)	35.2	0.000
DB1	0.0162 (114.5)	−0.0158 (46.7)	0.7739 (114.1)	0.0001 (1.1)	45.0	0.000
FME	0.0456 (85.0)	−0.0210 (34.8)	0.3839 (96.4)	0.0000 (0.0)	53.2	0.000
TUI	0.0054 (130.3)	−0.0095 (50.6)	1.3215 (127.0)	−0.0002 (2.7)	57.4	0.000
ALT	0.0224 (121.9)	−0.0144 (50.5)	0.5785 (142.4)	−0.0002 (2.1)	79.8	0.000
CBK	0.0016 (164.0)	−0.0048 (50.4)	2.4055 (152.0)	−0.0001 (2.0)	81.1	0.000
CONT	0.0131 (116.6)	−0.0168 (60.4)	0.8166 (139.6)	0.0002 (2.2)	85.8	0.000
ADS	0.0549 (113.2)	−0.0183 (38.1)	0.3528 (141.0)	−0.0002 (1.5)	118.2	0.000
BMW	0.0053 (173.8)	−0.0087 (69.6)	1.2029 (203.0)	−0.0001 (2.9)	173.7	0.000
TKA	0.0024 (148.6)	−0.0075 (61.0)	1.9075 (158.9)	0.0000 (1.3)	206.6	0.000
SCH	0.0106 (135.3)	−0.0101 (49.6)	0.8250 (168.8)	0.0000 (0.4)	232.6	0.000
RWE	0.0053 (212.2)	−0.0095 (86.2)	1.2460 (210.3)	0.0001 (3.3)	239.4	0.000
DTE	0.0002 (303.0)	−0.0010 (32.1)	5.0499 (232.7)	0.0000 (0.0)	292.8	0.000
IFX	0.0004 (196.7)	−0.0023 (45.6)	4.5335 (170.6)	0.0000 (0.4)	360.9	0.000
HVM	0.0015 (109.0)	−0.0043 (40.1)	2.8391 (130.9)	0.0000 (1.0)	363.8	0.000
VOW	0.0065 (21.2)	−0.0099 (17.7)	1.0472 (195.8)	0.0001 (0.2)	429.9	0.000
BAY	0.0024 (216.9)	−0.0046 (59.2)	1.6352 (225.8)	0.0000 (1.2)	458.2	0.000
BAS	0.0056 (219.9)	−0.0077 (77.4)	1.1206 (244.1)	0.0000 (1.1)	683.1	0.000
EOA	0.0060 (219.2)	−0.0070 (65.0)	1.0663 (252.7)	0.0000 (1.0)	1,011.3	0.000
DCX	0.0031 (258.2)	−0.0049 (65.4)	1.5638 (254.9)	0.0002 (7.2)	1,376.9	0.000
SAP	0.0370 (212.6)	−0.0147 (49.5)	0.5030 (237.4)	0.0006 (5.9)	1,609.9	0.000
MUV2	0.0196 (212.1)	−0.0106 (60.4)	0.6476 (246.9)	0.0001 (1.0)	2,101.9	0.000
DBK	0.0065 (248.7)	−0.0061 (57.7)	1.1517 (256.1)	0.0000 (0.9)	2,584.6	0.000
ALV	0.0187 (232.7)	−0.0080 (35.9)	0.6453 (294.4)	−0.0002 (4.5)	2,701.8	0.000
SIE	0.0052 (273.3)	−0.0039 (36.4)	1.1442 (297.3)	0.0001 (2.9)	3,827.8	0.000

2×4 quotes from the bid and ask side of the visible book are used to construct update and break even conditions derived from the zero marginal expected profit condition as in Såndas (2001). The numbers in parentheses are t-values. The fifth and sixth column report the GMM J statistic and the associated p-value. The stocks are sorted by ascending order of the J-statistic

break even conditions, eight update conditions, and the moment condition (6). Order sizes X_t are expressed in 1,000 shares.

Table 2 contains the first stage GMM results.[13]

We report parameter estimates, t-statistics and the value of the GMM J-statistic with associated p-values. Under the null hypothesis that the moment conditions are correctly specified, the J-statistic is asymptotically χ^2 with degrees of freedom equal to the number of moment conditions minus the number of estimated

[13] Two stage and iterated GMM estimates are similar and therefore not reported to conserve space. To compute the parameter standard errors and the J-statistic we employ the Bartlett Kernel with bandwidth equal to ten lags when computing the spectral density matrix. We have tested various lags and the results are robust with respect to bandwidth choice.

parameters. The estimation results based on the Xetra data are in line with the central findings reported by Såndas (2001). Only for two out of thirty stocks the model is not rejected at 1% significance level. In Såndas' (2001) application the model was rejected for all stocks. Like in Såndas' (2001) application, the transaction cost estimates (γ) are significantly negative, a result that is difficult to reconcile with the underlying theoretical model. Hence, even with a data generating process that corresponds very close to the theoretical framework, the model does not seem to fit the data very well.

Tables 3 and 4 report the results that are obtained when the modified moment conditions suggested in the previous section are used. As before, the first four

Table 3 First stage GMM results for the nonparametric specification

Ticker	α	γ	μ	$J(9)$	p-value
TUI	0.0040 (140.1)	−0.0091 (52.9)	−0.0001 (2.3)	0.7	1.000
LIN	0.0169 (122.1)	−0.0142 (49.8)	0.0003 (2.8)	3.4	0.945
DB1	0.0111 (110.2)	−0.0123 (39.0)	0.0001 (1.1)	11.1	0.270
HEN3	0.0301 (95.2)	−0.0120 (31.5)	−0.0002 (1.5)	19.7	0.020
ALT	0.0169 (129.8)	−0.0117 (47.9)	−0.0002 (1.8)	33.5	0.000
HVM	0.0012 (134.2)	−0.0058 (60.8)	0.0000 (0.8)	41.9	0.000
ADS	0.0403 (119.3)	−0.0125 (30.1)	−0.0002 (1.2)	68.3	0.000
MEO	0.0103 (141.2)	−0.0139 (64.3)	0.0000 (0.3)	72.8	0.000
FME	0.0299 (79.8)	−0.0119 (22.3)	0.0000 (0.2)	84.5	0.000
CONT	0.0094 (120.6)	−0.0127 (52.6)	0.0001 (2.1)	87.5	0.000
IFX	0.0003 (235.2)	−0.0032 (69.9)	0.0000 (0.2)	97.9	0.000
MAN	0.0076 (127.5)	−0.0125 (58.8)	0.0003 (5.0)	101.3	0.000
BMW	0.0039 (188.1)	−0.0072 (66.8)	−0.0001 (2.7)	112.9	0.000
LHA	0.0015 (162.9)	−0.0075 (68.5)	0.0001 (4.1)	165.3	0.000
VOW	0.0048 (28.4)	−0.0092 (21.9)	0.0001 (0.2)	169.4	0.000
SCH	0.0078 (147.3)	−0.0080 (45.9)	0.0000 (0.4)	173.0	0.000
DPW	0.0019 (168.3)	−0.0064 (53.0)	−0.0001 (2.9)	176.8	0.000
RWE	0.0038 (209.4)	−0.0073 (74.0)	0.0001 (3.2)	189.2	0.000
BAY	0.0018 (196.6)	−0.0034 (44.4)	0.0000 (1.3)	349.3	0.000
CBK	0.0013 (196.2)	−0.0060 (68.0)	0.0000 (2.0)	427.8	0.000
BAS	0.0041 (222.3)	−0.0057 (62.7)	0.0000 (0.8)	574.9	0.000
TKA	0.0019 (164.7)	−0.0077 (68.7)	0.0000 (1.1)	721.3	0.000
DCX	0.0022 (235.6)	−0.0036 (49.8)	0.0001 (5.8)	760.9	0.000
EOA	0.0043 (227.8)	−0.0047 (50.3)	0.0000 (0.7)	1,050.8	0.000
SAP	0.0274 (228.6)	−0.0099 (39.5)	0.0004 (4.9)	1,070.5	0.000
MUV2	0.0140 (219.1)	−0.0074 (47.7)	0.0000 (0.7)	1,508.2	0.000
DTE	0.0001 (380.6)	−0.0022 (73.5)	0.0000 (0.0)	1,514.1	0.000
DBK	0.0046 (265.3)	−0.0046 (49.6)	0.0000 (0.7)	1,633.5	0.000
ALV	0.0134 (241.2)	−0.0048 (22.7)	−0.0002 (4.0)	2,152.3	0.000
SIE	0.0039 (312.6)	−0.0032 (35.5)	0.0001 (2.6)	2,912.9	0.000

2×4 quotes from the bid and ask side of the visible book are used to construct update and break even conditions derived from the zero marginal expected profit condition as in Såndas (2001). For the construction of the moment conditions, the empirical distribution of the market order sizes is used instead of the exponential distribution. The stocks are sorted by ascending order of the J-statistic

Table 4 First stage GMM results based on average profit conditions instead of marginal profit conditions

Ticker	α	γ	λ	ξ	μ	J(8)	p-value	τ (%)
IFX	0.0004 (151.6)	0.0042 (24.3)	4.5335 (170.6)	−0.0051 (14.9)	0.0000 (0.2)	0.1	1.000	0.0157
DBK	0.0057 (102.7)	0.0137 (20.4)	1.1517 (256.1)	−0.0031 (19.4)	0.0000 (0.7)	0.3	1.000	0.0099
SAP	0.0286 (32.8)	0.0532 (9.7)	0.5030 (237.4)	−0.0058 (9.6)	0.0004 (4.1)	0.7	1.000	0.0109
DCX	0.0029 (145.8)	0.0059 (17.8)	1.5638 (254.9)	−0.0023 (17.9)	0.0002 (6.1)	0.8	0.999	0.0124
DB1	0.0152 (56.2)	0.0049 (2.0)	0.7739 (114.1)	−0.0016 (4.5)	0.0001 (1.1)	0.8	0.999	0.0251
SIE	0.0045 (117.0)	0.0133 (27.8)	1.1442 (297.3)	−0.0030 (23.7)	0.0001 (2.6)	0.8	0.999	0.0080
TUI	0.0053 (103.0)	0.0021 (2.8)	1.3215 (127.0)	−0.0016 (7.0)	−0.0002 (2.6)	0.9	0.999	0.0374
MUV2	0.0171 (89.0)	0.0234 (15.4)	0.6476 (246.9)	−0.0036 (15.7)	0.0001 (0.8)	1.1	0.998	0.0117
FME	0.0428 (24.1)	0.0101 (1.0)	0.3839 (96.4)	−0.0016 (1.7)	0.0000 (0.0)	1.5	0.992	0.0303
HVM	0.0015 (83.7)	0.0055 (11.7)	2.8391 (130.9)	−0.0032 (12.2)	0.0000 (0.9)	1.8	0.986	0.0218
ALT	0.0216 (83.1)	0.0055 (3.2)	0.5785 (142.4)	−0.0015 (6.2)	−0.0002 (2.1)	1.9	0.985	0.0259
BAS	0.0053 (161.4)	0.0037 (11.4)	1.1206 (244.1)	−0.0015 (16.6)	0.0000 (1.0)	3.1	0.930	0.0136
DTE	0.0002 (232.1)	0.0046 (49.9)	5.0499 (232.7)	−0.0061 (18.8)	0.0000 (0.0)	3.8	0.877	0.0064
ALV	0.0155 (80.3)	0.0280 (19.8)	0.6453 (294.4)	−0.0039 (18.9)	−0.0002 (3.8)	3.8	0.872	0.0099
ADS	0.0503 (31.8)	0.0252 (3.0)	0.3528 (141.0)	−0.0028 (3.4)	−0.0002 (1.4)	5.1	0.750	0.0191
VOW	0.0062 (21.8)	0.0020 (1.4)	1.0472 (195.8)	−0.0011 (3.7)	0.0001 (0.2)	6.5	0.595	0.0166
EOA	0.0053 (79.1)	0.0099 (12.0)	1.0663 (252.7)	−0.0027 (12.9)	0.0000 (0.8)	7.5	0.488	0.0107
CBK	0.0016 (139.6)	0.0020 (10.5)	2.4055 (152.0)	−0.0018 (13.5)	−0.0001 (1.8)	9.7	0.283	0.0248
BMW	0.0051 (143.8)	0.0018 (5.2)	1.2029 (203.0)	−0.0013 (13.4)	−0.0001 (2.8)	9.7	0.284	0.0176
BAY	0.0024 (168.5)	0.0037 (17.4)	1.6352 (225.8)	−0.0017 (17.6)	0.0000 (1.2)	12.9	0.117	0.0167
LIN	0.0232 (93.8)	−0.0107 (9.8)	0.5728 (134.6)	0.0004 (3.1)	0.0004 (3.1)	15.0	0.059	0.0305
DPW	0.0025 (131.8)	0.0012 (4.0)	1.8362 (150.0)	−0.0015 (8.7)	−0.0001 (3.2)	16.0	0.042	0.0252
HEN3	0.0405 (27.5)	−0.0004 (0.1)	0.3937 (114.0)	−0.0005 (0.6)	−0.0003 (1.9)	26.9	0.001	0.0241
SCH	0.0099 (92.5)	0.0052 (6.1)	0.8250 (168.8)	−0.0017 (9.4)	0.0000 (0.4)	35.1	0.000	0.0201
CONT	0.0132 (108.1)	−0.0105 (12.7)	0.8166 (139.6)	0.0009 (5.0)	0.0002 (2.2)	51.0	0.000	0.0338
MAN	0.0105 (105.0)	−0.0086 (10.2)	0.9445 (134.5)	0.0006 (3.4)	0.0004 (5.2)	53.0	0.000	0.0359

Table 4 (continued)

Ticker	α	γ	λ	ξ	μ	J(8)	p-value	τ (%)
LHA	0.0019 (137.1)	−0.0006 (2.0)	2.3210 (150.8)	−0.0008 (6.4)	0.0001 (4.3)	63.0	0.000	0.0307
MEO	0.0134 (115.8)	−0.0071 (9.3)	0.9066 (166.7)	0.0000 (0.1)	0.0000 (0.4)	99.1	0.000	0.0345
RWE	0.0051 (178.5)	0.0000 (0.1)	1.2460 (210.3)	−0.0009 (11.5)	0.0001 (3.3)	112.6	0.000	0.0188
TKA	0.0024 (139.4)	−0.0012 (5.0)	1.9075 (158.9)	−0.0002 (2.2)	0.0000 (1.3)	362.6	0.000	0.0285

2×4 quotes from the bid and ask side of the visible book are used to construct average update and average break even conditions. The exponential assumption on the distribution of the trade size is maintained. $\tau = \frac{\alpha \cdot \overline{m}}{\overline{P}}$, where $\overline{m}$ and $\overline{P}$ denote stock specific sample averages of the non-signed trade sizes (number of shares) and the midquotes, respectively. The stocks are sorted by ascending order of the J-statistic

quotes on each market side are used for the construction of break even and update conditions. Table 3 reports the estimation results for a specification that does not rely on a parametric assumption on the distribution of the market order size when constructing the marginal break even and the update conditions as described in section 3.2.1. The results reported in Table 4 are obtained when using the average break even and update conditions (11) and (12) for GMM estimation, while maintaining the parametric assumption (2) about the trade size distribution. For each parametric specification, the moment condition of Eq. (6) is employed, too. The full set of eight update conditions is exploited.

The estimates reported in Table 3 show that abandoning the parametric assumption concerning the market order size distribution improves the results only marginally. For four of the thirty stocks the model is not rejected at the 1% significance level. Hasbrouck's (2004) conjecture that the distributional assumption might be responsible for the model's empirical failure is therefore not supported. Generally, the estimates of the adverse selection components, transaction costs and drift parameters do not change dramatically compared to the baseline specification. The transaction cost estimates remain negative.

Maintaining the distributional assumption, but using average break even conditions instead of marginal break even conditions, considerably improves the empirical performance. Table 4 shows that we have model non-rejection for 22 out of the 30 stocks at the 1% significance level. With a single exception the estimates of the marginal transaction cost parameter γ are positive for those stocks for which the model is not rejected at 1% significance level. The size of the implied transaction cost estimates are broadly comparable with the relative realized spreads figures reported in Table 1. For example, the estimation results imply that transaction costs account for 0.013% of the euro value of a median sized DaimlerChrsyler trade. This value is quite comparable with the average relative realized spread which amounts to 0.010% (see Table 1).

Figure 1 shows graphically the improved empirical performance delivered by the revised methodology. The figure depicts means an medians of implied and observed ask side price schedules of four selected stocks. The results obtained from the baseline estimation which uses marginal moment conditions confirm the disturbing findings reported in Såndas' (2001). The price schedules implied by the model estimates are below the observed price schedules at all relevant volumes. The economically implausible negative price discount at small volumes is caused by the negative transaction costs estimates. This suggests that the model is not only rejected on the grounds of statistical significance, but that fundamentally fails to explain the data. The model does a bad job even in describing the 'average' state of the order book. However, Fig. 1 shows that when working with average break even and update conditions the empirical performance of the model is considerably improved. Especially the median observed price schedules correspond closely to those implied by the model.[14]

[14] We have also estimated a specification that combines the nonparametric approach towards trades sizes and average moment conditions, but the results (not reported) are not improved compared to the parametric version. In this version, the model is not rejected for 16 out of 30 stocks. The following analysis therefore focuses on the parametric specification using average moment conditions.

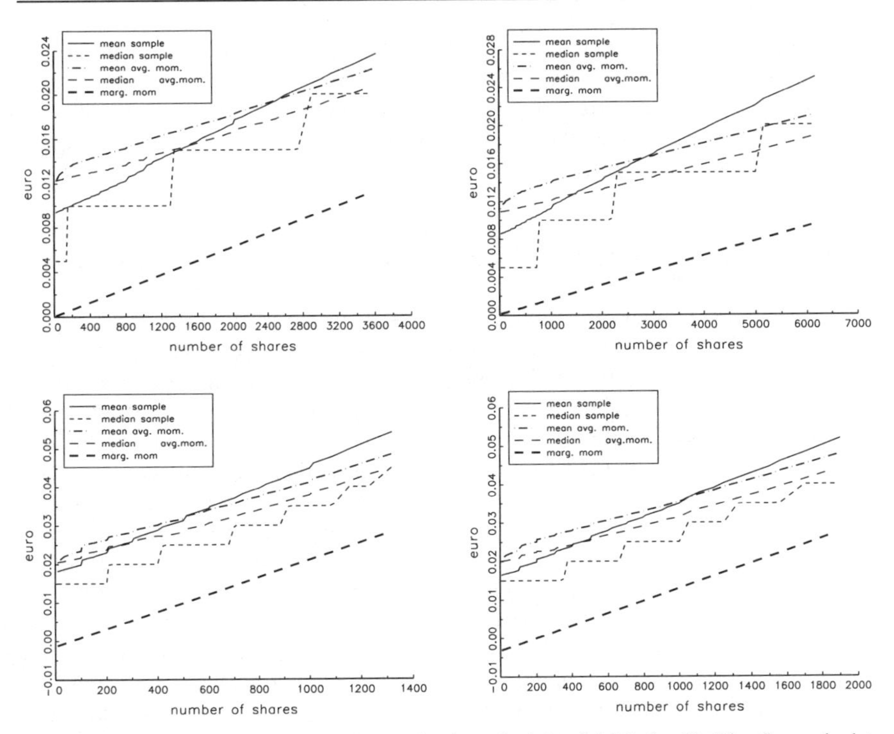

Fig. 1 Comparison of implied and observed price schedules (visible book). The figure depicts means and medians of implied and observed ask side price schedules of four selected stocks. In each figure the values on the horizontal axis show trade volumes (number of shares) up to the 0.9 quantile of the respective stock. The *vertical axis* show the per share price decrease that a sell trade of a given volume would incur if it were executed against the current book. The *solid line* depicts sample means and the *short dashed lines* sample medians computed by using all order book snapshots during the 3 month period. The *bold long-dashed lines* depict the mean slope implied by the estimation results reported in Table 4 (baseline model that uses marginal break even and update conditions). The *dash-dot lines* and the *long-dashed lines* are the mean and the median of the book slope as implied by the estimation results reported in Table 4 (revised specification which uses average break even and update restrictions). The stock in the *left upper panel* is DaimlerChrysler (DCX, from the largest trade volume quartile), the stock in the *right upper panel* is Bay. Hypo Vereinsbank (HVM, second volume quartile), the stock in the *left lower panel* is Altana (ALT, third volume quartile) and the stock in the *right lower panel* is Deutsche Boerse (DB1, fourth volume quartile)

4.2 Cross sectional analyses

Encouraged by the improved empirical performance of the revised methodology, this section uses the estimation results reported in Table 4 to conduct a cross sectional analysis of liquidity supply and adverse selection costs in the Xetra limit order book market. To ensure comparability across stocks, we follow a suggestion by Hasbrouck (1991) and standardize the adverse selection component α by computing

$$\tau = \frac{\alpha \cdot \overline{m}}{\overline{P}}, \tag{13}$$

where $\overline{m}$ is the average (non-signed), stock specific trade size expressed in number of shares. $\overline{P}$ is the sample average of the midquote of the respective stock. τ (times 100) approximates the percentage change of the stock price caused by a trade of (stock specific) 'average' size. This is a relative measure which is comparable across stocks. The τ estimates are reported in the last column of Table 4. In the following subsections we study the relation of τ and market capitalization, trading frequency, liquidity supply and alternative adverse selections measures.

4.2.1 Adverse selection effects, market capitalization and frequency of trading

In their seminal papers Hasbrouck (1991) and Easley et al. (1996) have reported empirical evidence that adverse selection effects are more severe for smaller capitalized stocks. Easley et al. (1996) use a formal model assuming a Bayesian market maker who updates quotes according to the arrival of trades while Hasbrouck (1991) estimates a vector autoregression (VAR) involving trade and midquote returns. Both methodologies have modest data requirements. To estimate the model by Easley et al. (1996) one only needs to count the number of buyer and seller initiated trades per trading day to estimate the probability of informed trading (PIN), the central adverse selection measure in this framework. As it allows a structural interpretation of the model parameter estimates the methodology is quite popular in empirical research. Hasbrouck's VAR methodology is not based on a formal model, but the reduced form VAR equations are compatible with a general class of microstructure models. The adverse selection measure is given by the cumulative effect of a trade innovation on the midquote return. To estimate the model, standard trade and quote data are sufficient.

Both methodologies are not specifically designed for limit order markets, but rather for market maker systems. Accordingly, their main applications have been to analyze NYSE and NASDAQ stocks. In the present paper, the data generating process, the theoretical background and the empirical methodology are quite different. However, we reach the same conclusion as Hasbrouck (1991) and Easley et al. (1996). The Spearman rank correlation of the market capitalization and the estimated standardized adverse selection component τ (using only the results for those 22 out of 30 stocks for which the model is not rejected at 1% significance level) is -0.928 (p-value <0.001). The correlation of τ and the daily number of trades is -0.946 (p-value <0.001), and the correlation of τ and the daily turnover is -0.966 (p-value <0.001). The estimation results thus confirm the previous evidence also for an open limit order market: Adverse selection effects are more severe for smaller capitalized stocks.

4.2.2 Adverse selection and liquidity supply

Many theoretical market microstructure models predict that liquidity supply and informed order flow are inversely related. As the standard framework for microstructure models is a stylized NYSE trading process with a single market maker quoting best bid and ask prices and associated depths (the 'inside market'), liquidity in those models is usually measured by the inside spread set by the specialist/market maker. Sequential trading models like Easley et al. (1996) and spread decomposition models like Glosten and Harris (1988) predict that liquidity (as measured by the spread) and informed order flow are inversely related. In the

Table 5 Correlation of standardized adverse selection component τ with liquidity indicators

Liquidity variable	Correlation	p-value
Quoted spread (%)	0.873	<0.0001
Effective spread (%)	0.794	<0.0001
Realized spread (%)	0.050	0.824

The table reports the cross sectional Spearman rank correlations of the standardized adverse selection component τ reported in Table 4 with average quoted, effective and realized spread reported in Table 1. To compute the correlations we include the stocks for which the model is not rejected at 1% significance level (22 out of 30 stocks). To obtain stock specific measures we take averages over all order book snapshots

presence of informed order flow, the market maker widens the spread in order to balance the losses that occur when trading with superiorly informed agents. More informed order flow thus implies reduced liquidity. Empirical analyses of specialist markets have confirmed this prediction. The results reported in Table 5 provide evidence that the inverse relation of inside spread and informed order flow also holds for open limit order book markets in which limit order traders, instead of specialists, determine the inside market. The table reports the cross sectional correlation (Spearman rank correlation) of the standardized adverse selection component τ and the effective, quoted, and the realized spreads. Effective and quoted spreads and τ are strongly positively correlated while the correlation with the realized spread and τ is not significantly different from zero. Given the interpretation of the realized spreads as a transaction costs measure which is purged of any informational effects, this is an expected result.[15]

4.2.3 Ad hoc versus model-based estimates of adverse selection effects in a limit order book market

In this subsection we investigate whether the adverse selection estimates obtained from the formal model and those delivered by the simple analysis of effective and realized spreads (see Section 2) point in the same direction. The two methodologies differ in two main aspects. First, the estimation of adverse selection components by taking the difference of effective and realized spread is not based on a specific theoretical model. The economic intuition behind the methodology, however, is quite clear, which explains the popularity of the approach. A large difference between effective and realized spread indicates informational content of the order flow as the midquote tends to move in the direction of the trade. If a market buy (sell) order initiates a trade at time t, then the midquote 5 min after the trade is on average above (below) the time t midquote. By contrast, the estimates of the standardized adverse selection component reported in Table 4 are based on a formal model assuming rational limit order traders who place their order submissions explicitly taking into account the amount of informational content of the order flow. Second, computation of price impacts by taking the difference of effective and realized spread only requires publicly available trade and (best) quote data.

[15] Huang and Stoll (1996) and DeJong et al. (1996) provide evidence for a negative correlation of realized spread and adverse selection costs.

To obtain the estimates in the formal framework considered in this paper, reconstructed order book data are needed. The latter methodology thus uses richer data, which are, however, more difficult to obtain.

But do the two different methodologies lead to the same conclusions? To address this question we compute the Spearman rank correlation between the standardized adverse selection components (τ) reported in Table 4 and the difference of effective and realized spread. The cross sectional correlation (using the 22 out of 30 stocks for which the model is not rejected at 1% significance level) is 0.95. The two different methodologies thus point in the same direction. This result indicates the robustness of the estimation results of the formal model and also provides a theoretical justification to use the popular ad hoc method for the analysis of adverse selection effects in limit order book markets.

5 Conclusion and outlook

An increasing number of financial assets trade in limit order markets. These markets can be characterized by the following keywords: Transparency, anonymity and endogenous liquidity supply. They are transparent, because a more or less unobstructed view on the liquidity supply is possible and anonymous, because prior to a trade the identity of none of the agents participating in the transaction is revealed. Liquidity supply is endogenous, because typically there are no dedicated market makers responsible for quoting bid and ask prices. The question how liquidity quality and price formation in such a trading design is affected by informed order flow is a crucial one, both from a theoretical and a practical point of view. Glosten (1994) has put forth a formal model that describes how an equilibrium order book emerges in the presence of potentially informed order flow. Såndas (2001) has confronted the Glosten model with real world data and reported quite discouraging results. His findings suggest that Glosten's model contains too many simplifying assumptions in order to provide a valid description of the intricate real world trading processes in limit order markets.

This paper shows that the ability of Glosten's basic framework to explain real world order book formation is greater than previously thought. We estimate the model using data produced by a DGP that closely corresponds to the Glosten's theoretical framework and confirm the previous finding that the baseline specification put forth by Såndas (2001) is generally rejected. However, relaxing the assumption about marginal zero profit order book equilibrium in favor of a weaker equilibrium condition considerably improves the empirical performance. The equilibrium condition proposed in this paper does not assume that traders immediately cancel a marginal order that shows non-positive expected profit. It also acknowledges the fact that competition between potential market makers will render the expected profit offered by the whole book ultimately to zero (after accounting for opportunity costs). Employing the revised econometric methodology, formal specification tests now accept the model in the vast majority of cases at conventional significance levels. A comparison of implied and observed order book schedules shows that the model estimated on the revised set of moment conditions fits the data quite well. We conclude that Glosten's theoretical framework can also be transferred into a quite useful empirical model.

On the other hand, the conjecture put forth by Hasbrouck (2004), which states that the distributional assumption regarding the market order sizes is responsible for the empirical model failure is not supported. The paper has developed a straightforward way to circumvent the restrictive distributional assumption and proposes a nonparametric alternative. However, this modification does not deliver an improved empirical performance.

Given the overall encouraging results, the empirical methodology is employed for an analysis of liquidity supply and adverse selection costs in a cross section of stocks traded in one of the largest European equity markets. The main results can be summarized as follows:

- We have provided new evidence, from a limit order market, that adverse selection effects are more severe for smaller capitalized, less frequently traded stocks. This corroborates the results of previous papers dealing with a quite different theoretical background, empirical methodology and market structure.
- The empirical results support one of the main hypothesis of the theory of limit order markets, namely that liquidity and adverse selection effects are inversely related.
- The adverse selection component estimates implied by the structural model and ad hoc measures of informed order flow which are based on a comparison of effective and realized spreads point in the same direction. This is a useful result, because it is not always possible to estimate the structural model, most often because of the lack of suitable data. The result also points towards the robustness of the structural model.

Avenues for further research stretch in various directions. The results reported in this paper have vindicated the empirical relevance of the Glosten type market order model. Practical issues in market design can thus be empirically addressed based on a sound theoretical framework. The revised methodology could be employed to evaluate changes in trading design on liquidity quality, with the advantage that the results can be interpreted on a sound theoretical basis. A comparison of (internationally) cross listed stocks seems also promising, especially after the NYSE's move towards adopting the key feature of an open limit order market, the public display of the limit order book. An interesting question would be to investigate whether the recently reported failures of cross listings (in terms of insufficient trading volume in the foreign markets) are due to market design features that aggravate potential adverse selection effects.

Second, a variety of methodological extensions could be considered. Såndas (2001) has already addressed the issue of state dependence of the model parameters. He used a set of plausible instruments to scale the model parameters. Recent papers on price impacts of trades point to alternative, powerful instruments that could be used, and which might improve the empirical performance and explanatory power. For example, Dufour and Engle (2000) have emphasized the role of time between trades within Hasbrouck's (1991) VAR framework. As the Glosten/Såndas type model considered in this paper is also estimated on irregularly spaced data, it seems natural to utilize their findings. Furthermore, the exogeneity of the market order flow is a restrictive assumption that should be relaxed. Gomber et al. (2004) and Coppejans et al. (2003) show that market order traders time their trades by submitting larger trade sizes at times when the book is relatively liquid.

Hence, using the liquidity state of the book as a scaling instrument for the expected order size parameter seems a promising strategy. As in many GMM applications, the number of moment conditions that are available is large, and the difficult task is to pick both relevant and correct moment conditions. Recent contributions by Andrews (1999) and Hall and Peixe (2003) could be utilized to base the selection of moment conditions on a sound methodological basis. Another direction of future research points to a further relaxation of the model's parametric assumptions. Specifically, the linear updating function A.1 could be replaced by a nonlinear relation of asset price and market order size. Combined with a conditional nonparametric distribution for the market order sizes this would provide a quite flexible modeling framework.

Acknowledgements We thank the participants for comments and suggestions. We benefited from discussions and collaborations with Helena Beltran, Michael Binder, Knut Griese, Alexander Kempf, Albert Menkveld, Patrik Såndas, Erik Theissen and Uwe Schweickert. The comments of two anonymous referees helped greatly to improve the quality of the exposition of the paper. We thank the German Stock Exchange for providing access to the Xetra order book data, and the CFR for financial support. The usual disclaimer applies.

Appendix: Derivation of revised moment conditions

This section outlines the background for the revised set of moment conditions describing order book equilibrium. We start by writing the zero expected profit condition for one unit of a limit sell order as

$$E(R_t - X_{t+1}) = 0, \tag{14}$$

where R_t denotes the net revenue (minus transaction costs) received from selling one unit of a limit order at price p_t to a market order trader who submitted a market buy order of size m_t.[16] X_{t+1} denotes the fundamental value of the stock after the arrival of a (buy) market order. X_{t+1} depends on the current value X_t and the signed market order size m_t, i.e. X_{t+1}=$g(m_t, X_t)$. For brevity of notation we henceforth omit the time t subscripts whenever it is unambiguous to do so.

The expected profit of the market order depends on the position of the limit order in the order queue and the distribution of market orders, i.e. we can write Eq. (14) as

$$\int_Q^\infty (R - g(m, X))f(m)dm = 0. \tag{15}$$

Q is the cumulated sell order volume standing in the book before the considered limit order unit and $f(m)$ denotes the probability density function of m. Alternatively, Eq. (15) can be written as

$$(R - E[g(m, X)|m \geq Q]) \cdot P(m \geq Q) = 0. \tag{16}$$

[16] The exercise is analogous for the bid side, but to conserve space, we focus on the sell side of the book.

Assuming the linear specification in Eq. (1) for $g(m, X)$, and dividing by the unconditional probability, $P(m \geq Q)$, Eq. (16) simplifies to

$$R - \alpha E[m|m \geq Q] - X - \mu \cdot = 0. \tag{17}$$

Eq. (17) highlights that the expected profit of a limit order trader depends on the upper tail expectation of the market order distribution.

Assuming exponentially distributed market order sizes as in Eq. (2) we have

$$E[m|m \geq Q] = Q + \lambda \tag{18}$$

Using $R=p-\gamma$ this yields

$$Q = \frac{p - X - \gamma - \mu}{\alpha} - \lambda, \tag{19}$$

which is a generalized form of Eq. (3). Without the distributional assumption, the equivalent of Eq. (19) is

$$E[m|m \geq Q] = \frac{p - X - \gamma - \mu}{\alpha} \tag{20}$$

Replacing $E[m \mid m \geq Q]$ by the conditional sample mean $\widehat{E}[m|m \geq Q]$, i.e. the observed upper tail market order distribution in the sample, one can construct update and break even moment conditions for GMM estimation which do not require a parametric assumption of market order sizes.

So far, the results are valid for an order book with a continuous price grid. We now focus on a specific offer side quote with price p_{+k} and corresponding limit order volume q_{+k}. Abstracting from the discreteness of limit order size shares and assuming that the execution probabilities for all units at the quote tick p_{+k} are identical, we calculate the expected profit of all limit orders with identical limit price p_{+k} by integrating the left hand side of equation, Eq. (17), viz[17]

$$\int_{Q_{+k-1}}^{Q_{+k}} (p_{+k} - \gamma - \alpha E[m|m \geq Q] - X - \mu) dQ \cdot P(m \geq Q_{+k-1}). \tag{21}$$

Assuming exponentially distributed order sizes and subtracting quote specific fixed execution costs ξ yields the total expected profit of the limit order volume at price p_{+k}. Dividing by the volume at quote q_{+k}, yields the average expected profit per share at the $+k$th quote,

$$\left(p_{+k} - X - \mu - \gamma - \frac{\xi}{q_{+k}} - \alpha\left(Q_{+k} + \lambda - \frac{q_{+k}}{2}\right)\right) \cdot P(m \geq Q_{+k-1}). \tag{22}$$

[17] The same result can be derived using the precise probabilities and a first-order Taylor approximation for the emerging exponential terms.

In the main text we discuss the implications of the situation that the average profit equals zero. This implies that

$$p_{+k} - X - \mu - \gamma - \frac{\xi}{q_{+k}} - \alpha\left(Q_{+k} + \lambda - \frac{q_{+k}}{2}\right) = 0. \tag{23}$$

Reordering Eq. (23) and replacing Q_{+k} by $Q_{+k-1}+q_k$ yields the average profit conditions Eq. (10) from which average break even and update conditions can be derived as described in the main text.

References

Ahn H-J, Bae K-H, Chan K (2001) Limit orders, depth, and volatility: evidence from the stock exchange of Hong Kong. J Finance 54:767–788

Andrews DWK (1999) Consistent moment selection procedures for generalized method of moments estimation. Econometrica 67:543–564

Bauwens L, Giot P (2001) Econometric modelling of stock market intraday activity (Kluwer)

Beltran H, Giot P, Grammig J (2004) Commonalities in the order book. Mimeo, CORE, Universities of Louvain, Namur, and Tübingen

Biais B, Hillion P, Spatt C (1995) An empirical analysis of the limit order book and the order flow in the Paris Bourse. J Finance 1655–1689

Biais B, Martimort D, Rochet J-C (2000) Competing mechanisms in a common value environment. Econometrica 68:799–837

Bloomfield R, O'Hara M, Saar G (2005) The make or take decision in an electronic market: evidence on the evolution of liquidity. J Financ Econ 75:165–199

Boehmer E (2004) Dimensions of execution quality: recent evidence for U.S. equity markets. EFA 2004 Working Paper Texas A&M University

Cao C, Hansch O, Wang X (2004) The informational content of an open limit order book. EFA 2004 Working Paper, Penn State University and Southern Illinois University

Cheung YC, de Jong F, Rindi B (2003) Trading European sovereign bonds: the microstructure of the mts trading platforms, Discussion paper EFA 2003 Annual Conference Paper, www.ssrn.com

Coppejans M, Domowitz I, Madhavan A (2003) Dynamics of liquidity in an electronic limit order book market, Discussion paper Department of Economics, Duke University

Degryse H, de Jong F, Ravenswaaij M, Wuyts G (2003) Aggressive orders and the resiliency of a limit order market, Working Paper, K.U. Leuven, Department of Economics

DeJong F, Nijman T, Roëll A (1996) Price effects of trading and components of the bid-ask spread on the Paris Bourse. J Empir Finance 193–213

Dufour A, Engle RF (2000) Time and the price impact of a trade. J Finance 55(6):2467–2499

Easley D, Kiefer NM, O'Hara M, Paperman J (1996) Liquidity, information, and less-frequently traded stocks. J Finance 51:1405–1436

Fernandes M, Grammig J (2005) Non-parametric specification tests for conditional duration models. J Econom 127:35–68

Foucault T (1999) Order flow composition and trading costs in a dynamic limit order market. J Financ Mark 2:99–134

Foucault T, Kadan O, Kandel E (2003) Limit order book as a market for liquidity, Mimeo, HEC

Glosten L (1994) Is the electronic limit order book inevitable? J Finance 49:1127–1161

Glosten LR, Harris LE (1988) Estimating the components of the bid-ask spread. J Financ Econ 21:123–142

Gomber P, Schweickert U, Theissen E (2004) Zooming in on liquidity, EFA 2004 Working Paper, University of Bonn

Grammig J, Heinen A, Reginfo E (2004) Trading activity and liquidity supply in a pure limit order book market, CORE Discussion Paper, 58 Université Catholique de Louvain

Hall A, Peixe F (2003) A consistent method for the selection of relevant instruments. Econ Rev 22:269–287

Hall A, Hautsch N, Mcculloch J (2003) Estimating the intensity of buy and sell arrivals in a limit order book market, Mimeo, University of Technology, Sidney and University of Konstanz

Hansen LP (1982) Large sample properties of generalized method of moments estimators. Econometrica 50:1029–1054
Hasbrouck J (1991) Measuring the information content of stock trades. J Finance 46:179–207
Hasbrouck J (2004) Lecture Notes for PhD Seminar in Empirical Market Microstructure, Stern School of Business, NYU
Huang R, Stoll H (1996) Dealer versus auction markets: a paired comparison of execution costs on NASDAQ and the NYSE. J Financ Econ 41:313–357
Kaniel R, Liu H (2001) So what orders do informed traders use? Discussion paper University of Texas at Austin
O'Hara M (1995) Market Microstructure Theory (Blackwell, Malden, MA)
Parlour C (1998) Price dynamics in limit order markets. R Financ Stud 11:789–816
Pascual R, Veredas D (2004) What pieces of limit order book information are informative? An empirical analysis of a pure order-driven market, Mimeo. Dept. of Business. Universidad de las Islas Baleares
Ranaldo A (2004) Order aggressiveness in limit order book markets. J Financ Mark 7(1):53–74
Såndas P (2001) Adverse selection and competitive market making: empirical evidence from a limit order market. Rev Financ Stud 14:705–734
SEC (2001) Report on the comparison of order executions across equity market structures, Office of Economic Analysis United States Securities and Exchange Commission Working Paper
Seppi D (1997) Liquidity provision with limit orders and a strategic specialist. Rev Financ Stud 10:103–150

Pierre Giot · Joachim Grammig

How large is liquidity risk in an automated auction market?

Abstract We introduce a new empirical methodology that models liquidity risk over short time periods for impatient traders who submit market orders. Using Value-at-Risk type measures, we quantify the liquidity risk premia for portfolios and individual stocks traded on the automated auction market Xetra. The specificity of our approach relies on the adequate econometric modelling of the potential price impact incurred by the liquidation of a portfolio. We study the sensitivity of liquidity risk towards portfolio size and traders' time horizon, and interpret its diurnal variation in the light of market microstructure theory.

1 Introduction

In economics and finance, the notion of liquidity is generally conceived as the ability to trade quickly a large volume with minimal price impact. In an attempt to grasp the concept more precisely, Kyle (1985) identifies three dimensions of liquidity: tightness (reflected in the bid–ask spread), depth (the amount of one-sided volume that can be absorbed by the market without causing a revision of the bid-ask prices), and resiliency (the speed of return to equilibrium). In modern automated auction markets, the liquidity supply solely depends on the state of the electronic order book which consists of previously entered, non-executed limit buy and sell orders. This set of standing orders determines the price-volume rela-

P. Giot
Department of Business Administration & CEREFIM, University of Namur,
Rempart de la Vierge, 8, 5000 Namur, Belgium

P. Giot
CORE, Université catholique de Louvain, Louvain, Belgium
E-mail: pierre.giot@fundp.ac.be

J. Grammig (✉)
Faculty of Economics, University of Tübingen, Mohlstrasse 36, 72074 Tübingen, Germany
E-mail: joachim.grammig@uni-tuebingen.de

J. Grammig
Centre for Financial Research, Cologne, Germany

tionship that a trader who requires immediacy of execution is facing.[1] If few limit buy or sell orders are present in the system or if many orders are present but for small trade sizes only, liquidity is low and marketable limit order trades may incur considerable price impacts. For example, Harris (2002) provides a complete taxonomy of the kinds of trades that can be submitted to exchanges and their impact on market liquidity. Broadly speaking and focusing solely on order books, liquidity providers (patient investors) submit non-aggressive limit orders, i.e. limit orders which do not face immediate execution but which provide liquidity to the system by filling the order book. Liquidity demanders (impatient traders) submit market orders which are executed against standing limit orders and which thus deplete the order book and decrease the overall liquidity.[2] Recent studies which focus on the interaction and dynamics of market orders vs limit orders in automated auctions include Biais et al. (1995), Handa and Schwartz (1996), Ahn et al. (2001) or Beltran et al. (2005a). Due to the interaction between limit and market orders, most studies conclude that there exists a dynamical equilibrium between limit order trading and transitory volatility. Examples of impatient traders include traders who wish to transact near the close of the trading session (so that the price of their trade is not far from the official closing price), see Cushing and Madhavan (2000), or momentum traders who are keen on entering immediate long or short positions (Keim and Madhavan 1997). In all cases, this behavior leads to increased volatility and trading costs.

Because of the price and time priority rules implemented at automated auction markets, the price impact of a buy (sell) side trade is an increasing (decreasing) function of the trade size. As in recent studies focusing on liquidity in automated auction markets (Gouriéroux et al. 1998; Irvine et al. 2000; Coppejans et al. 2004; Domowitz et al. 2005), we model the available liquidity by focusing on the unit price obtained by selling v shares at time t:

$$b_t(v) = \frac{\sum_k b_{k,t} v_{k,t}}{v} \tag{1}$$

where v is the volume executed at k different unique bid prices $b_{k,t}$ with corresponding volumes $v_{k,t}$ standing in the limit order book at time t. The unit price $a_t(v)$ of a buy of size v at time t can be computed analogously. Price impacts of buy and sell trades defined as in Eq. (1) provide important measures of ex-ante liquidity for impatient investors. As phrased in Irvine et al. (2000), "an ex-ante measure of liquidity is useful to investors, because it indicates the cost at which a trade can be immediately executed." The most obvious measure of ex-ante liquidity is the quoted inside spread. With full order book data, one can however do much better as the price impact of buy and sell trades for any given volume v (i.e. for volumes below or above the quoted inside depths) can be computed as in Eq. (1). In dealer based trading system (such as the over-the-counter trading in FOREX markets for

[1] That explains why market participants care about the whole information contained in the book and why order book modelling is of paramount importance. Pascual and Veredas (2004), Cao et al. (2004), Beltran et al. (2005b) or Beltran et al. (2005a), among others, focus on this issue.
[2] Several authors, such as Biais et al. (1995), have put forward a scale of aggressiveness for submitted orders. The most aggressive orders are market orders (also called marketable limit orders) that are fully matched with possibly many standing limit orders that sit on the other side of the book. On the other hand, the most patient investors submit limit orders that enter the order book below/above the best bid/ask prices.

example), ex-ante measures of liquidity are limited to quoted inside spread and quoted inside depths. It is also important to note that liquidity measures such as defined in Eq. (1) characterize committed liquidity as given by the standing limit orders only. With hybrid trading systems which mix characteristics of order book and dealer systems, unit prices $a_t(v)$ and $b_t(v)$ give an upper bound on the price to be paid for the trade as the additional participants can add liquidity prior to the execution of the trade, decreasing $a_t(v)$ or increasing $b_t(v)$.[3] This is also the case for automated auction markets which allow so-called hidden or iceberg orders (see below). In order book markets which feature hidden orders, ex-ante costs of trading measures and liquidity risk measures such as computed in this paper give an upper bound on these trading and liquidity costs (see also Beltran et al. 2005a).[4]

In this paper we will show that, with suitable data at hand, it is possible to quantify the liquidity risk over short term time horizons in automated auction markets. More precisely, we introduce liquidity risk measures that take into account the potential price impact of liquidating a portfolio. This approach is particularly relevant for short term impatient traders who submit market orders. The core of our methodology relies on the comparison of risk measures for so-called frictionless returns (i.e. no-trade returns) and actual returns (which take into account the actual trade price for a v-share trade). These actual returns are particularly relevant for short-term impatient traders who currently hold the stock and who are committed to shortly submit a marketable sell order. In contrast, frictionless returns refer to traders who hold the stock over the same time period, but do not intend to sell their shares. We rely on measures that originate from the Value-at-Risk methodology (see Section 3) to characterize the liquidity risk. It should however be stressed that our framework is not the usual 10-day VaR framework familiar to financial regulators. Hence, we use the VaR methodology to define our intraday risk measures, but in our paper these risk measures are meant to assess the intraday immediate liquidation risk faced by impatient traders. Consequently, we do not derive implications for financial regulators. In contrast, our approach is more similar to Andersen and Bollerslev (1997); Giot (2000, 2005) or Chanda et al. (2005) who characterize volatility on an intraday basis.

In contrast to a standard (frictionless) VaR approach, in which one uses prices based on mid-quotes, the *Actual VaR* approach pursued in this paper uses as inputs volume-dependent transaction prices. This takes into account the fact that buyer (seller) initiated trades incur increasingly higher (lower) prices per unit share as the trade volume increases. The liquidity risk component naturally originates from the volume dependent price impact incurred when the portfolio is liquidated. Our approach relies on the availability of intraday bid and ask prices valid for the immediate trade of any volume of interest. Admittedly, procuring such data from traditional market maker systems would be an extremely tedious task. However, the advent of modern automated auction systems offers new possibilities for em-

[3] Examples are a combination of a limit order book and market markets who bring additional liquidity (Euronext or Xetra, for non-actively traded stocks), or a combination of a limit order book, a specialist and floor traders (NYSE), see Sofianos and Werner (2000) or Venkatamaran (2001). Note that the German Stock Exchange recently adopted the price impact as defined in Eq. (1) as the key liquidity indicator for the automated auction system Xetra (see Gomber et al. 2002).

[4] Nevertheless, they do provide meaningful information as they characterize the worst-case scenarios.

pirical research.[5] Using a unique database (for three stocks traded on the Xetra platform) containing records of all relevant events occurring in an automated auction system, we construct real time order book histories over a three-month period and compute time series of potential price impacts incurred by trading a given portfolio of assets. Based on this data we estimate liquidity adjusted measures and liquidity risk premiums for portfolios and single assets. Our empirical results reveal a pronounced diurnal variation of liquidity risk which is consistent with predictions of microstructure information models. We show that, when assuming an impatient trader's perspective, accounting for liquidity risk becomes a crucial factor: the traditional (frictionless) measures severely underestimate the true risk of the portfolio.

The remainder of the paper is organized as follows: in Section 2, we provide background information about the Xetra system and describe our dataset. The empirical method is developed in Section 3. Results are reported in Section 4. Section 5 concludes and offers possible new research directions.

2 The dataset and the Xetra trading system

In our empirical analysis we use data from the automated auction system Xetra which is employed at various European trading venues, like the Vienna Stock Exchange, the Irish Stock Exchange and the European Energy Exchange.[6] Xetra was developed and is maintained by the German Stock Exchange and has operated since 1997 as the main trading platform for German blue chip stocks at the Frankfurt Stock Exchange (FSE). Whilst there still exist market maker systems operating parallel to Xetra—the largest of which being the Floor of the Frankfurt Stock Exchange—the importance of those venues has been greatly reduced, especially regarding liquid blue chip stocks. Similar to the Paris Bourse's CAC and the Toronto Stock Exchange's CATS trading system, a computerized trading protocol keeps track of entry, cancellation, revision, execution and expiration of market and limit orders. Until September 17, 1999, Xetra trading hours at the FSE extended from 8.30 A.M. to 5.00 P.M. CET. Beginning with September 20, 1999 trading hours were shifted to 9.00 A.M. to 5.30 P.M. CET. Between an opening and a closing call auction—and interrupted by another mid-day call auction—trading is based on a continuous double auction mechanism with automatic matching of orders based on clearly defined rules of price and time priority. Only round lot sized orders can be filled during continuous trading hours. Execution of odd-lot parts of an order (representing fractions of a round lot) is possible only in a call auction. During pre- and post-trading hours it is possible to enter, revise and cancel orders, but order executions are not conducted, even if possible.

[5] As mentioned above, this approach is valid for all order book markets. For automated auction markets which feature hidden orders, our approach delivers worst-case scenarios. This is however the best one can do as hidden orders are by definition not visible.

[6] Bauwens and Giot (2001) provide a complete description of order book markets and Biais et al. (1999) describe the opening auction mechanism used in order book markets and corresponding trading strategies. A lucid description of real world trading processes is found in Harris (2002). Further information about the organization of the Xetra trading process is provided in Deutsche Börse (1999).

Until October 2000, Xetra screens displayed not only best bid and ask prices, but the whole content of the order book to the market participants. This implies that liquidity supply and potential price impact of a market order (or marketable limit order) were exactly known to the trader. This was a great difference compared to e.g. Paris Bourse's CAC system where hidden orders (or 'iceberg' orders) may be present in the order book. As the name suggests, a hidden limit order is not visible in the order book. This implies that if a market order is executed against a hidden order, the trader submitting the market order may receive an unexpected price improvement. Iceberg orders have been allowed in Xetra since October 2000, heeding the request of investors who were reluctant to see their (potentially large) limit orders, i.e. their investment decisions, revealed in the open order book.

The transparency of the Xetra order book does not extend to revealing the identity of the traders submitting market or limit orders. Instead, Xetra trading is completely anonymous and dual capacity trading, i.e. trading on behalf of customers and principal trading by the same institution is not forbidden.[7] In contrast to a market maker system there are no dedicated providers of liquidity, like e.g. the NYSE specialists, at least not for blue chip stocks studied in this paper. For some small cap stocks listed in Xetra there may exist so-called Designated Sponsors—typically large banks—who are obliged to provide a minimum liquidity level by simultaneously submitting buy and sell limit orders.

The German Stock Exchange granted access to a database containing complete information about Xetra open order book events (entries, cancellations, revisions, expirations, partial-fills and full-fills of market and limit orders) that occurred between August 2, 1999 and October 29, 1999.[8] Due to the considerable amount of data and processing time to reconstruct the full order book in 'real time,' we had to restrict the number of assets we deal with in this study. Event histories were extracted for three blue chip stocks, DaimlerChrysler (DCX), Deutsche Telekom (DTE) and SAP. By combining these stocks we also form small, medium and large portfolios to compute the liquidity risk associated with trading portfolios. At the end of the sample period the combined weight of DaimlerChrysler, SAP and Deutsche Telekom in the DAX—the value weighted index of the 30 largest German stocks—amounted to 30.4 percent (October 29, 1999). Hence, the liquidity risk associated with the three stock portfolios is quite representative of the liquidity risk that an investor faces when liquidating the market portfolio of German Stocks.

Based on the event histories we perform a real time reconstruction of the order book sequences. Starting from an initial state of the order book, we track each change in the order book implied by entry, partial or full fill, cancellation and expiration of market and limit orders. This is done by implementing the rules of the Xetra trading protocol outlined in Deutsche Börse (1999) in the reconstruction program.[9] From the resulting real time sequences of order books snapshots at 10 and 30-min frequencies during the trading hours were taken. For each snapshot, the order book entries were sorted on the bid (ask) side in price descending (price ascending) order. Based on the sorted order book sequences we computed the unit

[7] Grammig et al. (2001) and Heidle and Huang (2002) have recently shown how the anonymity feature of automated auction systems can severely aggravate adverse selection effects.
[8] Note that during this period hidden orders were not allowed, and that trading hours shifted in the midst of the sample period.
[9] GAUSS programs for order book reconstruction are available from the authors upon request.

price $b_t(v)$, as defined in Eq. (1), implied by selling at time t volumes v of 1, 5,000, 20,000, and 40,000 shares, respectively. Our choice of volumes is motivated by the descriptive statistics and trading statistics for the three stocks (see below). Mid-quote prices were computed as the average of best bid and ask prices prevailing at time t. Of course these are equivalent to $b_t(1)$ and $a_t(1)$, respectively. If the trade volume v exceeds the depth at the prevailing best quote then $b_t(v)$ will be smaller than $b_t(1)$ (and $a_t(v) > a_t(1)$). By varying the trade volume v one can plot the slope of the instantaneous offer and demand curves. Because Xetra did not allow iceberg type orders during the time period under study, our reconstructed order book is the actual order book faced by market participants. This implies that the computed liquidity risk (see below) is the actual risk incurred by an impatient trader who submits aggressive buy or sell orders for a v-share volume.

Table 1 reports descriptive statistics on trading and liquidity supply activity for the three stocks. Trading activity is high, with 600 to 1,300 trades per day. The large number of (nonmarketable) limit order submissions, of which on average 60% are cancelled before execution, reflects active and competitive liquidity suppliers. The cumulative depth figures show that the order books can sustain large

Table 1 Data descriptives

	DCX	DTE	SAP
Trade descriptives			
Avg. no. of trades per day	1,297	922	661
Avg. transaction price (price per share in euros)	69.9	40.7	402.5
Avg. volume per trade (shares)	1,888	3,352	408
Median volume per trade (shares)	1,000	2,000	300
0.25 quantile volume per trade (shares)	500	900	100
0.75 quantile volume per trade (shares)	2,300	4,200	500
0.95 quantile volume per trade (shares)	5,000	9,900	1,000
Liquidity supply descriptives			
Avg. inside spread (euros)	0.076	0.069	0.732
Avg. volume (shares) at best ask	2,908	4,855	467
Avg. cumulated volume (shares) first two ask quotes	6,230	10,130	986
Avg. cumulated volume (shares) first three ask quotes	9,781	15,575	1,558
Avg. total ask side volume (shares)	350,063	317,725	32,300
Avg. volume (shares) at best bid	2,378	4,648	452
Avg. cumulated volume (shares) first two bid quotes	5,168	9,993	936
Avg. cumulated volume (shares) first three bid quotes	8,145	15,745	1,456
Avg. total bid side volume (shares)	346,334	322,311	33,885
Daily avg. no. submitted non-marketable limit orders per day	3,139	2,311	3,038
Daily avg. no. limit order cancelations per day	1,968	1,395	2,346

The table reports descriptives of the liquidity demand reflected in trade events and liquidity supply reflected in the Xetra order book for Daimler Chrysler (DCX), Deutsche Telekom (DTE) and SAP. The sample period extends from August 2, 1999 to October 29, 1999 and comprises 65 trading days. The descriptives are computed using observations from the continuous trading hours. Until September 17, 1999 the continuous trading period extended from 8.30 A.M. to 5.00 P.M. CET. Beginning with September 20, 1999 the trading hours were shifted to 9.00 A.M. to 5.30 P.M. CET. Except for the daily figures we compute averages over the trade events using the snapshots of the order book immediately before the trade for the liquidity supply variables

trades, also those hypothetical trade sizes that we consider in this paper. Looking at the distribution of trade sizes one can see that the majority of the trades is relatively small, and that the distribution is right-skewed. The hypothetical trade sizes that we consider for the liquidity adjusted VaR method proposed in this paper can thus be considered as large trades, i.e. beyond the 95% quantile of the trade size distribution. This is quite intended as we want to assess the risk associated with liquidating large positions.

3 Methodology

In this section, we detail the econometric methodology used to characterize the liquidity risk faced by impatient traders. We first present an existing model that focuses on liquidity risk in the VaR framework. Then we put forward our new methodology that capitalizes on the ex-ante known state of the order book to derive actual measures of liquidity risk. Throughout this section, we focus on the mean component of the liquidity risk and on the 'uncertainty' component (the standard deviation combined with the quantile), with both measures being combined in a single number akin to a VaR measure.

VaR type market risk measures can be traced back to the middle of the 1990s. First, the 1988 Basel Accord specified the total market risk capital requirement for a financial institution as the sum of the requirements of the equities, interest rates, foreign exchange and gold and commodities positions.[10] Secondly, the 1996 Amendment proposed an alternative approach for determining the market risk capital requirement, allowing the use of an internal model in order to compute the maximum loss over 10 trading days at a 1% confidence level. This set the stage for Value-at-Risk models which take into account the statistical features of the return distribution to quantity the market risk.[11]

Although the use of VaR models was a breakthrough with respect to the much cruder models used earlier, the notion of liquidity risk has been conspicuously absent from the VaR methodology until the end of the 1990s. Subramanian and Jarrow (2001) characterize the liquidity discount (the difference between the market value of a trader's position and its value when liquidated) in a continuous time framework. Empirical models incorporating liquidity risk are developed in Jorion (2000), or Bangia et al. (1999), but none of the methods does explicitly take into account the price impact incurred when liquidating a portfolio of assets. Bangia et al. (1999), henceforth referred to as BDSS, suggest a liquidity risk correction procedure for the VaR framework. BDSS relate the liquidity risk component to the distribution of the inside half-spread. In the first step of the procedure, the VaR is computed as the α percent quantile of the mid-quote return distribution (assuming normality). This quantile is then increased by a factor based on the excess kurtosis of the returns. In a second step, liquidity costs are allowed for by taking as inputs the historical average half-inside-spread and its volatility. This adjusts the VaR for the fact that buy and sell orders are not executed at the quote mid-point, but that

[10] This sum is a major determinant of the eligible capital of the financial institution based on the 8% rule.

[11] Further general information about VaR techniques and regulation issues are available in Dowd (1998), Jorion (2000), Saunders (2000) or at the Bank of International Settlement website http://www.bis.org.

(extreme) variations in the spread may occur. BDSS assume a perfect correlation between the frictionless VaR and the exogenous cost of liquidity. This yields the total VaR being equal to the sum of the market VaR and liquidity cost. Switching from returns to price levels, BDSS express the VaR at level α (including liquidity costs) as:

$$P_t = \frac{a_t(1) + b_t(1)}{2} \left[\left(1 - e^{\mu + Z_\alpha \sigma}\right) + \frac{1}{2} \left(\mu_S + Z'_\alpha \sigma_S\right) \right] \quad (2)$$

where μ and σ are the mean and volatility of the market (mid-quote) returns, μ_S and σ_S are the mean and volatility of the relative spread, Z_α and Z'_α are the α percent quantiles of the distribution of market returns and spread respectively and P_t is the VaR at level α (expressed as a price) taking into account market risk and liquidity costs.

The BDSS procedure offers the possibility to allow for VaR liquidity risk when only the best bid and best ask prices are available. This is, for example, the case when using the popular TAQ data supplied by NYSE. A volume dependent price impact is, quite deliberately, not taken into account as such information is not available from such standard databases. However, a more precise way to allow for liquidity risk becomes feasible with richer data at hand. The approach pursued in this paper relies on the availability of time series of intraday bid and ask prices valid for the immediate trade of a given volume (thanks to the procedure detailed in Section 2). In a market maker setting this requires a time series of quoted bid and ask prices for a given volume. In an automated auction market, unit bid and ask prices can be computed according to Eq. (1) using open order book data. Obtaining such data for a market maker system will be almost impossible. As market makers are obliged to quote only best bid and ask prices with associated depths, quote driven exchanges can and will at best supply this limited information set for financial market research. As a matter of fact, this is the situation where the BDSS approach adds the greatest value in correcting VaR for liquidity risk. In a computerized auction market much richer data can be exploited. As the automated trading protocol keeps track of and records all events occurring in the system it is possible to reconstruct real time series of limit order books from which the required unit bid prices $b_t(v)$ can be straightforwardly computed. Furthermore and as stressed above, the Xetra trading system we work with did not authorize hidden orders at that time; thus, our reconstructed order book and unit prices computed from Eq. (1) actually give the transaction prices relevant for impatient investors.

In order to compute the liquidity risk measures to be introduced below, econometric specifications for two return processes are required. First, for mid-quote returns (referred to as frictionless returns) which are defined as the log ratio of consecutive mid-quotes:[12]

$$r_{mm,t} = \ln \frac{a_t(1) + b_t(1)}{a_{t-1}(1) + b_{t-1}(1)}.$$

[12] Frictionless returns are, somewhat misleadingly, referred to as trading returns in the BDSS framework.

Second, for actual returns which are defined as the log ratio of mid-quote and consecutive unit bid price valid for selling a volume v shares at time t:

$$r_{mb,t}(v) = \ln \frac{b_t(v)}{0.5(a_{t-1}(1) + b_{t-1}(1))}.$$

In the market microstructure framework discussed at the beginning of the paper, the $r_{mb,t}(v)$ returns are relevant for short-term impatient traders who currently (i.e. at time $t-1$) hold the stock and who are committed to submit a marketable sell order for v shares at time t; in contrast, the $r_{mm,t}$ returns refer to a no-trade outcome at time t. For the analysis of liquidity risk associated with a portfolio consisting of $i=1,\ldots,N$ assets with volumes v^i, actual returns are obtained by computing the log ratio of the market value when selling the portfolio at time t, $\sum_{i=1}^{N} b_t(v^i)v^i$, and the value of the portfolio evaluated at time $t-1$ mid-quote prices. To compute frictionless portfolio returns, the portfolio is evaluated at mid-quote prices both at t and $t-1$.

The computation of the liquidity risk faced by impatient traders relies on the individual computation and then comparison of two VaR measures that pertain to the $r_{mm,t}$ and $r_{mb,t}(v)$ returns.[13] For both types of returns the VaR is estimated in the standard way, namely as the one-step ahead forecast of the α percent return quantile. We refer to the VaR computed on the $\{r_{mb,t}(v)\}_{t=1}^{T}$ returns sequence as the *Actual VaR*. Our econometric specifications of the return processes build on previous results on the statistical properties of intraday spreads and return volatility. Two prominent features of intraday return and spread data have to be accounted for. First, spreads feature considerable diurnal variation (see e.g. Chung et al. 1999). Microstructure theory suggests that inventory and asymmetric information effects play a crucial role in procuring these variations. Information models predict that liquidity suppliers (market makers, limit order traders) widen the spread in order to protect themselves against potentially superiorly informed trades around alleged information events, such as the open. Second, as shown by e.g. by Andersen and Bollerslev (1997), conditional heteroskedasticity and diurnal variation of return volatility have to be taken into account. When specifying the conditional mean of the actual return processes we therefore allow for diurnal variations in actual returns, since these contain, by definition, the half-spread. We adopt the specification of Andersen and Bollerslev (1997) to allow for volatility diurnality and conditional heteroskedasticity in the actual return process. Furthermore, diurnal variations in mean returns and return volatility are assumed to depend on the trade volume, as suggested by Gouriéroux et al. (1999). For convenience of notation we suppress the volume dependence of actual returns and write the model as:

$$r_{mb,t} = \psi_t + \delta_0 + \sum_{i=1}^{r} \delta_i r_{mb,t-1} + u_t, \tag{3}$$

[13] See Giot (2005) for an application of VaR type market risk measures to high-frequency returns.

where

$$u_t = \sqrt{\phi_t h_t \varepsilon_t}, \tag{4}$$

$$h_t = \omega + \sum_{i=1}^{q} \alpha_i u_{t-1}^2 + \sum_{i=1}^{p} \beta_i h_{t-1}. \tag{5}$$

The innovations ε_t are assumed to be independently identically Student distributed with ν_1 degrees of freedom. The functions ψ_t and ϕ_t account for diurnal variation in the level of actual returns and return volatility, respectively. We have suppressed the volume dependence of actual returns only for brevity of notation. The discerning reader will recognize that in a more extensive notation all Greek letters would have to be written with a volume index v. Accordingly, the model parameters are estimated for each volume dependent actual return process. We employ a four-step procedure that is described as follows.

First, the diurnal component ψ_t is estimated by a non-parametric regression approach. Given returns available at the s-minute sampling frequency, we subdivide the trading day into s-minute bins, compute the average actual return (over all days in the sample) by bin and smooth the resulting time series using the Nadaraya–Watson estimator.[14] In the second step, a time series of diurnally adjusted returns is obtained by subtracting the estimate $\hat{\psi}_t$ from the actual return $r_{mb,t}$. The resulting time series is used to estimate the AR-parameters process by OLS. The sequence of AR residuals provides the input for modelling actual return volatility. In the third step, the diurnal volatility function ϕ_t is estimated nonparametrically by applying the Nadaraya–Watson estimator to the estimated squared AR residuals, $\{\hat{u}_t^2\}$ which are sorted in s-minute bins. In step four, the squared AR residuals are divided by the estimates $\hat{\phi}_t$. The resulting series is finally used to estimate the GARCH parameters by conditional Maximum Likelihood.[15]

This specification implies that the conditional standard deviation of the actual return at time t, $\sigma_t(r_{mb,t}(v))$, evolves as:

$$\sigma_t(r_{mb,t}) = \sqrt{\phi_t h_t}. \tag{6}$$

[14] See for example Gouriéroux et al. (1998).

[15] As the trading hours shifted during the sample period, the diurnal functions ψ) and ϕ_t are estimated separately for each sub-sample. When estimating the parameters of the autoregressive model components, we have to account for non-trading periods (overnight, weekends) and prevent that end-of-day observations shape the dynamics of the start-of-day returns. For this purpose, we adopt a procedure proposed by Engle and Russell (1998) and re-initialize the AR process at the start of each day. Sample average returns are used as initial values. We do not consider joint estimation which, although feasible, would impose a considerable computational burden. Recent research on related models of intraday price processes has shown that the joint estimation results are quite similar to multi-step procedures like the one outlined above (see Engle and Russell 1998, or Martens et al. 2002).

Correspondingly, the Actual VaR at time t–1 for the actual return at time t given confidence level α is then given by:

$$VaR_{mb,t} = \mu_{mb,t} + t_{\alpha,\nu_1}\sigma_t(r_{mb,t}), \tag{7}$$

where

$$\mu_{mb,t} = \psi_t + \delta_0 + \sum_{i=1}^{r} \delta_i(r_{mb,t-1} - \psi_{t-1}))$$

and t_{α,ν_1} is the α-percent quantile of the student distribution with ν_1 degrees of freedom.[16] We refer to $\mu_{mb,t}$ as the mean component and to $t_{\alpha,\nu_1}\sigma_t(r_{mb,t})$ as the volatility component of the Actual VaR. With respect to the liquidity risk faced by an impatient investor (over the [t–1, t] time interval), $VaR_{mb,t}$ thus provides one number that both summarizes the expected cost (shaped by the mean component $\mu_{mb,t}$) and the volatility cost (determined by $t_{\alpha,\nu_1}\sigma_t(r_{mb,t})$). This duality is central to our methodology as it stresses that the liquidity risk for impatient investors involves both an expected immediacy cost (which is strongly determined by the volume dependent spread) and a second cost that is related to the uncertainty of the trade price for the given traded volume over the [t–1, t] time interval.

In a second step, the computation of the relative (i.e. for an actual trade of v shares vs a no trade situation) liquidity risk premium measures which will be discussed below also requires a VaR estimate based on mid-quote returns (referred to as frictionless VaR). The econometric specification corresponds to the Actual VaR with the exception that there is no need to account for a diurnal variation in mean returns.[17] For notational convenience let us use the same greeks as for the actual return specification:

$$\begin{aligned} r_{mm,t} &= \mu_{mm,t} + u_t, \\ \mu_{mm,t} &= \delta_0 + \sum_{i=1}^{r} \delta_i r_{mm,t-1}, \\ u_t &= \sqrt{\phi(t) h_t \varepsilon_t}, \\ h_t &= \omega + \sum_{i=1}^{q} \alpha_i u_{t-1}^2 + \sum_{i=1}^{p} \beta_i h_{t-1}. \end{aligned} \tag{8}$$

The innovations ε_t are assumed to be independently identically Student distributed with ν_2 degrees of freedom. ϕ_t accounts for diurnal variation in frictionless return volatility. Parameter estimation is performed along the same lines as outlined above. The frictionless VaR at α percent confidence level is:

$$VaR_{mm,t} = \mu_{mm,t} + t_{\alpha,\,\nu_2}\sigma_t(r_{mm,t}). \tag{9}$$

[16] For notational simplicity we suppress the dependence of the VaR measures on α as we do not vary the significance level in the empirical analysis.

[17] In a weakly efficient market, frictionless returns are serially uncorrelated. However, for ultra-high frequency time horizons, some small statistically significant autocorrelations can be expected (see Campbell et al. 1997, or Engle 2000).

where $\sigma_t(r_{mm,t}) = \sqrt{\phi(t)h_t}$. As above we refer to the two terms on right hand side of Eq. (9) as mean and volatility component of the frictionless VaR.

Besides the computation of $VaR_{mb,t}$ type liquidity risk measures, we thus also propose to quantify the relative liquidity risk by comparing the frictionless and Actual VaR. More precisely, two relative liquidity risk premium measures are used, one based on the difference, the other on the ratio of frictionless and Actual VaR:

$$\Lambda_t = VaR_{mm,t} - VaR_{mb,t}, \tag{10}$$

$$\lambda_t = \frac{\Lambda_t}{VaR_{mm,t}}. \tag{11}$$

Omitting the economically negligible mean component of the frictionless VaR, we can rearrange Eq. (11) and write the relative liquidity risk premium as:

$$\lambda_t = \left[\frac{\mu_{mb,t}}{t_{\alpha,\nu_2}\sigma_t(r_{mm,t})}\right] + \left[\frac{t_{\alpha,\nu_1}\sigma_t(r_{mb,t})}{t_{\alpha,\nu_2}\sigma_t(r_{mm,t})} - 1\right]. \tag{12}$$

Equation (12) shows that λ_t can be conceived as the sum of two terms naturally referred to as mean and volatility component of the relative liquidity risk premium. Note that in a more extensive notation, one would write the dependence of both liquidity premiums on portfolio size v and confidence level α.

We want to stress two points before applying the relative liquidity measures Λ_t and λ_t to the three stocks in our dataset. First, λ_t is, by definition, a relative conditional measure. If the VaR horizon is short and the volatilities of both actual and frictionless returns are relatively small and of comparable size then the mean component of the Actual VaR will be the most important determinant of λ_t. Ceteris paribus, the importance of the actual return mean component will increase with trade volume and so will both relative liquidity risk premium measures. At longer VaR horizons, both actual and frictionless return volatility will naturally increase due to non-liquidity related market risk (i.e. the common component that both shapes the best bid and ask prices in the order book and the bid and ask prices achievable for large volume trades).

This reduces the relative importance of the λ-mean component as the denominator of the first term in Eq. (12) grows. If both actual and frictionless return volatility increase by the same factor then the relative liquidity risk premium is expected to approach zero at longer VaR horizons. Second, when studying intraday variations of the relative liquidity risk premium the difference measure Λ_t is more appropriate. Both spreads and volatility of intraday returns are expected to exhibit diurnal variation. By construction, small changes in the intraday frictionless return volatility may exert a considerable impact on the diurnal variation of λ_t, whilst Λ_t is robust against such fluctuations.

4 Empirical results

4.1 Parameter estimates

The Nadaraya–Watson, OLS and Maximum likelihood estimates are obtained using GAUSS procedures written by the authors. Table 2 reports the estimates of the parametric model (AR-GARCH parameters) based on 10-minute returns of equal volume stock (EVS) portfolios and individual stocks. Table 4 (deferred to the Appendix) contains the half-hour frequency results.[18] An EVS portfolio contains the same number of shares for each stock in the portfolio. We will henceforth generally use the notion *portfolio* both for EVS portfolios and single stocks, conceiving the latter as a portfolio containing only a single stock. We report results on small volume (v=5,000), medium volume (v=20,000) and big volume (v=40,000) portfolios. Based on the Schwarz–Bayes-Criteria and Ljung–Box statistics, AR(1)-GARCH(1,1) (half-hour frequency) and AR(3)-GARCH(1,1) (10-min frequency) specifications were selected. Whilst the AR parameter estimates and the Ljung–Box statistics computed on raw returns indicate only small autocorrelations of frictionless returns (as expected in an at least weakly efficient market), the serial dependence of the actual returns on the medium and big portfolios is more pronounced. This results holds true especially at the higher frequency and indicates that persistence in spreads increase with trade volume. After accounting for mean diurnality and serial dependence in actual returns the AR residuals do not display significant autocorrelations. Comparing the Ljung–Box statistic before and after accounting for conditional heteroskedasticity and volatility diurnality reveals that the model does a good job in reducing serial dependence in squared returns. The GARCH parameter estimates and degrees of freedom are quite stable across portfolio sizes and their order of magnitude is comparable to what is found when estimating GARCH models on intraday returns (see Andersen and Bollerslev 1997). Based on the estimation results we compute the frictionless and Actual VaR at α=0.05 as well as the sequence of relative and difference liquidity measures Λ_t and λ_t.

4.2 The diurnal variation of liquidity risk

For the purpose of studying the intraday variation of the relative liquidity risk we take sample averages of the difference measure Λ_t by time of day t, smooth the resulting series by applying the Nadaraya–Watson estimator, and investigate its diurnal variation during trading hours. Figure 1 displays the considerable diurnal variation of the relative liquidity risk premium especially for big portfolios. Liquidity risk is highest at the start of the trading day and sharply declines during the next two hours whilst remaining at a constant level throughout the remainder of the trading day. This pattern is stable across both sample sub-periods with different trading hours. A trader who plans to sell large volumes at the start of the trading day

[18] Diurnal variation in actual returns and volatility diurnality is taken into account employing nonparametric regression techniques, hence no estimation results are presented regarding the diurnal functions ψ_t and ϕ_t. Our GAUSS procedures library is available upon request. The MAXLIK or CML module is required.

Table 2 Estimation results at the 10-min frequency

	Frictionless	ν=5,000	ν=20,000	ν=40,000
EVS portfolios				
δ_0	3.9E−05 (3.1E−05)	0.0E+00 (3.1E−05)	0.0E+00 (3.0E−05)	1.0E−06 (3.2E−05)
δ_1	0.015 (0.018)	0.020 (0.018)	0.041 (0.018)	0.108 (0.018)
δ_2	0.011 (0.018)	0.013 (0.018)	0.020 (0.018)	0.068 (0.018)
δ_3	0.015 (0.018)	0.040 (0.018)	0.055 (0.018)	0.090 (0.018)
ω	0.054 (0.027)	0.061 (0.032)	0.072 (0.032)	0.104 (0.038)
α_1	0.051 (0.015)	0.049 (0.015)	0.050 (0.014)	0.055 (0.014)
β_1	0.868 (0.050)	0.863 (0.055)	0.850 (0.052)	0.807 (0.057)
ν	9.360 (1.437)	9.996 (1.625)	10.567 (1.784)	11.346 (1.995)
Q	0.06 [1.26]	0.06 [3.35]	0.18 [17.76]	1.05 [154.91]
Q^2	4.60 [23.81]	2.49 [14.32]	0.69 [24.58]	0.17 [188.48]
DCX				
δ_0	3.0E−06 (3.7E−05)	−2.0E−06 (3.6E−05)	−1.0E−06 (3.6E−05)	−2.0E−06 (4.0E−05)
δ_1	0.025 (0.018)	0.033 (0.018)	0.082 (0.018)	0.189 (0.018)
δ_2	0.001 (0.018)	0.004 (0.018)	0.020 (0.018)	0.069 (0.018)
δ_3	0.011 (0.018)	0.024 (0.018)	0.036 (0.018)	0.079 (0.018)
ω	0.114 (0.036)	0.118 (0.051)	0.164 (0.049)	0.126 (0.048)
α_1	0.105 (0.021)	0.089 (0.023)	0.098 (0.020)	0.081 (0.019)
β_1	0.667 (0.079)	0.694 (0.104)	0.628 (0.089)	0.718 (0.085)
ν	5.249 (0.507)	5.666 (0.577)	6.312 (0.683)	7.230 (0.861)
Q	0.02 [2.96]	0.06 [5.65]	0.03 [34.57]	0.34 [258.85]
Q^2	1.56 [58.49]	1.25 [42.07]	3.49 [47.48]	2.78 [228.73]
DTE				
δ_0	7.6E-05 (4.9E−05)	2.0E-06 (4.9E−05)	2.0E-06 (4.9E−05)	2.0E-06 (5.2E−05)
δ_1	0.013 (0.018)	0.033 (0.018)	0.062 (0.018)	0.121 (0.018)
δ_2	0.030 (0.018)	0.029 (0.018)	0.034 (0.018)	0.067 (0.018)
δ_3	0.005 (0.018)	0.015 (0.018)	0.038 (0.018)	0.070 (0.018)
ω	0.052 (0.021)	0.040 (0.021)	0.056 (0.022)	0.082 (0.027)
α_1	0.055 (0.013)	0.041 (0.013)	0.057 (0.014)	0.075 (0.015)
β_1	0.841 (0.046)	0.883 (0.045)	0.843 (0.046)	0.789 (0.051)
ν	5.657 (0.578)	6.009 (0.657)	6.704 (0.784)	7.497 (0.917)
Q	0.19 [4.36]	0.26 [11.94]	0.46 [32.41]	0.27 [132.27]
Q^2	12.42 [82.20]	12.55 [67.59]	6.52 [74.93]	4.55 [176.03]
SAP				
δ_0	7.3E−05 (4.7E−05)	1.0E−06 (4.7E−05)	2.0E−06 (4.8E−05)	4.0E−06 (5.1E−05)
δ_1	0.017 (0.018)	0.050 (0.018)	0.109 (0.018)	0.204 (0.018)
δ_2	−0.001 (0.018)	0.009 (0.018)	0.028 (0.018)	0.074 (0.018)
δ_3	−0.011 (0.018)	0.008 (0.018)	0.038 (0.018)	0.071 (0.018)
ω	0.099 (0.028)	0.086 (0.029)	0.073 (0.027)	0.125 (0.047)
α_1	0.089 (0.016)	0.067 (0.014)	0.065 (0.014)	0.093 (0.020)
β_1	0.673 (0.067)	0.752 (0.063)	0.792 (0.057)	0.685 (0.089)

Table 2 (continued)

	Frictionless	v=5,000	v=20,000	v=40,000
ν	4.123 (0.338)	4.748 (0.421)	5.486 (0.543)	6.029 (0.634)
Q	0.14 [0.10]	0.15 [8.44]	0.07 [62.64]	2.72 [310.22]
Q^2	3.80 [73.24]	7.10 [66.27]	6.26 [129.33]	4.66 [505.71]

Section 3 provides details of the estimation procedure. The first column gives estimation results using frictionless returns. The other columns report estimation results based on actual returns. Parameter standard errors are given in parentheses. The Q-rows report the Ljung–Box Q-statistic computed on the AR residuals. For the computation of the Ljung–Box statistics the number of autocorrelations included is equal to 2, and observations of different trading days where excluded from the estimation of autocovariances. In brackets the Ljung–Box statistic of the raw return data is reported. The Q^2 rows reports the Ljung–Box statistic of the GARCH residuals. The figures in brackets are Ljung–Box statistic of squared raw returns

is expected to incur a significant price impact, i.e. has to be ready to pay a considerable liquidity risk premium. The diversification effect smoothes the intraday variation, but the pronounced liquidity risk premium during the first half hours after the open cannot be diversified away.

Figure 2 details this finding by displaying the intraday variation of the mean and volatility components of the Actual VaR, i.e. both components of the liquidity

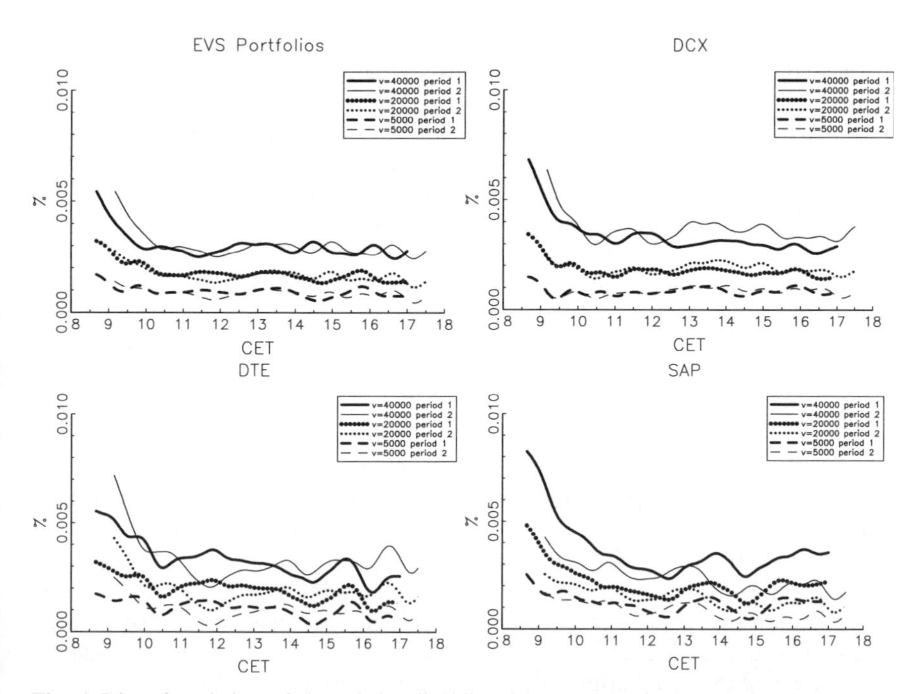

Fig. 1 Diurnal variation of the relative liquidity risk premium Λ_t for equal volume stock portfolios and individual stocks. Period 1 refers to the first half of the sample period, August, 2, 1999–September 17, 1999 when the Xetra continuous trading hours extended from 8.30 A.M. CET —5.00 P.M. CET. Period 2 refers to the second half of the sample period, September 20, 1999–October 29, 1999 when Xetra continuous trading hours extended from 9.00 A.M. CET—5.30 P.M. CET. The numbers have been multiplied by 100 to represent percentages. α=0.05

risk for impatient traders, for the big EVS portfolio (v=40,000).[19] Both mean and volatility component contribute to the diurnal variation of the Actual VaR and hence to the time-of-day pattern of the liquidity risk premium. In the afternoon, NYSE pre-trading exerts an effect on the volatility component of both frictionless and Actual VaR, but as both VaR measures are affected by the same order of magnitude, the relative liquidity risk premium is not affected.

The intraday pattern of the relative liquidity risk premium and Actual VaR provides additional empirical support for the information models developed by Madhavan (1992) and Foster and Viswanathan (1994). Madhavan (1992) considers a model in which information asymmetry is gradually resolved throughout the trading day implying higher spreads at the opening. In the Foster and Viswanathan (1994) model, competition between informed traders leads to high return volatility and spreads at the start of trading. Analyzing NYSE intraday liquidity patterns using the inside spread, Chung et al. (1999) have argued that the high level of the spread at the NYSE opening and its subsequent decrease provides evidence for the information models à la Madhavan and Foster/Viswanathan. Accordingly, the diurnal variation of liquidity risk is consistent with the predictions implied by those models. Due to alleged information asymmetries, liquidity suppliers are initially cautious, i.e. the liquidity risk premium is large. As the information becomes gradually incorporated during the trading process, the liquidity risk premium decreases with increasing liquidity supply.

4.3 Unconditional liquidity risk premiums from traders' perspective

The aftermath of the LTCM debacle showed that disregard of liquidity risk associated with intraday trading of large volumes can lead to devastating results even from a macroeconomic perspective. Let us assess the importance of short term liquidity risk in the present sample. The relative liquidity risk measure λ as well as the difference measure Λ are defined as *conditional* measures given information at time t–1. One can estimate the *unconditional* liquidity risk premium $\lambda = E(\lambda_t)$ by taking sample averages:

$$\overline{\lambda} = T^{-1}\sum_{t-1}^{T}\lambda_t = T^{-1}\sum_{t=1}^{T}\frac{\mu_{mb,t}}{t_{\alpha,\nu_2}\sigma_t(r_{mm,t})} + T^{-1}\sum_{t=1}^{T}\frac{t_{\alpha,\nu_1}\sigma_t(r_{mb,t})}{t_{\alpha,\nu_2}\sigma_t(r_{mm,t})} - 1 \tag{13}$$

and study the dependence of the unconditional liquidity risk premium on the size of the portfolio to be liquidated. Equation 13 shows that the decomposition of the relative liquidity risk premium λ_t into mean and volatility component remains valid for the unconditional liquidity risk premium.

Table 3 reports the estimated unconditional liquidity risk premium $\overline{\lambda}$. The decomposition into mean and volatility component is contained in Table 5. The results show that taking account of liquidity risk at the intraday level is quite crucial. Even at medium portfolio size, the liquidity risk premium is considerable. At half-hour horizon the underestimation of the VaR of the medium EVS portfolio amounts to 34%. For the big EVS portfolio the VaR is underestimated at half-hour

[19] As above we take sample averages by time of day for each component and apply the Nadaraya–Watson smoother.

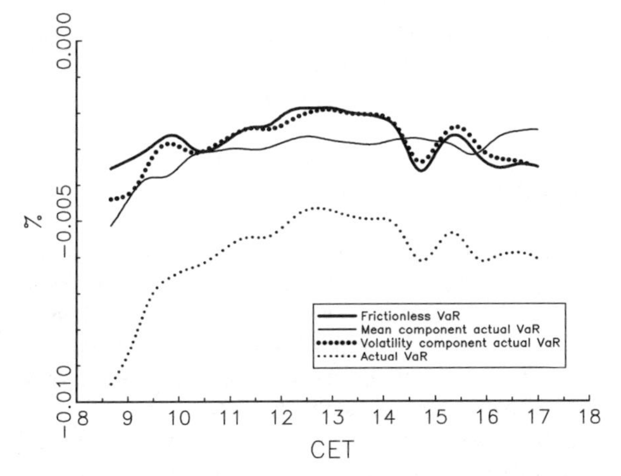

Fig. 2 Decomposition of relative liquidity premium $\Lambda_t = VaR_{mm,t} - VaR_{mb,t}$. The figure displays for the big EVS portfolio (v=40,000) and for the first half of the sample period (August, 2, 1999–September 17, 1999) the diurnal variation of the components of Λ_t. The Actual VaR ($VaR_{mb,t}$) is the sum of the mean component, $\mu_{mb,t}$, and the volatility component, $t_{\alpha,v_1}\sigma_t(r_{mb,t})$ (see Eq. (7)). The frictionless VaR ($VaR_{mm,t}$) is defined in Eq. (9). The numbers are multiplied by 100 to represent percentages. α=0.05

horizon by 61%. At the shorter horizon the underestimation becomes even more severe with the medium EVS portfolio's VaR being underestimated by 68%. The decomposition exercise shows that when increasing portfolio size the volatility component of the liquidity risk premium remains small relative to the mean component since the price impact incurred by trading a large portfolio becomes the dominating factor. The relative liquidity risk premium decreases, ceteris paribus, at the longer VaR horizon. The reason is that both the frictionless and the Actual VaR's volatility components obey the square root of time rule, and increase (in absolute terms) by about the same order of magnitude. This reduces the sig-

Table 3 Unconditional relative liquidity risk premium $\overline{\lambda}$ estimates at different VaR time horizons

VaR horizon	v=5,000	v=20,000	v=40,000
EVS portfolios			
10-min	0.35	0.68	1.20
half-hour	0.17	0.34	0.61
DCX			
10-min	0.30	0.64	1.21
half-hour	0.15	0.32	0.60
DTE			
10-min	0.28	0.51	0.86
half-hour	0.11	0.22	0.39
SAP			
10-min	0.29	0.50	0.83
half-hour	0.12	0.23	0.38

Table 5 (Appendix) contains the decomposition $\overline{\lambda}$ into mean and volatility components

nificance of the mean component while the order of magnitude of the volatility component remains approximately constant. In other words, market risk compared to liquidity risk becomes more important at longer horizons.

5 Conclusion and outlook

This paper quantified liquidity risk in an automated auction market by employing a new empirical technique which extends the classical frictionless VaR methodology. The notion of an Actual VaR measure was introduced which takes into account the potential price impact of liquidating a portfolio. This Actual VaR measure is particularly relevant for impatient investors who submit market orders. Indeed it provides one number that summarizes the expected cost of immediacy and the volatility cost due to the market risk. This duality is central to our methodology as it stresses that the liquidity risk for impatient investors over short time periods involves both a spread dependent expected cost and a second cost that is related to the variance of the price process (for the given traded volume) over the short time interval being studied. Relative liquidity risk measures were also put forward that are defined on the difference and ratio of Actual and standard (frictionless) VaR. It was argued that automated auction markets, in which a computer manages order entry and matching, provide suitable and accurate data for the task of estimating the Actual VaR. In order to measure the liquidity risk using the methodology pursued in this paper one has to provide a real time limit order book reconstruction from which price impacts of trading a given volume can be computed. Using data from the automated auction system Xetra, liquidity risk was quantified both for portfolios and for individual stocks. The dependence of liquidity risk premiums on time-of-day, trade volume and VaR time horizon was outlined.

The analysis revealed a pronounced diurnal variation of the liquidity risk premium. The peak of the liquidity risk premium at the open and its subsequent decrease is consistent with the predictions of the microstructure information models considered by Madhavan (1992) and Foster and Viswanathan (1994). The results show that when assuming a trader's perspective, accounting for liquidity risk becomes a crucial factor: the traditional (frictionless) measures severely underestimate the risk faced by impatient investors who are committed to trade. This result is the more pronounced the bigger the portfolio size and the shorter the time horizon.

Avenues for further research stretch in several directions. In the classification of Dowd (1998), this paper has focused on the normal liquidity risk in contrast to crisis liquidity risk. The latter refers to situations where "a market can be very liquid most of the time, but lose its liquidity in a major crisis" (Dowd 1998). The methodology applied in this paper could be readily used to evaluate crisis liquidity risk using intraday data specific to such crisis periods where liquidity dried up for a few days/weeks (for example during the Asian crisis of the summer months of 1997 or close to the LTCM debacle in September 1998). Second, in a cross section study for a larger collection of stocks one could relate liquidity risk premiums to firm-specific characteristics like e.g. the equity–debt ratio. This would facilitate investigating the question whether corporate financing decisions or ownership structure affect firm specific liquidity risk. Thirdly, it would be interesting to look at how investors adapt to the prevailing liquidity risk. Indeed, traders could decide to split their orders into smaller chunks to avoid prohibitive trading costs. This paves the way for dynamic strategies where market participants are interested in dynamic risk measures which take into account the fact that large orders can be split into smaller orders. To our knowledge, this however asks for much more complicated models that are quite different from the literature on high-frequency volatility models such as considered in this paper.

Acknowledgements We are grateful to Deutsche Börse AG for providing access to the limit order data and to Kai-Oliver Maurer and Uwe Schweickert who offered invaluable experise regarding the Xetra trading system. Helpful comments and suggestions were offered by Erik Theissen, Michael Genser, Rico von Wyss and seminar participants at the Sfb 386 workshop on stochastic volatility and risk management at Munich University, December 2002, the 2003 EEA/ESEM congress in Stockholm, the 54th ISI session in Berlin 2003 and at the CREST finance seminar in October 2003. We thank Helena Beltran-Lopez for her cooperation in the preparation of the datasets and Bogdan Manescu for research assistance. Finally we thank the two anonymous referees for useful suggestions and comments.

1 Appendix

1.1 Additional tables

1.1.1 Table 4 Estimation results at the 30-min frequency

	Frictionless	v=5,000	v=20,000	v=40,000
EVS portfolios				
δ_0	1.2E−04 (9.3E−05)	−3.0E−06 (9.2E−05)	−3.0E−06 (9.0E−05)	−2.0E−06 (9.0E−05)
δ_1	0.052 (0.031)	0.065 (0.031)	0.067 (0.031)	0.081 (0.031)
ω	0.029 (0.019)	0.038 (0.024)	0.035 (0.021)	0.045 (0.026)
α_1	0.029 (0.013)	0.031 (0.014)	0.030 (0.013)	0.027 (0.013)
β_1	0.935 (0.031)	0.923 (0.037)	0.926 (0.033)	0.916 (0.039)
ν	9.386 (10.048)	20.167 (10.548)	16.230 (7.030)	13.687 (5.048)
Q	0.20 [3.46]	0.20 [4.45]	0.19 [4.87]	0.32 [7.69]
Q^2	0.26 [4.99]	0.66 [1.13]	0.66 [1.62]	0.81 [2.34]
DCX				
δ_0	2.9E−02 (1.9E−02)	−7.0E−06 (1.1E−04)	−6.0E−06 (1.1E−04)	−7.0E−06 (1.1E−04)
δ_1	0.029 (0.013)	0.012 (0.031)	0.021 (0.031)	0.061 (0.031)
ω	0.122 (0.071)	0.129 (0.073)	0.194 (0.141)	0.274 (0.185)
α_1	0.060 (0.024)	0.058 (0.023)	0.069 (0.031)	0.087 (0.037)
β_1	0.744 (0.123)	0.744 (0.121)	0.647 (0.220)	0.535 (0.271)
ν	6.590 (1.396)	7.058 (1.585)	7.536 (1.756)	8.666 (2.159)
Q	2.94 [3.14]	3.03 [3.13]	2.40 [1.86]	1.96 [3.70]
Q^2	0.60 [16.69]	0.09 [13.30]	0.12 [10.11]	0.14 [6.93]
DTE				
δ_0	2.3E−04 (1.5E−04)	−2.0E−06 (1.5E−04)	−1.0E−06 (1.5E−04)	3.0E−06 (1.5E−04)
δ_1	0.036 (0.031)	0.042 (0.031)	0.054 (0.031)	0.070 (0.031)
ω	0.051 (0.023)	0.053 (0.024)	0.054 (0.025)	0.058 (0.027)
α_1	0.029 (0.012)	0.030 (0.012)	0.029 (0.012)	0.028 (0.013)
β_1	0.881 (0.042)	0.877 (0.042)	0.876 (0.045)	0.872 (0.049)
ν	5.889 (1.156)	5.944 (1.171)	5.631 (1.072)	5.766 (1.107)

Table 4 (continued)

	Frictionless	ν=5,000	ν=20,000	ν=40,000
Q	0.09 [1.52]	0.04 [2.05]	0.23 [3.53]	1.36 [6.93]
Q^2	5.38 [5.42]	3.14 [2.30]	3.54 [1.66]	4.47 [1.57]
SAP				
δ_0	2.2E−04 (1.4E−04)	4.0E−06 (1.4E−04)	4.0E−06 (1.4E−04)	5.0E−06 (1.4E−04)
δ_1	0.056 (0.031)	0.058 (0.031)	0.081 (0.031)	0.107 (0.031)
ω	0.233 (0.133)	0.233 (0.123)	0.252 (0.116)	0.304 (0.163)
α_1	0.075 (0.034)	0.057 (0.028)	0.065 (0.029)	0.061 (0.031)
β_1	0.592 (0.200)	0.626 (0.175)	0.595 (0.161)	0.528 (0.223)
ν	7.677 (1.771)	8.524 (2.157)	8.826 (2.324)	8.448 (2.134)
Q	1.99 [3.39]	1.96 [5.32]	1.84 [10.47]	1.67 [20.75]
Q^2	2.16 [11.23]	0.97 [8.94]	0.39 [10.17]	0.46 [15.58]

See Table 2 for explanations.

1.1.2 Table 5 Decomposition of the unconditional relative liquidity risk premium $\overline{\lambda}$ at different VaR time horizons

The table reports the three parts of $\overline{\lambda}=T^{-1}\sum_{t=1}^{T}\frac{\mu_{mb,t}}{t_{\alpha,\nu_2}\sigma_t(r_{mm,t})}+T^{-1}\sum_{t=1}^{T}\frac{t_{\alpha,\nu_1}\sigma_t(r_{\mathrm{mb},t})}{t_{\alpha,\nu_2}\sigma_t(r_{\mathrm{mm},t})}-1$. Because of rounding errors, the sum of mean and volatility components might differ from the figures reported in Table 3.

VaR horizon	ν=5,000	ν=20,000	ν=40,000
EVS portfolios			
10-min	0.36+0.99−1	0.70+0.98−1	1.16+1.04−1
half-hour	0.19+0.98−1	0.38+0.96−1	0.64+0.97−1
DCX			
10-min	0.29+1.01−1	0.62+1.03−1	1.08+1.13−1
half-hour	0.17+0.99−1	0.35+0.98−1	0.60+1.00−1
DTE			
10-min	0.27+1.01−1	0.49+1.02−1	0.77+1.08−1
half-hour	0.12+0.98−1	0.25+0.98−1	0.40+0.99−1
SAP			
10-min	0.28+1.02−1	0.47+1.03−1	0.73+1.10−1
half-hour	0.13+0.99−1	0.23+1.00−1	0.37+1.02−1

References

Ahn H, Bae K, Chan K (2001) Limit orders, depth and volatility: evidence from the stock exchange of Hong Kong. J Finance 56:767–788

Andersen TG, Bollerslev T (1997) Intraday periodicity and volatility persistence in financial markets. J Empir Finance 4:115–158

Bangia A, Diebold FX, Schuermann T, Stroughair JD (1999) Modeling liquidity risk, with implications for traditional market risk measurement and management, The Wharton Financial Institutions Center WP 99-06

Bauwens L, Giot P (2001) Econometric modelling of stock market intraday activity. Kluwer Academic Publishers

Beltran H, Durré A, Giot P (2005a) Volatility regimes and the provision of liquidity in order book markets, CORE Discussion Paper 2005/12

Beltran H, Giot P, Grammig J (2005b) Commonalities in the order book, CORE Discussion Paper 2005/11

Biais B, Hillion P, Spatt C (1995) An empirical analysis of the limit order book and the order flow in the Paris Bourse. J Finance 50:1655–1689

Biais B, Hillion P, Spatt C (1999) Price discovery and learning during the preopening period in the Paris Bourse. J Polit Econ 107:1218–1248

Campbell JY, Lo AW, MacKinlay AC (1997) The econometrics of financial markets. Princeton University Press, Princeton

Cao C, Hansch O, Wang X (2004) The information content of an open limit order book, Mimeo, available at http://ssrn.com/abstract=565324

Chanda A, Engle RF, Sokalska ME (2005) High frequency multiplicative component GARCH, Mimeo

Chung KH, Van Ness BF, Van Ness RA (1999) Limit orders and the bid-ask spread. J Financ Econ 53:255–287

Coppejans M, Domowitz I, Madhavan A (2004) Resiliency in an automated auction, Working Paper, ITG Group

Cushing D, Madhavan A (2000) Stock returns and trading at the close. J Financ Mark 3:45–67

Deutsche Börse AG (1999) Xetra Market Model Release 3 Stock Trading, Technical Report

Domowitz I, Hansch O, Wang X (2005) Liquidity commonality and return comovement. Forthcoming in Journal of Financial Markets

Dowd K (1998) Beyond Value-at-Risk. Wiley

Engle RF (2000) The econometrics of ultra high frequency data. Econometrica 68:1–22

Engle RF, Russell J (1998) Autoregressive conditional duration; a new model for irregularly spaced transaction data. Econometrica 66:1127–1162

Foster F, Viswanathan S (1994) Strategic trading with asymmetrically informed investors and long-lived information. J Financ Quant Anal 29:499–518

Giot P (2000) Time transformations, intraday data and volatility models. J Comput Financ 4:31–62

Giot P (2005) Market risk models for intraday data. Eur J Financ 11:309–324

Gomber P, Schweickert U, Deutsche Börse AG (2002) Der Market Impact: Liquiditätsmass im elektronischen Handel. Die Bank 7:185–189

Gouriéroux C, Le Fol G, Meyer B (1998) Analyse du carnet d'ordres. Banque et Marchés 36:5–20

Gouriéroux C, Le Fol G, Jasiak J (1999) Intraday market activity. J Financ Mark 2:193–226

Grammig J, Schiereck D, Theissen E (2001) Knowing me, knowing you: trader anonymity and informed trading in parallel markets. J Financ Mark 4:385–412

Handa P, Schwartz RA (1996) Limit order trading. J Finance 51:1835–1861

Harris L (2002) Trading and exchanges. Oxford University Press

Heidle HG, Huang RD (2002) Information-based trading in dealer and auction markets: an analysis of exchange listings. J Financ Quant Anal 37:391–424

Irvine P, Benston G, Kandel E (2000) Liquidity beyond the inside spread: measuring and using information in the limit order book, Mimeo, Goizueta Business School, Emory University, Atlanta

Jorion P (2000) Value-at-Risk. McGraw-Hill

Keim DB, Madhavan A (1997) Transaction costs and investment style: an interexchange analysis of institutional equity trades. J Financ Econ 46:265–292

Kyle AS (1985) Continuous auctions and insider trading. Econometrica 53:1315–1335

Madhavan A (1992) Trading mechanisms in securities markets. J Finance 47:607–642

Martens M, Chang y, Taylor SJ (2002) A comparison of seasonal adjustment methods when forecasting intraday volatility. J Financ Res 25:283–299

Pascual R, Veredas D (2004) What pieces of limit order book information are informative? An empirical analysis of a pure order-driven market, WP Stern Business School FIN-04-003

Saunders A (2000) Financial institutions management. McGraw-Hill

Sofianos G, Werner IM (2000) The trades of NYSE floor brokers. J Financ Mark 3:139–176

Subramanian A, Jarrow RA (2001) The liquidity discount. Math Financ 11:447–474

Venkatamaran K (2001) Automated versus floor trading: an analysis of execution costs on the Paris and New York Stock Exchange. J Finance 56:1445–1485

Anthony D. Hall · Nikolaus Hautsch

Order aggressiveness and order book dynamics

Abstract In this paper, we study the determinants of order aggressiveness and traders' order submission strategy in an open limit order book market. Applying an order classification scheme, we model the most aggressive market orders, limit orders as well as cancellations on both sides of the market employing a six-dimensional autoregressive conditional intensity model. Using order book data from the Australian Stock Exchange, we find that market depth, the queued volume, the bid-ask spread, recent volatility, as well as recent changes in both the order flow and the price play an important role in explaining the determinants of order aggressiveness. Overall, our empirical results broadly confirm theoretical predictions on limit order book trading. However, we also find evidence for behavior that can be attributed to particular liquidity and volatility effects.

Keywords Open limit order book · Aggressive market orders · Aggressive limit orders and cancellations · Multivariate intensity

JEL Classification G14 · C32 · C41

1 Introduction

Limit order book data provide the maximum amount of information about financial markets at the lowest aggregation level. A theme in the recent literature is to obtain a better understanding of all of the aspects of a trader's fundamental decision problem: when to submit an order; which type of order to submit; and, on which side of the market to submit the order.

A. D. Hall
School of Finance and Economics, University of Technology, Sydney, Australia

N. Hautsch (✉)
Department of Economics, University of Copenhagen, Studiestraede 6, 1455 Copenhagen-K, Denmark
E-mail: Nikolaus.Hautsch@econ.ku.dk

In this paper, we study traders' order aggressiveness in an open limit order book market. Applying an order categorization scheme, we model the arrival rate of most aggressive market orders, limit orders as well as cancellations on both sides of the market in dependence of the state of the book. The six-dimensional point process implied by the random and irregular occurrence of the different types of orders is modelled in terms of the (multivariate) intensity function, associated with the contemporaneous instantaneous arrival rate of an order in each dimension. The intensity function is a natural concept to overcome the difficulties associated with the asynchronous arrival of individual orders and allows for a continuous-time modelling of the simultaneous decision of when and which order to submit given the state of the market.

In the previous literature on order aggressiveness, the trader's decision problem has typically been addressed by applying the order classification scheme proposed by Biais et al. (1995). In this classification scheme, orders are categorized according to their implied price impact and their implied execution probability determined by their position in the book. The major advantage of this approach is its ease of application since all of the information on order aggressiveness is encapsulated into a (univariate) variable which permits modelling the degree of aggressiveness using a standard ordered probit model with explanatory variables that capture the state of the order book.[1] However, there are three major drawbacks of this model. First, it is not a dynamic model, so any dynamics within the individual processes as well as all interdependencies between the processes are ignored. Ignoring multivariate dynamics and spill-over effects can induce misspecifications and biases. Second, Coppejans and Domowitz (2002) show that with respect to particular order book variables, trades behave quite differently from limit orders and cancellations. This raises the question as to whether it is reasonable to treat these events as the ordered realizations of the same (single) variable.[2] Third, modelling order aggressiveness based on an ordered response model ignores the timing of orders. Thus, the trader's decision is modelled conditional on the fact that there *is* a submission of an order at a particular point in time while the question of *when* to place the order is ignored.

Our study avoids these difficulties and extends the existing approaches by Coppejans and Domowitz (2002), Ranaldo (2004), and Pascual and Veredas (2004) in several directions. First, the use of a multivariate autoregressive intensity model explicitly accounts for order book dynamics and interdependencies between the individual processes. Second, as we model them as individual processes, we allow for the possibility that market orders, limit orders and cancellations behave differently in their dependence on particular order book variables. Instead of trying to capture order aggressiveness in terms of a single variable, we account for the multiple dimensions of the decision problem. Third, the concept of the intensity function implies a natural continuous-time measurement of a trader's degree of aggressiveness. As the multivariate intensity function provides the instantaneous order arrival probability per time at each instant and in each dimension, it naturally

[1] See e.g. Al-Suhaibani and Kryzanowski (2000), Griffiths et al. (2000), Hollifield et al. (2002), Ranaldo (2004) or Pascual and Veredas (2004).

[2] For this reason, Pascual and Veredas (2004) consider the decision process as a sequential process with two steps. In the first step, the trader chooses between a market order, limit order and a cancellation, while in the second step, he decides the exact order placement.

addresses the question of where and when it is likely that an order will be placed given the current state of the market.

In order to reduce the impact of noise in the data and to allow for a better identification of systematic relationships, we explicitly focus on most aggressive market orders, limit orders and cancellations. On top of the common classification scheme proposed by Biais et al. (1995), we select only those orders whose volumes are substantially larger than the average order volume.[3] According to previous studies in this field, these are the most aggressive and interesting orders. Note that this classification scheme also applies to cancellations, and we define a cancellation as aggressive whenever a large volume is cancelled. In this sense, our approach can be seen as an extension of the study by Coppejans and Domowitz (2002) who also focus on the arrival rate of trades, limit orders and cancellations. However, they do not explicitly study high volume orders but consider all incoming orders. Moreover, as they analyze the individual processes separately using a generalized version of Engle and Russell's (1998) (univariate) autoregressive conditional duration (ACD) model, their framework does not allow for any multivariate interdependencies between the individual processes.

In this setting, we state the following research questions: (1) Can we confirm previous results regarding the determinants of order aggressiveness and traders' order submission strategies when the multivariate dynamics of limit order books are fully taken into account? (2) How strong are the (dynamic) interdependencies between the individual processes and how important is it to account for the order book dynamics? (3) After modelling the multivariate dynamics, what is the additional explanatory power of order book variables? (4) Can we confirm theoretical predictions regarding the impact of order book variables on traders' order submission strategies?

Our analysis is based on order book data from the five most liquid stocks traded on the Australian Stock Exchange (ASX) during the period July–August 2002. By replicating the electronic trading at the ASX, we reconstruct the complete order book at each instant of time. The order arrival intensities are modelled using a six-dimensional version of the autoregressive conditional intensity (ACI) model introduced by Russell (1999), where we include explanatory variables that capture the current state of the order book as well as recent changes in the book.

It turns out that market depth, the queued volume, the bid-ask spread, recent volatility, as well as recent changes in both the order flow and the price play an important role in explaining the determinants of order aggressiveness. We show that the impact of these variables is quite stable over a cross-section of stocks. Moreover, these results hold irrespective of the specification of the model dynamics. Confirming the results of Coppejans and Domowitz (2002) we also observe that the arrival rates of market orders and limit orders can behave quite differently in their dependence of the state of the order book. Therefore, a limit order should not necessarily be considered simply as a less aggressive version of a market order. This finding motivates modelling the individual processes in a multivariate setting. Moreover, we find clear evidence for multivariate dynamics and interdependencies between the individual processes.

It is also shown that the inclusion of order book variables clearly improves the goodness-of-fit of the model. In addition, we demonstrate that a model that

[3] For more details, see Sect. 4.3.

includes order book variables, but excludes dynamics, outperforms a dynamic specification without covariates. This result clearly indicates that traders' order aggressiveness and order submission strategy is affected by the state of the book.

Regarding the impact of order book variables on order aggressiveness, our results broadly confirm the theoretical results on traders' optimal order submission strategies as derived by Parlour (1998) and Foucault (1999). In particular, the impact of depth on order aggressiveness can be explained by "crowding out" effects as discussed in Parlour (1998). Moreover, our findings provide evidence for the notion that traders use the order book information to infer expected future price movements. Nevertheless, we also observe behavior that is not consistent with predictions implied by theoretical dynamic equilibrium models. For instance, we find evidence for liquidity driven order submissions after mid-quote changes in the recent past. Furthermore, no support is found for the hypothesis that the current volatility affects the mix between aggressive market and limit orders. Rather, we observe that a rise in volatility increases the overall order submission activity in the market.

The remainder of the paper is organized in the following way: In Sect. 2, we discuss economic hypotheses on the basis of recent theoretical research on limit order book trading. Section 3 presents the econometric approach. In Sect. 4, we describe the data as well as descriptive statistics characterizing the limit order books of the individual stocks traded at the ASX. The empirical results are reported and discussed in Sect. 5 and Sect. 6 concludes.

2 Economic hypotheses

The desire for a deeper understanding of market participants' order submission strategies in a limit order book market has inspired a wide range of theoretical and empirical research.[4] In a limit order book market investors must choose between limit orders and market orders and as a result traders face a dilemma. The advantage of a market order is that it is executed immediately. However, with a limit order, while traders have the possibility of improving their execution price, they face the risk of non-execution as well as the risk of being "picked off". The latter arises from the possibility that, as a result of new information entering the market, a limit order can become mispriced. These economic principles form the basis of numerous theoretical approaches in this area.

Parlour (1998) proposes a dynamic equilibrium model in which traders with different valuations for an asset arrive randomly in the market. The endogenous execution probability of a limit order then depends both on the state of the book and how many market orders will arrive over the remainder of the day. She shows that both the past, through the state of the book, and the future, through the expected order flow, affect the placement strategy and cause systematic patterns in transaction and order data. The major underlying idea is the mechanism of a "crowding out" of market sell (buy) orders after observing market buy (sell) orders. This is due to the effect that after a buy (sell) market order, a limit order at the ask

[4] See e.g. Glosten (1994), Handa and Schwartz (1996), Harris and Hasbrouck (1996), Seppi (1997), Harris (1998), Bisière and Kamionka (2000), Griffiths et al. (2000), Lo and Sapp (2003), Cao et al. (2003), or Ranaldo (2004) among others.

(bid) has a higher execution probability. Since the payoff to limit orders increases with the probability of execution, a trader who wants to sell (buy) is now more likely to submit a sell (buy) limit order instead of a sell (buy) market order. Because of this crowding out of market orders on the opposite side, buy (sell) market orders are less frequent after sell (buy) market orders than after buy (sell) market orders. This trading behavior has been confirmed by the empirical work of Biais et al. (1995), Griffiths et al. (2000) and Ranaldo (2004). Parlour's model also predicts that the probability of observing a limit buy order after the arrival of a limit buy order is smaller than the probability of observing a limit buy order after any other transaction. This is due to the fact that a lengthening of the queue at one level decreases the execution probability of further limit orders at the same level and thus makes them less attractive. Applying this theoretical underpinning, the "crowding out" argument implies testable relationships between changes in the depth of the book volume and their impact on traders' incentive to post market orders, limit orders or cancellations. As a result, we can formulate the following hypothesis:

(1) An increase of the depth on the ask (bid) side

- increases the aggressiveness of market trading on the bid (ask) side,
- decreases the aggressiveness of limit order trading on the ask (bid) side,
- increases the probability of cancellations on the ask (bid) side.

A traders' order submission strategy does not only depend on the current state of the book but also on recent movements in the price. Positive price movements during the recent past indicate an aggressiveness in buy limit orders and buy market trading leading to a relative decline of the ask depth compared to the bid depth. Applying the crowding out concept from Parlour's model we can formulate Hypothesis (2) as follows:

(2) Past price movements are

- negatively (positively) correlated with the aggressiveness of market trading on the bid (ask) side,
- positively (negatively) correlated with the aggressiveness of limit order trading on the ask (bid) side,
- negatively (positively) correlated with the probability of cancellations on the ask (bid) side.

Foucault (1999) proposes a dynamic equilibrium model to explain traders' choice between limit orders and market orders as a function of the asset's volatility. In this model, investors' valuations of shares differ and traders' order placement strategies depend on their valuations as well as the best offers in the book. Foucault (1999) shows that the volatility of the asset is a determinant of the mix between market and limit orders. Since higher volatility increases the pick-off risk, this increases the reservation prices of limit order traders, widening spreads and increasing the cost of market trading. As a result traders' incentive to post limit orders (market orders) increases (decreases) leading to Hypothesis (3):

(3) A higher volatility decreases the aggressiveness in market order trading and increases the aggressiveness in limit order trading.

An important determinant of liquidity is the inside spread between the best ask and bid price. The bid-ask spread determines the cost of crossing the market and thus the cost of utilizing market orders. Cohen et al. (1981) show that the existence of a bid-ask spread is a result of the "gravitational pull" of a limit order and is an equilibrium property of the market. Handa et al. (2003) demonstrate that the size of the spread increases with the degree of adverse selection and the difference in valuation between low and high valuation investors. However, while these studies focus on the existence and properties of the spread, Foucault's model provides testable implications regarding the impact of the spread on the aggressiveness in market trading and limit order trading, and this is formulated as Hypothesis (4):

(4) The higher the bid-ask spread, the lower the aggressiveness in market trading and the higher the aggressiveness in limit order trading.

These formulated hypotheses underpin the rationale for the construction of appropriate explanatory variables in Sect. 5.

3 The econometric approach

The arrival of aggressive market orders, limit orders and cancellations is modelled as a multivariate (financial) point process. The econometric literature on the modelling of financial point processes was originated by the seminal paper by Engle and Russell (1998) who introduced the class of autoregressive conditional duration (ACD) models. While this model was successfully applied to univariate duration processes,[5] it is not easily extended to a multivariate framework. The reason is that in a multivariate context the individual processes occur asynchronously, which is difficult to address in a discrete time duration model.

A natural way to model multivariate point processes is to specify the (multivariate) intensity function leading to a continuous-time framework. In this paper, we apply a six-dimensional version of the autoregressive conditional intensity (ACI) model proposed by Russell (1999).[6] Following the notation of Hall and Hautsch (2004), let t denote the calendar time and define t_i^k, $k=1,\ldots,K$, as the arrival times of a K-dimensional point process. Let $N^k(t) = \sum_{i\geq 1} \mathbb{1}_{\{t_i^k \leq t\}}$ and $M^k(t) = \sum_{i\geq 1} \mathbb{1}_{\{t_i^k < t\}}$ respectively represent the right- and left-continuous counting functions associated with the k-type process. Correspondingly, $M(t) =$

[5] This model has been extended in several directions, see e.g. Bauwens and Giot (2000), Lunde (2000), Dufour and Engle (2000), Grammig and Maurer (2000), Zhang et al. (2001), Fernandes and Grammig (2001), Coppejans and Domowitz (2002) or Bauwens and Veredas (2004) among others. For an overview, see Hautsch (2004).

[6] An interesting alternative would be the latent factor intensity (LFI) model proposed by Bauwens and Hautsch (2003), where the key idea is to allow for a common latent component which jointly drives the individual processes. Even though such a specification would be particularly interesting for the modelling of limit order book processes, its estimation requires substantial computational effort. As the computational burden for a six-dimensional process with included order book variables is already quite high, we leave the application of the LFI model in this context to future research.

$\sum_{i\geq 1} \mathbb{1}_{\{t_i<t\}}$ is the left-continuous counting function regarding the pooled process consisting of *all* individual points.

Define the multivariate intensity function as

$$\lambda(t;\mathcal{F}_t) = \left(\lambda^1(t;\mathcal{F}_t), \lambda^2(t;\mathcal{F}_t), \ldots, \lambda^K(t;\mathcal{F}_t)\right),$$

where

$$\lambda^k(t;\mathcal{F}_t) = \lim_{\Delta\downarrow 0} \frac{1}{\Delta} \Pr\left[\left(N^k(t+\Delta) - N^k(t)\right) > 0|\mathcal{F}_t\right],\ k = 1,\ldots,K, \tag{1}$$

denotes the conditional intensity function associated with the counting process $N^k(t)$, given the information set $\mathcal{F}_t$ consisting of the history of the complete order and trading process up to t. In this framework $\lambda^k(t;\mathcal{F}_t)$ corresponds to the instantaneous arrival rate of an aggressive order or cancellation, and thus is a natural continuous-time measure for the degree of order aggressiveness at each instant.

Russell (1999) proposes parameterizing $\lambda^k(t;\mathcal{F}_t)$ in terms of a proportional intensity structure

$$\lambda^k(t;\mathcal{F}_t) = \Psi^k_{M(t)}\lambda^k_0(t)s^k(t), \quad k = 1,\ldots,K, \tag{2}$$

where Ψ_i^k is a function capturing the dynamics of the k-type process, $\lambda^k_0(t)$ denotes a k-type baseline intensity component that specifies the deterministic evolution of the intensity until the next event and $s^k(t)$ is a k-type seasonality component that may be specified using a spline function. The basic idea of the ACI model is to specify the dynamic component Ψ_i^k in terms of an autoregressive process. Assume that Ψ_i^k is specified in log-linear form, i.e.

$$\Psi^k_i = \exp\left(\widetilde{\Psi}^k_i + z'_{i-1}\gamma^k\right), \tag{3}$$

where z_i denotes the vector of explanatory variables capturing the state of the market at arrival time t_i and γ^k the corresponding parameter vector associated with process k. Then, the ACI(1,1) model is obtained by parameterizing the $(K\times 1)$ vector $\widetilde{\Psi}_i = \left(\widetilde{\Psi}^1_i, \widetilde{\Psi}^2_i, \ldots, \widetilde{\Psi}^K_i\right)$ in terms of a VARMA type specification,

$$\widetilde{\Psi}_i = \sum_{k=1}^{K} \left(A^k e_{i-1} + B\widetilde{\Psi}_{i-1}\right) y^k_{i-1}, \tag{4}$$

where $A^k=\{\alpha^j_k\}$ denotes a $(K\times 1)$ innovation parameter vector and $B=\{\beta^{ij}\}$ is a $(K\times K)$ matrix of persistence parameters. Moreover, y_i^k defines an indicator variable that takes the value 1 if the i-th point of the pooled process is of type k.

The innovation term e_i is computed from the integrated intensity function associated with the most recently observed process. Hence,

$$e_i = \sum_{k=1}^{K} \left(1 - \int_{t^k_{N^k(t_i)-1}}^{t^k_{N^k(t_i)}} \lambda^k(s; \mathcal{F}_s) ds \right) y^k_i. \tag{5}$$

Under fairly weak assumptions, the integrated intensity function corresponds to an i.i.d. standard exponential variate.[7] Therefore, e_i is a random mixture of exponential variates. For this reason, weak stationarity of the model depends on the eigenvalues of the matrix B. If these lie inside the unit circle, the process $\widetilde{\Psi}_i$ is weakly stationary.

Since the intensity function has left-continuous sample paths, Ψ_i also has to be a left-continuous function and predetermined, at least instantaneously *before* the arrival of a new event. Therefore, Ψ_i is known instantaneously after the occurrence of t_{i-1} and does not change until t_i. Then $\lambda^k(t; \mathcal{F}_t)$ changes between t_{i-1} and t_i only as a deterministic function of time according to the functions $\lambda^k_0(t)$ and $s^k(t)$.

The baseline intensity function $\lambda^k_0(t)$ is specified in terms of the backward recurrence times $x^k(t) = t - t^k_{M^k(t)}$, $k=1,\ldots,K$, of all processes and may be specified using a Weibull-type parameterization depending on the parameters ω^k and p^k_r,

$$\lambda^k_0(t) = \exp\left(\omega^k\right) \prod_{r=1}^{K} x^r(t)^{p^k_r - 1}, \quad \left(p^k_r > 0\right). \tag{6}$$

Moreover, the seasonality functions $s^k(t)$ are parameterized as linear spline functions given by[8]

$$s^k(t) = 1 + \sum_{j=1}^{S} \nu^k_j \left(t - \tau_j\right) \cdot \mathbb{1}_{\{t > \tau_j\}}, \tag{7}$$

where τ_j, $j=1,\ldots, S$, denote the S nodes within a trading day and ν^k_j the corresponding parameters.

As a result, by denoting W as the data matrix consisting of all points and explanatory variables and denoting θ as the parameter vector of the model, the log-likelihood function of the multivariate ACI model is given by

$$\ln \mathcal{L}(W; \theta) = \sum_{k=1}^{K} \sum_{i=1}^{n} \left\{ -\int_{t_{i-1}}^{t_i} \lambda^k(s; \mathcal{F}_s) ds + y^k_i \ln \lambda^k\left(t_i; \mathcal{F}_{t_i}\right) \right\}, \tag{8}$$

where t_0=0 and n denotes the number of points of the pooled process. Under correct specification of the model, the resulting k-type ACI residuals

$$\hat{\varepsilon}^k_i = \int_{t^k_{i-1}}^{t^k_i} \hat{\lambda}^k(s; \mathcal{F}_s) ds$$

[7] See Brémaud (1981) and Bowsher (2002).

[8] In order to identify the constant ω^k, $s^k(t)$ is set to one at the beginning of a trading day.

should be distributed as i.i.d. unit exponential. As a result, model diagnostics can be undertaken by evaluating the dynamical and distributional properties of these residuals. Engle and Russell (1998) have proposed a test against excess dispersion based on the asymptotically normal test statistic $\sqrt{n^k/8}(\hat{\sigma}^2_{\varepsilon^k} - 1)$, where $\hat{\sigma}^2_{\varepsilon^k}$ is the empirical variance of the k-type residual series and n^k denotes the number of points observed for process k.

4 Data and descriptive statistics

4.1 Trading at the ASX

The Australian Stock Exchange (ASX) is a continuous double auction electronic market. After a pre-opening period followed by a staggered sequence of opening call auctions, normal trading takes place continuously between 10:00 A.M. and 16:00 P.M. The market is closed with a further call auction and a late trading period. For more details regarding the daily market schedule of the ASX, see Hall and Hautsch (2004). During normal trading, orders can be entered as market orders which will execute immediately and limit orders which enter the queues. On the ASX, as orders are not allowed to walk up (down) the book, a market order with a large quoted volume will first be matched with the pending volume on the first level of the opposite queue. Trades will be generated and traded orders deleted until there is no more order volume at the posted price. The remaining part of the order enters the queue as a corresponding limit order. When a market order is executed against several pending limit orders, a trade record for each market order—limit order pair is generated. Since these multiple trades are generated by a single market order, we aggregate them into a single trade record.

Limit orders are queued in the buy and sell queues according to a strict price-time priority order. During normal trading, pending limit orders can be modified or cancelled without restrictions.[9] All trades and orders are visible to the public. Orders with a total value exceeding $200,000 can be entered with a hidden volume. However, sufficient information is available to unambiguously reconstruct all transactions.

4.2 Descriptive statistics

Our empirical analysis is based on the order book data from the five most liquid stocks traded at the ASX during the period 1 July to 30 August 2002, namely Broken Hill Proprietary Limited (BHP), National Australia Bank (NAB), News Corporation (NCP), Telstra (TLS) and Woolworths (WOW). The samples are extracted from the Stock Exchange Automated Trading System and contain time stamped prices, volumes and identification attributes of all orders as well as information about opening and closing auctions. By replicating the execution engine of the ASX, with explicit consideration of all trading rules, we can fully reconstruct the individual order books at any time. Our resulting samples consist of

[9] Clearly, modifying the order volume or the order price can affect the order priority. For more details, see Hall and Hautsch (2004).

Table 1 Order book characteristics

	BHP		NAB		NCP		TLS		WOW	
	Mean	S.D.	Mean	S.D.	Mean	S.D.	Mean	S.D.	Mean	S.D.
sprd	1.075	0.311	2.175	1.516	1.277	0.599	1.006	0.079	1.485	0.922
trvol	2,746.7	7,688.5	1,174.1	2,901.3	1,113.7	4,131.5	8,877.5	39,414.2	1,646.7	4,906.6
buyvol	5,692.4	9,783.9	2,830.6	4,016.9	4,817.5	7,408.2	2,1233.8	57,399.5	3,707.5	7,184.2
sellvol	8,336.2	12,190.6	2,902.7	3,902.7	5,358.7	7,789.6	21,816.1	60,877.4	4,410.3	6,785.4
qvol	8,303.4	38,223.4	3,534.3	4,858.5	5,235.3	6,238.4	35,513.0	68,337.0	4,832.7	7,329.7
qavol	9,704.9	12,842.4	3,621.1	5,169.3	5,209.8	6,058.5	35,077.7	68,038.8	5,123.8	7,046.5
qbvol	7,212.8	49,667.3	3,453.4	4,549.0	5,264.2	6,436.2	35,927.5	68,618.7	4,585.3	7,553.3
cvol	11,172.2	81,984.5	4,308.8	6,063.3	4,533.8	5,999.2	61,007.7	97,236.4	6,021.8	9,469.3
cavol	11,595.3	17,307.7	4,348.3	6,392.8	4,315.7	5,438.5	64,477.8	100,813.0	5,793.7	8,761.2
cbvol	10,778.7	112,671.2	4,271.5	5,736.1	4,774.1	6,553.5	58,166.4	94,120.8	6,211.8	10,018.1
d_askp	5.842	86.084	6.407	24.454	7.247	52.743	3.198	11.616	4.818	16.021
d_bidp	4.662	16.628	8.543	71.853	6.131	14.328	2.377	7.798	4.104	14.611
amq	0.115	0.316	0.423	0.809	0.101	0.319	0.040	0.197	0.230	0.518
avol	1,452.4	343.7	265.1	62.1	920.6	256.4	5,409.2	1,558.8	501.8	185.0
bvol	1,376.9	665.7	222.9	78.7	720.3	216.3	6,237.1	1,647.8	344.5	230.7
adiff_1	0.897	0.671	2.018	1.926	1.271	0.939	0.577	0.268	1.376	1.323
adiff_2	1.308	1.007	2.880	2.752	1.884	1.319	0.643	0.357	2.001	1.931

Table 1 (continued)

	BHP		NAB		NCP		TLS		WOW	
	Mean	S.D.	Mean	S.D.	Mean	S.D.	Mean	S.D.	Mean	S.D.
	BHP		NAB		NCP		TLS		WOW	
	Mean	S.D.	Mean	S.D.	Mean	S.D.	Mean	S.D.	Mean	S.D.
adiff_5	2.717	2.116	5.829	5.403	3.712	2.340	0.866	0.565	4.024	3.415
bdiff_1	0.898	0.650	1.851	1.700	1.155	0.869	0.582	0.273	1.187	1.032
bdiff_2	1.269	0.954	2.523	2.381	1.633	1.204	0.656	0.371	1.546	1.409
bdiff_5	2.488	1.948	4.806	4.639	3.077	2.078	0.937	0.612	2.781	2.610
adep_1	22.257	10.725	2.404	1.870	11.055	7.154	102.651	35.792	6.352	4.686
adep_2	34.429	21.732	3.614	3.262	15.621	12.257	196.338	77.341	9.862	8.665
adep_5	45.662	40.078	4.577	5.093	18.572	16.865	420.273	216.059	12.729	14.358
bdep_1	20.477	13.596	2.093	1.697	9.203	5.613	117.936	39.229	4.516	4.183
bdep_2	31.411	23.779	3.228	2.924	13.874	10.131	223.766	85.101	7.479	7.465
bdep_5	42.927	38.160	4.293	4.507	17.876	15.783	455.945	240.004	11.290	12.538

Means and standard deviations of various order book characteristics based on the BHP, NAB, NCP, TLS and WOW stock traded at the ASX. The samples contain all market and limit orders of the individual stocks traded at the ASX during July–August 2002, corresponding to 45 trading days. All prices are measured in cents. The order book characteristics are: Bid-ask spread (*sprd*), traded volume (*trvol*), traded buy/sell volume (*buyvol, sellvol*), quoted volume (*qvol*), quoted ask/bid volume (*qavol, qbvol*), cancelled volume (*cvol*), cancelled ask/bid volume (*cavol, cbvol*), difference between quoted ask price and the current best ask price (*d_askp*), difference between current best bid quote and quoted bid price (*d_bidp*), absolute mid-quote change (*amq*), as well as cumulated ask/bid volume (*avol, bvol*, in units of 1,000 shares). Furthermore, $adiff\_x = p_{x,a} - mq$, where $p_{x,a}$ denotes the price associated with the x%-quantile of the cumulated ask volume and mq denotes the mid-quote. Correspondingly, $bdiff\_x = mq - p_{x,b}$, where $p_{x,b}$ denotes the price associated with the x%-quantile of the cumulated bid volume. Moreover, $adep\_x = (x/100) \cdot avol/(p_{x,a} - mq)$ and $bdep\_x = (x/100) \cdot bvol/(mq - p_{x,b})$, measured in units of 1,000 shares

data covering the normal trading period, where we remove data from the opening and closing call auctions as well as all market crossings and off-market trades. The resulting samples consist of 147,552, 107,595, 252,009, 97,804, and 59,519 observations for BHP, NAB, NCP, TLS and WOW, respectively.

Table 1 shows descriptive statistics characterizing the order books of the five individual stocks during the sample period. We observe an average bid-ask spread ranging between 1.0 ticks for TLS and 2.2 ticks for NAB. For all stocks, the average sell volume is slightly higher than the buy volume which may be explained by a slightly down market during the period of analysis. However, comparing the average posted as well as cancelled ask and bid volumes, there are no systematic differences. The variables *d_askp* and *d_bidp* measure the difference between the current posted price of a limit order and the current best ask and bid price, respectively. Thus, we observe that on average limit orders are placed within a distance of about five ticks to the current best ask and bid price. Again, smaller spreads are set for TLS (around two to three ticks), whereas NAB and NCP reveal relatively wide spreads of around six to eight ticks. The variables *adiff_x* and *bdiff_x* represent the price difference between the mid-quote and the price associated with the x-th quantile of the standing ask and bid volume, respectively.[10] Therefore, they reflect the average piecewise steepness of the bid and ask reaction curves. For most stocks, the average shape of both curves is relatively symmetric, but for WOW we observe a slightly higher average depth on the bid side. Finally, the variables *adep_x* measure the market depth in terms of the ratio of the volume associated with the x-th quantile and the corresponding implied price impact. For instance, *adep_5*=45.662 for BHP means that up to the 5%-quantile we observe on average a standing ask volume of 45,662 shares per tick. Again, we observe relatively symmetric shapes of the individual bid and ask queues.

4.3 Order aggressiveness at the ASX

Biais et al. (1995) propose a scheme that classifies orders according to their implied price impact and their position in the order book. By setting the limit price, the limit volume and attributes associated with specific execution rules traders implicitly determine the aggressiveness of their order, and thus influence both the execution probability and the implied price impact. Generally, the most aggressive order is a market order which is allowed to be matched with several price levels on the opposite side, i.e. an order which is allowed to "walk up" or "down" the book. Accordingly, the least aggressive order is a cancellation where a pending limit order is removed from the book. As explained above ASX market orders are not allowed to walk up or down the ask or bid queues, respectively. Hence, at the ASX the most aggressive order is an order which has a volume that exceeds the standing volume on the first level of the opposite queue and results in an immediately executed market order for the matched volume and a limit order for the remaining volume. Accordingly, we define a "normal" market order as a buy or sell order whose volume can be fully matched with pending limit orders. With respect to limit orders, we apply the scheme proposed by Biais et al. (1995) and classify ask and bid limit orders according to the distance between the posted limit price and the

[10] Table 1 contains an exact definition of the variables.

current best bid and ask price. Thus, we distinguish between "most aggressive" limit orders (whose price undercuts or overbids the current best ask or bid limit price, respectively), "aggressive" limit orders (which are placed directly in the current first level of the ask or bid queue), and "normal" limit orders (which enter the higher levels of the order book). Finally, cancelled limit orders are regarded as the least aggressive orders. This classification is shown in Table 2.

Table 3 shows the average numbers of the different types of orders. We observe that the proportions of the particular order types are not stable across stocks and show clear variations. For instance, the percentage of aggressive market orders varies between 2 and 7%. Accordingly, the proportion of most aggressive limit order varies between 0.1 and 4%. On average, we observe a higher proportion of the most aggressive ask limit orders than that of the corresponding bid limit orders. Furthermore, it turns out that on average around 5% of all limit orders are cancelled. An exceptionally high proportion of cancellations of around 11% is observed for NCP.

In our empirical analysis, we focus on the three most interesting groups of orders: aggressive buy and sell orders, most aggressive ask and bid limit orders as well as ask and bid cancellations. Moreover, we introduce a further criterion for order aggressiveness which goes beyond the scheme shown in Table 2. In particular, we exclusively select only those orders which have a quoted volume which is significantly above average. This additional selection criteria is applied for two reasons: First, order aggressiveness is naturally linked to the size of the posted volume. For economic significance, it makes a difference whether a small or a high volume is quoted. For high volumes, the economic trade-off between the costs of immediacy and the pick-off risk is much more relevant than for small orders. Second, focussing exclusively on the big trades should reduce the noise in the data and should help to identify distinct patterns and relationships. Since there is a natural trade-off between focusing on orders whose order volumes are on the one hand significantly above the average volume but, on the other hand, still

Table 2 Classification of order aggressiveness at the ASX

Aggressive buy order	Quoted volume exceeds the first level of the standing ask volume
Normal buy order	Quoted volume does not exceed the first level of the standing ask volume
Most aggressive ask order	Limit price undercuts the current best ask price
Aggressive ask order	Limit price is at the current best ask price
Normal ask order	Limit price is above the current best ask price
Cancelled ask order	Cancellation of a standing ask order
Aggressive sell order	Quoted volume exceeds the first level of the standing bid volume
Normal sell order	Quoted volume does not exceed the first level of the standing bid volume
Most aggressive bid order	Limit price overbids the current best bid price
Aggressive bid order	Limit price is at the current best bid price
Normal bid order	Limit price is below the current best bid price
Cancelled bid order	Cancellation of a standing bid order

Table 3 Descriptive statistics of trade and limit order arrival processes at the ASX

	BHP		NAB		NCP		TLS		WOW	
	Num.	Prop.	Num.	Prop.	Num.	Prop.	Num.	Prop.	Num.	Prop.
Total number	147,552		107,595		252,009		97,804		59,519	
Aggr. Buys, $v \geq 75\%$	1,946	0.013	2,657	0.025	2,385	0.009	522	0.005	838	0.014
Aggr. buys	5,998	0.041	8,057	0.075	7,700	0.031	1,761	0.018	3,359	0.056
Normal buys	28,349	0.192	14,303	0.133	21,402	0.085	19,142	0.196	10,125	0.170
Most aggr. asks, $v \geq 75\%$	1,511	0.010	1,763	0.016	2,058	0.008	241	0.002	795	0.013
Most aggr. asks	4,092	0.028	4,695	0.044	6,879	0.027	909	0.009	2,460	0.041
Aggr. asks	15,065	0.102	11,632	0.108	14,544	0.058	12,595	0.129	6,787	0.114
Normal asks	10,306	0.070	7,963	0.074	50,769	0.201	9,565	0.098	4,106	0.069
Canc. asks, $v \geq 75\%$	1,029	0.007	842	0.008	528	0.002	716	0.007	273	0.005
Canc. asks	6,689	0.045	5,912	0.055	30,390	0.121	4,572	0.047	2,763	0.046
Aggr. sells, $v \geq 75\%$	2,338	0.016	2,867	0.027	2,372	0.009	551	0.006	892	0.015
Aggr. sells	6,133	0.042	8,016	0.075	7,711	0.031	1,747	0.018	3,559	0.060
Normal sells	15,455	0.105	13,100	0.122	17,249	0.068	17,707	0.181	7,329	0.123
Most aggr. bids, $v \geq 75\%$	1,611	0.011	1,455	0.014	1,174	0.005	353	0.004	348	0.006
Most aggr. bids	5,772	0.039	4,626	0.043	5,903	0.023	1,249	0.013	2,109	0.035
Aggr. bids	17,441	0.118	12,967	0.121	15,333	0.061	13,848	0.142	8,010	0.135
Normal bids	14,900	0.101	8,504	0.079	42,375	0.168	9,125	0.093	5,594	0.094
Canc. bids, $v \geq 75\%$	881	0.006	879	0.008	625	0.002	719	0.007	292	0.005
Canc. bids	7,329	0.050	6,265	0.058	27,624	0.110	5,584	0.057	3,318	0.056
All aggr. Orders, $v \geq 75\%$	9,316		10,463		9,142		3,102		3,438	

Descriptive statistics of order arrival processes of the BHP, NAB, NCP, TLS and WOW stock. The order categories are defined in Table 2. "$v \geq 75\%$" means that the quoted volume is equal or higher than the 75%-quantile of all order volumes. The table shows the number of orders in the individual categories as well as their corresponding percentage with respect to the complete sample of the individual stock. The sample contains all market and limit orders of the individual stocks traded on the ASX during July and August 2002, corresponding to 45 trading days

retaining sufficient observations in the individual dimensions, we use the 75%-quantile as a natural compromise for the choice of the selection threshold. Hence, we select those aggressive orders whose posted volume is equal or higher than the 75%-quantile of all order volumes. Correspondingly, aggressive cancellations are cancellations of pending limit orders with a volume equal or higher than the 75%-quantile of all order volumes. This selection rule leads to a significant reduction of the sample size resulting in 9,316, 10,463, 9,142, 3,102, and 3,438 observations for BHP, NAB, NCP, TLS, and WOW respectively (see Table 3).

5 Empirical results

As the estimation of six-dimensional ACI processes is a challenging task requiring the estimation of a large number of parameters, we estimate restricted ACI specifications. In order to reduce the number of parameters, we specify the baseline intensity functions in terms of a Weibull parameterization, where we do not allow for interdependencies between the individual functions, i.e. $p_r^k=1 \forall k \neq r$. This restriction is motivated by the fact that the consideration of interdependencies would require estimating 30 additional parameters without significantly improving the model's goodness-of-fit in terms of the BIC. Similar arguments hold for the specification of spill-over effects in the persistence terms, where we restrict the matrix B to be specified as a diagonal matrix. In order to account for deterministic intra-day seasonality patterns, we specify three linear spline functions for the processes of aggressive market orders, limit orders, and cancellations based on 1 h nodes.[11] To ease the numerical optimization of the log-likelihood function, we standardize the time scale by the average duration of the pooled process.

To test the economic hypotheses formulated in Sect. 2, we define several explanatory variables to capture the state of the market. The market depth on the ask side is measured by the (log) ratio between the current 5% ask volume quantile and the corresponding price impact, formally given by $AD=\ln[0.05 \cdot avol/(p_{0.05,a}-mq)]$, where *avol* denotes the aggregated volume pending on the ask queue, $p_{0.05,a}$ is the limit price associated with the 5% ask volume quantile and mq denotes the mid-quote. Correspondingly, the bid depth is given by $BD=\ln[0.05 \cdot bvol/(mq-p_{0.05,b})]$. The choice of the 5% quantile is driven by the trade-off between a parsimonious specification[12] and an appropriate measurement of market depth. However, recent studies (see e.g. Pascual and Veredas 2004 or Hall and Hautsch 2004) show that traders' order submission is dictated by the depth in the lower sections of the book. Therefore, we presume that the impact of market depth is well approximated by the volume–price relation over the 5% volume quantile. In order to account not only for the volume–price ratio solely, but also for the volume level itself, we include $AV=\ln(avol)$ and $BV=\ln(bvol)$ as separate regressors. Furthermore, we capture the (signed) cumulative changes in the logarithmic aggregated ask volume (*DAV*), the logarithmic aggregated bid volume (*DBV*) as well as in the mid-quote (*MQ*) process during the past 5 min. Finally, we include the current volatility (*VL*), measured by the average squared mid-quote changes during the past 5 min as well as the current bid-ask spread (*SP*).

In order to analyze the importance of order book dynamics and the information provided by the open limit order book for the goodness-of-fit and the explanatory power of the model, we estimate three different specifications. Table 4 reports the estimation results based on an ACI model including both dynamic variables as well as order book variables. Table 5 is based on a specification which includes order book information, but does not account for any dynamics in the multivariate process. Hence, in this specification, $\tilde{\Psi}_i$ is set to zero. Finally, Table 6 gives the results of a specification which accounts for dynamic structures but excludes any order book covariates.

[11] However, motivated by the results by Hall and Hautsch (2004), we assume identical seasonality patterns on the ask and bid side.
[12] Note that for each regressor six parameters have to be estimated.

Table 4 Fully specified ACI models

	BHP	NAB	NCP	TLS	WOW		BHP	NAB	NCP	TLS	WOW
Constants and backward recurrence parameters											
ω^1	−1.059***	−0.336***	−0.294***	−1.085***	−1.015***	p^1	0.844***	0.797***	0.820***	0.832***	0.737***
ω^2	−0.301***	−0.285***	−0.157	−1.207***	−0.672***	p^2	0.845***	0.778***	0.798***	0.836***	0.731***
ω^3	−0.793***	−0.782***	−0.490***	−0.722***	−0.859***	p^3	0.834***	0.905***	0.878***	0.825***	0.810***
ω^4	−0.497***	−1.348***	−1.081***	−0.420*	−1.637***	p^4	0.818***	0.862***	0.834***	0.700***	0.838***
ω^5	−1.352***	−1.397***	−1.576***	−1.025***	−1.674***	p^5	0.742***	0.749***	0.873***	0.653***	0.789***
ω^6	−1.704***	−1.442***	−1.410***	−1.363***	−1.980***	p^6	0.762***	0.741***	0.859***	0.671***	0.780***
Innovation parameters											
α_1^1	0.115***	0.114***	0.076***	0.235***	0.176***	α_2^1	0.003	0.029**	0.018**	0.046	0.081***
α_1^2	−0.010	0.013	0.004	0.179***	0.055***	α_2^2	0.120***	0.115***	0.085***	0.244***	0.177***
α_1^3	0.052***	0.038***	0.012	0.160	0.082***	α_2^3	−0.010	−0.009	0.013*	−0.007	0.054**
α_1^4	−0.033**	−0.009	−0.007	0.076	0.059*	α_2^4	0.018	0.042***	0.003***	0.263*	0.058*
α_1^5	0.031**	0.037**	0.064***	0.074**	0.016	α_2^5	0.046***	0.049***	0.022*	−0.030	0.145***
α_1^6	0.044***	0.027*	0.042***	0.049**	0.199***	α_2^6	0.010	0.042***	0.041***	0.030*	0.138***
α_3^1	0.042***	−0.002	−0.015*	0.085	0.074***	α_4^1	0.028**	0.053***	0.005	−0.079*	0.071***
α_3^2	0.026*	−0.007	0.018*	−0.159**	0.070***	α_4^2	0.064***	0.048***	0.013	0.046	0.021
α_3^3	0.097***	0.073***	0.028***	0.438***	0.122***	α_4^3	0.012	0.008	0.038***	−0.393**	0.084***
α_3^4	0.076***	−0.004	0.026***	0.010	0.175***	α_4^4	0.116***	0.083***	0.033***	−0.042	0.215***
α_3^5	0.051***	0.034**	0.022*	0.063	0.132***	α_4^5	0.031*	0.056**	0.029*	0.017	0.058
α_3^6	0.056***	0.032**	0.008	−0.038	0.029	α_4^6	0.022	0.010	0.010	−0.014	0.201***
α_5^1	−0.008	−0.017	−0.010	−0.019	0.118***	α_6^1	0.002	−0.054**	0.012	0.127***	0.083***
α_5^2	0.023	0.013	−0.004	0.185***	0.133***	α_6^2	0.017	−0.024	0.012	−0.056***	0.013
α_5^3	−0.007	0.017	−0.063***	0.186*	0.022	α_6^3	0.000	0.014	0.024	−0.451**	−0.021
α_5^4	0.058*	0.015	0.010	0.129	0.104*	α_6^4	0.001	−0.002	0.012	−0.088	0.058
α_5^5	0.002	0.097***	0.025	0.068**	0.122***	α_6^5	0.005	0.030	−0.006	0.029	−0.185***
α_5^6	0.094***	0.002	−0.001	0.030	−0.141**	α_6^6	0.067***	0.119***	0.017	0.019	0.068*

Table 4 (continued)

	BHP	NAB	NCP	TLS	WOW		BHP	NAB	NCP	TLS	WOW
Persistence parameters											
β^{11}	0.980***	0.961***	0.993***	0.922***	0.994***	β^{44}	0.950***	0.984***	0.998***	−0.251***	0.995***
β^{22}	0.969***	0.968***	0.980***	0.820***	0.993***	β^{55}	0.980***	0.980***	0.993***	0.948***	0.994***
β^{33}	0.955***	0.991***	0.995***	0.481***	0.994***	β^{66}	0.979***	0.982***	0.994***	−0.973***	0.991***
Seasonality parameters											
$\nu^{12}_{11:00}$	−0.673***	−0.646***	−1.216***	8.226***	−1.355***	$\nu^{34}_{11:00}$	0.160	−0.334	−0.831***	3.955**	−1.183***
$\nu^{12}_{12:00}$	0.138	0.266	0.961***	−14.730***	1.228***	$\nu^{34}_{12:00}$	−1.159**	0.183	0.492	−7.780**	0.672
$\nu^{12}_{13:00}$	−0.981***	−1.341***	−0.937***	2.870*	−0.570*	$\nu^{34}_{13:00}$	−0.438***	−1.551***	−0.957***	3.113***	0.179
$\nu^{12}_{14:00}$	3.161***	3.553***	2.620***	10.182***	2.222***	$\nu^{34}_{14:00}$	2.956	3.548***	2.661***	5.563**	1.526***
$\nu^{12}_{15:00}$	−0.286	−0.521*	−0.721***	−6.032**	−0.034	$\nu^{34}_{15:00}$	0.025***	−1.032**	−0.269	−4.674	−0.422
$\nu^{12}_{16:00}$	0.468	0.516	1.509***	6.824**	4.069***	$\nu^{34}_{16:00}$	−0.237	0.652	−0.986**	1.557	0.896
$\nu^{56}_{11:00}$	−0.572*	−1.387***	−1.309***	−0.264	−1.894***	$\nu^{56}_{14:00}$	3.755***	2.814***	2.770***	4.863***	1.495***
$\nu^{56}_{12:00}$	0.266	1.555***	1.163**	0.359	1.817***	$\nu^{56}_{15:00}$	−1.313***	−1.044***	−0.827*	−2.817***	0.184
$\nu^{56}_{13:00}$	−1.531***	−1.605***	−1.135***	−2.265***	−0.349	$\nu^{56}_{16:00}$	1.071*	0.799*	0.273	3.798***	0.516
Explanatory variables											
AD^1	−7.287***	−2.318***	−3.390***	−18.057***	−2.130***	BD^1	3.974***	1.903***	2.575***	8.315***	1.036**
AD^2	4.400***	1.340***	2.405***	4.035***	0.828**	BD^2	−7.144***	−1.247***	−3.381***	−16.624***	−0.586
AD^3	−7.708***	−2.072***	−3.448***	−28.527***	−3.019***	BD^3	0.148	−0.185	−0.390	4.385***	−1.137**
AD^4	0.709*	−0.302	0.596	2.784**	−2.468***	BD^4	−8.672***	−2.400***	−4.005***	−26.667***	−2.832***
AD^5	4.822***	6.074***	7.804***	2.902***	4.521***	BD^5	1.516***	−0.048	−1.026	1.510**	0.904
AD^6	0.584	0.177	0.270	−1.085	0.937	BD^6	6.410***	6.437***	5.877***	4.674***	5.545***
AV^1	3.108***	1.579***	1.129	10.325***	1.952	DAV^1	−0.649**	−0.411**	−0.974***	−2.109***	−1.097***
AV^2	0.477	−0.006	0.044	1.283	−4.123***	DAV^2	−0.698**	−0.161	−0.899***	−1.309***	−1.558***
AV^3	8.480***	5.378***	8.408***	23.421***	3.956***	DAV^3	0.172	−0.456*	−0.376	−0.662	−0.963***
AV^4	11.203***	−10.071***	−12.398***	−5.250***	−5.426*	DAV^4	−0.910**	−0.232	−0.868**	−0.938***	0.841*
AV^5	−0.602	0.635	3.546***	5.198***	−0.008	DAV^5	1.341***	1.811***	1.585***	−0.390	0.993*

Table 4 (continued)

	BHP	NAB	NCP	TLS	WOW		BHP	NAB	NCP	TLS	WOW
AV^6	−3.195***	−7.523***	−4.154***	−4.753***	−6.877***	DAV^6	0.363	−0.968***	−0.851*	0.524*	−0.269
BV^1	0.379	−1.429**	−0.241	−2.643*	−1.420	DBV^1	−0.440**	−0.629***	−0.351**	−1.498***	−0.774***
BV^2	1.869***	−0.271	0.751	10.025***	3.522**	DBV^2	−0.290	−0.549***	−0.368*	−1.588***	−0.465**
BV^3	−4.407***	−5.105***	−6.919***	−6.776***	−1.855	DBV^3	−0.447*	−0.228	−0.243	−0.991*	−0.244
BV^4	15.485***	11.177***	13.989***	21.712***	8.300**	DBV^4	0.539**	−0.167	0.605	−0.105	0.342
BV^5	−4.124***	−5.123***	−9.011***	−8.142***	−4.821**	DBV^5	−1.576***	−0.289	−0.599*	0.112	−0.604**
BV^6	−1.756**	2.774***	−0.647	2.399**	2.235	DBV^6	0.543*	1.046***	1.017***	−0.980***	1.522***
MQ^1	−1.849***	−0.305***	−0.596***	−0.925**	−0.239**	VL^1	3.773***	0.242***	2.901***	0.124	0.977***
MQ^2	1.551***	0.248***	0.329***	1.836***	0.256***	VL^2	4.790***	0.232***	2.498***	0.073	1.018***
MQ^3	−0.257*	−0.010	−0.217**	−0.770*	−0.261***	VL^3	3.936***	0.042	1.361***	0.055	0.153
MQ^4	0.467***	−0.016	−0.247**	0.911*	−0.161	VL^4	3.921***	0.144***	0.695**	−0.090	0.414*
MQ^5	−1.098***	−0.096	−0.839***	0.608	−0.112	VL^5	5.712***	0.266***	2.992***	0.055	1.125***
MQ^6	1.788***	−0.062	0.424**	0.440	0.698***	VL^6	5.848***	0.273***	1.551***	0.069	1.188***
SP^1	−1.925***	−0.386***	−1.130***	−1.594	−0.471***						
SP^2	−1.699***	−0.297***	−0.892***	−3.223*	−0.273***						
SP^3	0.639***	0.199***	0.416***	1.661***	0.250***						
SP^4	0.549***	0.173***	0.459***	1.753***	0.235***						
SP^5	−0.857***	0.044	−0.279***	−1.218**	0.086						
SP^6	−0.969***	0.124***	−0.136*	−0.887*	−0.169**						
Diagnostics											
Obs	9,316	10,463	9,142	3,102	3,438						
LL	−20,145	−23,836	−19,894	−6,343	−6,527						
BIC	−20,721	−24,419	−20,468	−6,850	−7,040						
Aggressive buy orders							Aggressive sell orders				
Mean of $\widehat{\varepsilon}_i$	1.014	1.001	0.993	0.952	0.995		0.994	1.000	0.991	1.042	1.002
S.D. of $\widehat{\varepsilon}_i$	1.159	1.067	1.095	1.190	1.114		1.139	1.058	1.088	1.241	1.100

Table 4 (continued)

	BHP	NAB	NCP	TLS	WOW	BHP	NAB	NCP	TLS	WOW
LB(20) of $\widehat{\varepsilon}_i$	13.039	18.503	25.243	14.846	13.610	14.905	28.017	27.049	26.972	17.114
Exc. disp.	5.379***	2.533**	3.443***	3.369***	2.477**	5.106***	2.295**	3.179***	4.501***	2.216**
Aggressive ask limit orders						Aggressive bid limit orders				
Mean of $\widehat{\varepsilon}_i$	1.020	0.999	0.991	0.978	0.992	1.015	0.997	0.979	1.032	0.959
S.D. of $\widehat{\varepsilon}_i$	1.095	1.058	1.086	1.088	1.050	1.091	1.035	1.097	1.083	1.025
LB(20) of $\widehat{\varepsilon}_i$	10.806	22.876	13.522	29.364*	32.165**	20.823	13.765	11.254	10.760	11.538
Exc. disp.	2.740***	1.796*	2.891***	1.009	1.040	2.714***	0.985	2.472**	1.151	0.339
Aggressive ask cancellations						Aggressive bid cancellations				
Mean of $\widehat{\varepsilon}_i$	0.991	1.004	1.011	1.029	1.031	1.006	1.023	1.005	1.005	1.083
S.D. of $\widehat{\varepsilon}_i$	0.975	0.963	0.932	0.975	1.045	0.932	0.972	0.992	0.905	1.095
LB(20) of $\widehat{\varepsilon}_i$	17.313	20.241	11.585	31.632**	22.407	12.733	27.965	10.571	37.679***	15.504
Exc. disp.	0.549	0.744	1.067	0.464	0.542	1.368	0.578	0.133	1.701*	1.202

Maximum likelihood estimates of six-dimensional ACI(1,1) models for intensity processes of (1) aggressive buy orders, (2) aggressive sell orders, (3) aggressive ask limit orders, (4) aggressive bid limit orders, (5) aggressive cancellations of ask orders, (6) aggressive cancellations of bid orders. Backward recurrence functions are specified in terms of individual univariate Weibull parameterizations. The persistence vectors A^k are fully parameterized, whereas B is parameterized as diagonal matrix. Three spline functions are specified for market orders ($\nu_{\cdot}^{12}$), limit orders ($\nu_{\cdot}^{34}$) and cancellations ($\nu_{\cdot}^{56}$) based on 1 h nodes between 10 A.M. and 4 P.M. The exact definition of the covariates is found in Sect. 5. All covariates except *VOL* are scaled by 10. Standard errors are computed based on OPG estimates. The time series are re-initialized at each trading day

Diagnostics: Log Likelihood (*LL*), Bayes Information Criterion (*BIC*) and diagnostics (mean, standard deviation, Ljung–Box statistics and excess dispersion test) of ACI residuals $\widehat{\varepsilon}_i^k$

Table 5 ACI models without dynamics

	BHP	NAB	NCP	TLS	WOW		BHP	NAB	NCP	TLS	WOW
Constants and backward recurrence parameters											
ω^1	−0.970***	−0.304***	−0.256**	−0.578*	−1.849***	p^1	0.828***	0.772***	0.779***	0.806***	0.557***
ω^2	−0.250*	−0.243**	−0.088	−0.814**	−0.569***	p^2	0.830***	0.759***	0.777***	0.827***	0.575***
ω^3	−0.671***	−0.845***	−0.517***	−0.611**	−0.639***	p^3	0.837***	0.865***	0.849***	0.795***	0.701***
ω^4	−0.450***	−1.463***	−1.131***	−0.440*	−1.827***	p^4	0.804***	0.837***	0.811***	0.700***	0.614***
ω^5	−1.343***	−1.427***	−1.583***	−0.718**	−0.987***	p^5	0.713***	0.706***	0.823***	0.649***	0.604***
ω^6	−1.771***	−1.552***	−1.393***	−1.030***	−3.317***	p^6	0.727***	0.714***	0.791***	0.668***	0.627***
Seasonality parameters											
$\nu^{12}_{11:00}$	−0.779***	−0.817***	−1.134***	5.225***	−1.216***	$\nu^{34}_{11:00}$	−0.403*	−0.403	−0.812***	4.292**	−0.571*
$\nu^{12}_{12:00}$	0.247	0.424	0.861***	−9.370***	0.998***	$\nu^{34}_{12:00}$	−0.360	0.245	0.594*	−8.839**	−0.240
$\nu^{12}_{13:00}$	−0.867***	−1.119***	−0.951***	0.947	−0.257	$\nu^{34}_{13:00}$	−0.415	−1.408***	−1.149***	4.130**	0.862*
$\nu^{12}_{14:00}$	2.917***	3.231***	2.764***	8.350***	1.551***	$\nu^{34}_{14:00}$	2.459***	3.449***	2.791***	5.828***	1.174*
$\nu^{12}_{15:00}$	−0.474**	−0.909***	−1.008***	−4.896***	−1.108**	$\nu^{34}_{15:00}$	−0.129	−1.453***	−0.210	−5.695*	−1.713**
$\nu^{12}_{16:00}$	−0.072	0.584*	1.315***	4.353**	1.640***	$\nu^{34}_{16:00}$	−0.147	0.741	−1.208**	2.312	2.539***
$\nu^{56}_{11:00}$	−0.414	−1.361***	−1.125***	−0.220	−1.224***	$\nu^{56}_{14:00}$	3.742***	2.795***	3.108***	4.911***	1.569**
$\nu^{56}_{12:00}$	0.042	1.531***	1.018***	0.352	0.920	$\nu^{56}_{15:00}$	−1.188***	−1.263***	−0.898	−2.894***	−1.288
$\nu^{56}_{13:00}$	−1.498***	−1.592***	−1.353**	−2.334***	0.047	$\nu^{56}_{16:00}$	0.233	0.685*	−0.396	3.788***	0.932
Explanatory variables											
AD^1	−5.852***	−2.001***	−2.542***	−18.482***	−1.714***	BD^1	3.915***	1.971***	2.843***	7.732***	0.465
AD^2	4.520***	1.436***	2.372***	3.827***	1.153***	BD^2	−6.109***	−1.208***	−3.027***	−17.238***	−0.745*
AD^3	−6.738***	−1.523***	−3.776***	−27.456***	−1.939***	BD^3	0.113***	0.024	−0.621*	3.950***	−1.365***
AD^4	0.578	0.008	−0.225	3.221**	−0.477	BD^4	−7.338***	−2.499***	−4.289***	−26.201***	−3.493***
AD^5	5.127***	6.341***	7.808***	2.694***	5.165***	BD^5	1.746***	−0.108	−0.974	0.881	−0.589
AD^6	1.036**	0.333	0.570	−1.004	0.942	BD^6	6.696***	6.540***	6.174***	4.977***	4.212***

Table 5 (continued)

	BHP	NAB	NCP	TLS	WOW		BHP	NAB	NCP	TLS	WOW
AV^1	1.717***	0.639	−2.788***	12.936***	0.091	DAV^1	−0.415	−0.261	−0.648***	−1.970***	0.587*
AV^2	1.510***	0.455	0.802*	0.909	1.200*	DAV^2	−0.615**	−0.210	−0.796***	−0.990***	−0.437*
AV^3	7.577***	4.381***	8.662***	21.358***	3.234***	DAV^3	0.367	−0.566**	−0.306	−0.499	0.164
AV^4	−10.311***	−10.219***	−9.006***	−5.775***	−15.237***	DAV^4	−0.837**	−0.302	−0.660*	−1.029***	1.491***
AV^5	−0.774	0.157	1.334	3.986***	3.446***	DAV^5	1.587***	1.868***	1.630***	−0.346	1.719***
AV^6	−3.138***	−6.224***	−5.912***	−4.956***	−12.886***	DAV^6	0.442	−1.120***	−0.642	0.598**	0.314
BV^1	0.787*	−0.651*	3.028***	−3.281*	2.461***	DBV^1	−0.234	−0.514***	−0.301*	−0.932***	−0.374*
BV^2	0.084	−0.759**	−0.244	10.528***	−1.222*	DBV^2	−0.143	−0.284***	−0.284	−1.270***	0.310*
BV^3	−4.146***	−4.487***	−6.735***	−5.331***	−1.634*	DBV^3	−0.227	−0.218	−0.140	−1.055**	−0.410**
BV^4	13.824***	11.335***	11.352***	21.499***	18.504***	DBV^4	0.440	0.082	0.734*	−0.125	0.635**
BV^5	−4.307***	−4.653***	−6.657***	−6.256***	−7.446***	DBV^5	−1.502***	−0.129	−0.784**	0.181	0.329
BV^6	−2.250***	1.421*	0.830	2.351**	10.951***	DBV^6	0.551*	1.270***	1.271***	−0.937***	1.600***
MQ^1	−1.642***	−0.310***	−0.678***	−0.856***	−0.183**	VL^1	3.989***	0.261***	2.613***	0.068	0.279*
MQ^2	1.493***	0.177***	0.380***	1.547***	0.289***	VL^2	4.621***	0.256***	2.442***	0.001	0.265*
MQ^3	−0.260*	−0.026	−0.253***	−0.782	−0.035	VL^3	3.772***	0.029	1.405***	0.039	−0.021
MQ^4	0.482***	−0.049	−0.109	1.034**	0.029	VL^4	3.892***	0.114***	0.975***	−0.073	−0.342
MQ^5	−1.321***	−0.171**	−0.856***	0.411	−0.399**	VL^5	5.922***	0.271***	2.889***	0.026	0.483*
MQ^6	1.870***	−0.080	0.433***	0.205	0.586***	VL^6	5.942***	0.273***	1.547***	0.083	0.755***
SP^1	−1.951***	−0.389***	−1.148***	−3.472	−0.665***	SP^4	0.535***	0.188***	0.493***	1.745***	0.193***
SP^2	−1.740***	−0.299***	−0.917***	−2.502	−0.377***	SP^5	−0.898***	0.051*	−0.311***	−1.514**	−0.019
SP^3	0.639***	0.221***	0.416***	1.787***	0.229***	SP^6	−1.010***	0.129***	−0.157**	−1.332*	−0.196**
Diagnostics											
Obs	9,316	10,463	9,142	3,102	3,438						
LL	−20,426	−24,192	−20,210	−6,422	−7,322						
BIC	−20,809	−24,581	−20,593	−6,759	−7,664						

Table 5 (continued)

	BHP	NAB	NCP	TLS	WOW	BHP	NAB	NCP	TLS	WOW
Aggressive buy orders						Aggressive sell orders				
Mean of $\widehat{\varepsilon}_i$	1.011	1.004	1.004	0.984	1.027	1.001	1.004	1.006	1.018	1.039
S.D. of $\widehat{\varepsilon}_i$	1.227	1.061	1.121	1.274	1.136	1.155	1.075	1.102	1.219	1.108
LB(20) of $\widehat{\varepsilon}_i$	138.654***	228.349***	328.073***	28.760*	875.175***	152.729***	380.321***	189.016***	40.895***	1,168.166***
Exc. disp.	7.904***	2.312**	4.458***	5.039***	2.976**	5.719***	2.952**	3.724***	4.052***	2.421**
Aggressive ask limit orders						Aggressive bid limit orders				
Mean of $\widehat{\varepsilon}_i$	1.002	1.001	1.001	0.993	0.993	0.988	0.995	0.987	1.017	0.948
S.D. of $\widehat{\varepsilon}_i$	1.087	1.077	1.138	1.126	1.070	1.078	1.038	1.132	1.068	1.073
LB(20) of $\widehat{\varepsilon}_i$	59.655***	179.270***	98.016***	36.974**	687.721***	169.180***	192.434***	57.125***	9.576	457.756***
Exc. disp.	2.502**	2.384**	4.756***	1.470	1.454	2.312**	1.068	3.428***	0.947	1.000
Aggressive ask cancellations						Aggressive bid cancellations				
Mean of $\widehat{\varepsilon}_i$	1.003	0.995	1.002	1.008	1.013	1.005	1.002	1.003	1.012	1.037
S.D. of $\widehat{\varepsilon}_i$	1.004	0.964	0.952	0.954	1.161	0.936	0.960	1.034	0.908	1.168
LB(20) of $\widehat{\varepsilon}_i$	37.028**	83.534***	41.465***	48.019***	233.517***	46.825***	80.124***	61.366***	38.402***	110.162***
Exc. disp.	0.107	0.718	0.747	0.851	2.032**	1.287	0.813	0.611	1.654*	2.207**

Maximum likelihood estimates of six-dimensional ACI(1,1) models without dynamics for intensity processes of (1) aggressive buy orders, (2) aggressive sell orders, (3) aggressive ask limit orders, (4) aggressive bid limit orders, (5) aggressive cancellations of ask orders, (6) aggressive cancellations of bid orders. Backward recurrence functions are specified in terms of individual univariate Weibull parameterizations. The autoregressive matrices A^k and B are set to zero. The spline functions are specified for market orders ($\nu_{\cdot}^{12}$), limit orders ($\nu_{\cdot}^{34}$) and cancellations ($\nu_{\cdot}^{56}$) based on one hour nodes between 10 A.M. and 4 P.M. The exact definition of the covariates is found in Sect. 5. All covariates except *VOL* are scaled by 10. Standard errors are computed based on OPG estimates. The time series are re-initialized at each trading day

Diagnostics: Log Likelihood (*LL*), Bayes Information Criterion (*BIC*) and diagnostics (mean, standard deviation, Ljung–Box statistics and excess dispersion test) of ACI residuals $\widehat{\varepsilon}_i^k$

5.1 Statistical results

The ACI models were estimated by maximum likelihood using the MAXLIK-procedure of GAUSS. It should be stressed that despite of the high-dimensionality of the processes and the large number of parameters, the processes converged smoothly and without numerical difficulties. The following statistical results can be summarized. First, for all processes, we find significantly declining backward recurrence functions as revealed by the estimated parameters $p^k<1$. Thus, the event arrival rates decline with the length of the spell which is a well known result for financial duration data. However, as indicated by the residual diagnostics, the specification of the backward recurrence function is not sufficient in all of the models to completely capture the distributional properties of the processes. In particular, the specifications of Tables 4 and 5 reveal significant excess dispersion, which is not the case for the models reported in Table 6. This implies that the inclusion of order book variables makes it more difficult to capture the distributional properties of the data.

Second, as indicated by the significantly positive estimates of α_k^j, we find evidence for positive autocorrelation in the individual processes. This implies that the individual arrival rates are clustered, and this is true for all of the individual processes including cancellations. Nearly all processes show quite high persistence, as revealed by parameter estimates of β^{ij} close to unity.[13] Moreover, the estimates of the parameters α_k^j indicate significant spill-over effects between the individual processes. The strongest interactions seem to exist between the individual sides of the market which is particularly true for market orders,[14] but it also holds for limit orders. Weaker interdependencies, but in most cases still significant, are also found between the arrival rates of aggressive market orders and aggressive limit orders. It is interesting that these spill-overs are primarily positive, so we do not find evidence for the fact that a high aggressiveness on one side of the market negatively influences the aggressiveness on the opposite side. These results suggest that interdependencies between the particular order arrival processes are obviously not driven by individual trading behavior as predicted by the economic underpinnings outlined in Sect. 2. Rather, such effects seem to be driven by an underlying process of general market activity which simultaneously affects all individual processes. Similar results are also found for the ask and bid cancellation intensity. In most cases, the parameters α_i^5 and α_i^6 are significantly positive, indicating that a higher market activity also increases the intensity of order cancellations. This is particularly true for the impact of aggressive market trading on the cancellation intensity. However, since in most cases the parameters α_5^i and α_6^i are insignificant, it turns out that the market and limit order processes do not seem to be affected by the arrival rate of order cancellations.

Third, the estimates of the seasonality functions provide evidence of distinct deterministic intra-day patterns in the intensities of aggressive market trading, limit order trading and cancellations. The estimated individual intra-day seasonality

[13] The only exceptions are found for TLS. Here, some of the persistence parameters are even negative. These somewhat peculiar results might be explained by the fact that the dynamics in the order and cancellation processes for TLS are relatively weak and obviously interfere with the covariate processes.

[14] This finding is consistent with the results of Hall and Hautsch (2004) who find similar results when analyzing the continuous buy–sell pressure at the ASX.

Table 6 ACI models without covariates

	BHP	NAB	NCP	TLS	WOW		BHP	NAB	NCP	TLS	WOW
Constants and backward recurrence parameters											
ω^1	−0.893***	−0.724***	−0.635***	−1.204***	−1.109***	p^1	0.822***	0.787***	0.803***	0.797***	0.705***
ω^2	−0.742***	−0.658***	−0.544***	−1.153***	−1.014***	p^2	0.798***	0.766***	0.781***	0.843***	0.721***
ω^3	−0.967***	−1.138***	−0.783***	−1.665***	−0.740***	p^3	0.760***	0.870***	0.821***	0.671***	0.784***
ω^4	−0.986***	−1.385***	−1.330***	−1.259***	−1.705***	p^4	0.736***	0.816***	0.780***	0.595***	0.779***
ω^5	−1.449***	−1.600***	−1.910***	−1.210***	−1.959***	p^5	0.706***	0.734***	0.832***	0.665***	0.746***
ω^6	−1.604***	−1.537***	−1.876***	−1.216***	−1.851***	p^6	0.755***	0.730***	0.862***	0.672***	0.749***
Innovation parameters											
α_1^1	0.049***	0.124***	0.063***	0.147***	0.134***	α_2^1	0.046***	0.056***	0.028***	0.047**	0.049***
α_1^2	0.027**	0.034**	0.012	0.124***	0.064***	α_2^2	0.085***	0.133***	0.078***	0.096***	0.097***
α_1^3	−0.008	0.010	−0.018*	0.037	0.078***	α_2^3	0.013	0.010	0.038***	0.043	0.050**
α_1^4	0.037***	0.017*	0.062***	0.036	0.107***	α_2^4	−0.007	−0.013	−0.025	−0.057*	−0.002
α_1^5	−0.014	0.041*	0.021	0.033	−0.039*	α_2^5	0.095***	0.154***	0.055***	0.085***	0.133***
α_1^6	0.167***	0.091***	0.063***	0.088	0.151***	α_2^6	−0.033	0.038**	0.053***	0.027	0.081***
α_3^1	0.025**	−0.008	−0.005***	0.092**	0.056***	α_4^1	0.016	0.017	0.007	−0.016	0.086***
α_3^2	0.017	−0.009	0.018*	0.049	0.054***	α_4^2	0.007	0.041**	0.004	−0.053	0.037*
α_3^3	0.110***	0.084***	0.096***	0.182***	0.108***	α_4^3	−0.003	−0.005	0.019	0.008	0.051*
α_3^4	0.005	−0.037***	−0.029*	0.129***	0.080***	α_4^4	0.151***	0.110***	0.186***	0.046	0.224***
α_3^5	−0.006	0.056**	0.040**	0.034	0.083***	α_4^5	−0.019	0.032	−0.001	−0.008	0.073**
α_3^6	0.022	0.026	−0.024	0.045	0.005	α_4^6	−0.003	−0.005	0.001	0.035	0.125***
α_5^1	−0.029*	−0.007	−0.021	0.004	0.087***	α_6^1	0.039**	−0.050*	0.026	0.036	0.044***
α_5^2	0.025*	0.037	0.006	0.034	0.088***	α_6^2	0.003	−0.038	0.001	−0.003	0.006***
α_5^3	−0.001	0.013	−0.011	−0.024	0.025	α_6^3	0.013	−0.003	−0.027	0.025	−0.001***
α_5^4	−0.058**	0.017	−0.050	0.018	0.032	α_6^4	0.002	−0.016	−0.008	0.018	0.088**
α_5^5	0.045**	0.134***	0.055*	0.001	0.056	α_6^5	0.049**	0.031	0.022	0.025	−0.014***
α_5^6	0.122***	0.031	0.007	0.013	−0.164***	α_6^6	0.149***	0.094***	0.042	0.001	0.102

Table 6 (continued)

	BHP	NAB	NCP	TLS	WOW		BHP	NAB	NCP	TLS	WOW
Persistence parameters											
β^{11}	0.987***	0.945***	0.994***	0.986***	0.994***	β^{44}	0.997***	0.998***	0.990***	0.992***	0.995***
β^{22}	0.987***	0.951***	0.984***	0.985***	0.995***	β^{55}	0.987***	0.961***	0.993***	0.983***	0.997***
β^{33}	0.997***	0.994***	0.997***	0.994***	0.994***	β^{66}	0.961***	0.972***	0.993***	0.979***	0.994***
Seasonality parameters											
$\nu^{12}_{11:00}$	−1.257***	−1.223***	−1.665***	−0.793***	−1.553***	$\nu^{34}_{11:00}$	−1.449***	−1.291***	−1.495***	−0.913***	−1.660***
$\nu^{12}_{12:00}$	0.791***	0.912***	1.499***	−0.137	1.339***	$\nu^{34}_{12:00}$	0.989***	1.326***	1.505***	−0.189	1.245***
$\nu^{12}_{13:00}$	−0.577***	−0.929***	−0.736***	−0.005	−0.362	$\nu^{34}_{13:00}$	−0.172	−1.098***	−1.110***	1.040*	0.191
$\nu^{12}_{14:00}$	2.330***	2.666***	2.046***	2.603***	1.666***	$\nu^{34}_{14:00}$	1.675***	2.189***	2.151***	1.692***	1.003***
$\nu^{12}_{15:00}$	−0.311	−0.411*	−0.579***	−1.491***	−0.134	$\nu^{34}_{15:00}$	−0.975***	−0.755***	−0.336	−2.316***	−0.632
$\nu^{12}_{16:00}$	0.286	0.355	1.291***	1.402**	2.487***	$\nu^{34}_{16:00}$	0.698**	0.174	−0.695**	3.150***	1.044*
$\nu^{56}_{11:00}$	−0.884***	−1.224***	−1.506***	−0.430	−1.612***	$\nu^{56}_{14:00}$	3.110***	3.064***	2.309***	4.714***	1.505***
$\nu^{56}_{12:00}$	0.502	1.249***	1.304***	0.598	1.227**	$\nu^{56}_{15:00}$	−0.646	−0.870**	−0.586	−2.483***	−0.127
$\nu^{56}_{13:00}$	−1.110***	−1.572***	−0.860***	−2.250***	−0.040	$\nu^{56}_{16:00}$	0.173	0.924	0.569	3.274***	1.130
Diagnostics											
Obs	9,316	10,463	9,142	3,102	3,438						
LL	−22,351	−25,144	−21,277	−7,538	−6,967						
BIC	−22,680	−25,477	−21,605	−7,828	−7,260						
Aggressive buy orders							Aggressive sell orders				
Mean of $\widehat{\varepsilon}_i$	0.989	0.999	0.992	0.958	0.988		0.992	0.996	1.001	1.002	1.012
S.D. of $\widehat{\varepsilon}_i$	1.016	1.020	1.051	0.930	1.087		1.027	1.044	1.058	0.957	1.084
LB(20) of $\widehat{\varepsilon}_i$	13.858	18.542	18.892	27.911	20.361		12.845	27.835	20.016	17.917	17.768
Exc. disp.	0.515	0.760	1.831*	1.079	1.869*		0.934	1.712	2.075**	0.698	1.864*
Aggressive ask limit orders							Aggressive bid limit orders				
Mean of $\widehat{\varepsilon}_i$	0.983	1.009	1.015	0.987	0.987		0.997	0.986	0.964	0.966	0.946
S.D. of $\widehat{\varepsilon}_i$	1.036	1.048	1.070	0.953	1.001		1.012	1.033	1.020	0.921	0.992

Table 6 (continued)

	BHP	NAB	NCP	TLS	WOW	BHP	NAB	NCP	TLS	WOW
LB(20) of $\widehat{\varepsilon}_i$	15.927	14.314	11.126	31.249*	17.930	19.197	21.943	17.132	17.175	18.180
Exc. disp.	1.015	1.488	2.333**	0.494	0.037	0.344	0.918	0.495	1.003	0.093
Aggressive ask cancellations						Aggressive bid cancellations				
Mean of $\widehat{\varepsilon}_i$	0.995	1.002	1.012	1.008	1.008	1.010	1.009	1.003	1.010	1.037
S.D. of $\widehat{\varepsilon}_i$	0.935	0.962	0.950	0.906	0.975	0.929	0.942	0.978	0.876	1.047
LB(20) of $\widehat{\varepsilon}_i$	23.337	25.336	16.429	20.911	19.263	11.344	24.900	13.915	24.142	8.333
Exc. disp.	1.411	0.749	0.787	1.681*	0.288	1.428	1.169	0.374	2.200**	0.584

Maximum likelihood estimates of six-dimensional ACI(1,1) models for intensity processes of (1) aggressive buy orders, (2) aggressive sell orders, (3) aggressive ask limit orders, (4) aggressive bid limit orders, (5) aggressive cancellations of ask orders, (6) aggressive cancellations of bid orders. Backward recurrence functions are specified in terms of individual univariate Weibull parameterizations. The persistence vectors A^k are fully parameterized, whereas B is parameterized as diagonal matrix. Three spline functions are specified for market orders ($\nu_{\cdot}^{12}$), limit orders ($\nu_{\cdot}^{34}$), and cancellations ($\nu_{\cdot}^{56}$) based on one hour nodes between 10 A.M. and 4 P.M. Standard errors are computed based on OPG estimates. The time series are re-initialized at each trading day

Diagnostics: Log Likelihood (*LL*), Bayes Information Criterion (*BIC*) and diagnostics (mean, standard deviation, Ljung–Box statistics and excess dispersion test) of ACI residuals $\widehat{\varepsilon}_i^k$

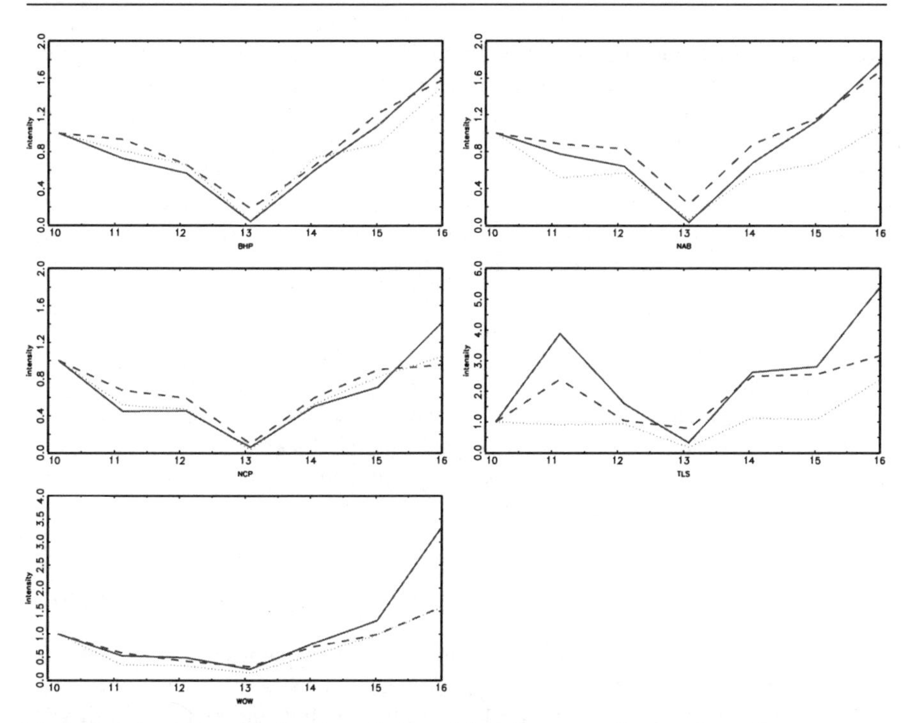

Fig. 1 Intraday seasonality functions. Estimated intraday seasonality functions of the processes of aggressive market orders (*solid line*), aggressive limit orders (*broken line*), and aggressive cancellations (*dotted line*) for the BHP, NAB, NCP, TLS and WOW stock traded at the ASX. The estimates are based on the ACI specifications in Table 4

functions, reported in Fig. 1, reveal similar patterns in the form of the well known U-shape. It appears that the intra-day seasonalities are induced by the general level of trading activity implying relatively high intensities after the opening, a significant decline around noon and a distinct increase before the closure of the market.

Fourth, as indicated by the Ljung–Box statistics based on the ACI residuals, the dynamic specifications (i.e. Tables 4 and 6) seem to appropriately capture the dynamic properties of the data. Moreover, for four out of five stocks, the inclusion of dynamics is absolutely essential in order to capture the serial dependence in the data. This is illustrated by the diagnostics for the non-dynamic specification (i.e. Table 5) where the Ljung–Box statistics reveal significant dynamic misspecifications.[15]

Fifth, by comparing the BIC values of the three different specifications reported in Tables 4, 5 and 6, we can conclude that the inclusion of both dynamic variables as well as limit order book variables clearly improves the goodness-of-fit of the model. However, for four out of five stocks, the specifications where order book variables are omitted (i.e. Table 6) clearly underperform the non-dynamic models (i.e. Table 5) in terms of explanatory power. This finding suggests that a

[15] An exception is TLS. A shown by Table 5, even for a non-dynamic ACI model, the corresponding Ljung–Box statistics associated with the ACI residuals are already quite low indicating a reasonable goodness-of-fit.

specification which includes order book information has a significantly higher explanatory power than a pure dynamic model without covariates. Nevertheless, as indicated by the BIC values in Table 4, in four of five cases the inclusion of dynamics on top of order book variables leads to a further increase of the BIC. Summarizing these findings, we conclude that the state of the order book plays a particularly important role in explaining the degree of order aggressiveness in the individual processes. Nonetheless, in addition order book dynamics have to be taken into account in order to obtain a well-specified model.

5.2 Economic results

A particular important finding is that for most of our order book covariates, we find a remarkable robustness over the cross-section of stocks with no systematic differences between the individual stocks. Regarding our economic hypotheses, we can summarize the following findings.

5.2.1 The impact of market depth

Our estimation results show a clear confirmation of Hypothesis (1). In fact, an increase of the depth on the ask side (*AD*) increases the aggressiveness in sell market order trading, decreases it in sell limit order trading and increases it in ask cancellations. The converse is true for the depth on the bid side (*BD*) leading to a rising intensity of aggressive buys as well as bid cancellations and a declining intensity for aggressive buy limit orders. This finding clearly confirms the crowding out concept as discussed in Parlour (1998). Furthermore, we also find a significantly negative relation between the depth on a certain side of the market and traders' preference to post aggressive market orders on that side. Hence, trader's preference for aggressive buys (sells) increases when the ask (bid) depth declines. This finding cannot be solely explained by a pure crowding out effect but supports the idea that traders use information from the book to infer price expectations. As discussed in Hall and Hautsch (2004), greater ask (bid) depth indicates that a relatively higher proportion of volume is to be sold (bought) at a comparably low (high) price. This induces a negative (positive) price signal which increases traders' preference to post sell (buy) market orders. Interestingly, we do not observe a clear-cut and significant impact of changes in the depth on the limit order and cancellation activity on the opposite side of the market.

5.2.2 The impact of the conditional cumulated volume

The variables *AD* and *BD* measure market depth by the log volume-price ratio associated with the 5% volume quantile. In addition, the cumulated (log) ask and bid volume (*AV* and *BV*) control for the amount of volume pending in the queue. Hence, the regressors associated with *AV* and *BV* reflect the impact of changes in the cumulated volume given the corresponding market depth (as measured by *AD* and *BD*). We find a positive impact of the cumulated ask (bid) volume on the intensity of aggressive sell (buy) limit orders. Since we condition on the volume-price ratio, an increase of the volume must also imply a higher price impact leading to a higher dispersion of limit prices and thus an increase of the execution probability on that side of the market. Therefore, this result is consistent with the theoretical predictions of the Parlour (1998) model and Hypothesis (1). However, we also find slight evidence for a positive impact on market order aggressiveness and a negative impact on the limit order aggressiveness at the same side of the market. Hence, obviously a higher ask (bid) volume accompanied with a higher dispersion of limit prices indicates that market participants expect increasing (decreasing) prices leading to a rise (decline) in the preference for immediacy. Overall, we observe a significant reduction of the order submission and cancellation activity on the opposite side of the market.

Concerning the influence of changes in the cumulated volume during the recent past we find that an increase in the cumulated volume on one market side during the past 5 min decreases the intensity of market and limit orders on both sides of the market, and simultaneously increases the cancellation intensity on the same side. Hence, after periods in which substantial (one-sided) volume has been accumulated in the queues, we observe a type of mean reversion effect causing a reduction of the overall order flow and an increase in the tendency to remove pending orders.

5.2.3 The impact of recent mid-quote movements

Mid-quote movements during the past 5 min have a significant impact on traders' preference to post aggressive market and limit orders. Interestingly, we find a clear rejection of Hypothesis (2). In particular, positive price movements decrease (increase) traders' overall activity on the ask (sell) side. This finding is not explained by crowding out effects or the informational content of the book. A possible reason for this effect could be that, after significant price movements, traders' order submission strategies are dominated by liquidity considerations. This might be explained by the fact that a movement of the mid-quote accompanied by the absorption of a substantial part of the pending volume on one side leads to an increase of the costs of aggressive trading on that side. Then, traders obviously become reluctant to post further aggressive orders on that market side. Moreover, it also turns out that traders significantly reduce the cancellation intensity.

5.2.4 The impact of past volatility

No clear-cut confirmation of Hypothesis (3) is found. We observe an increase of the overall order submission and cancellation activity in periods of a higher mid-quote volatility. These results are highly significant and consistent over all stocks. These results contradict the implications of Foucault (1999) since the notion of a crowding out of market trading towards limit order trading is clearly rejected. Rather, we find evidence that a higher volatility is accompanied by a higher limit order book activity. A possible reason for this finding could be the well known positive relation between volatility and order volumes. Since we explicitly focus on large orders, our results could be driven by the fact that a higher volatility increases the overall arrival rate of higher order volumes.

5.2.5 The impact of the bid-ask spread

In this instance we find a clear confirmation of Hypothesis (4). Traders' preference for aggressive market trading significantly decreases when the bid-ask spread rises. Conversely, the aggressiveness of limit order trading increases. Hence, the higher the bid-ask spread, the lower traders' incentive to cross the market and to post a market order on the opposite side. In this case, market agents are willing to bear execution risk by posting limit orders. Interestingly, we find weak evidence for the fact that a widening of the bid-ask spread also leads to a decline of the cancellation intensity on both sides of the market. Hence, the decrease of the risk of non-execution induced by higher spreads reduces traders' willingness to cancel pending orders.

5.3 Summarizing the results

Overall, we find clear evidence that the arrival rate of aggressive market orders, limit orders, and cancellations is affected by the state of the order book and that the inclusion of order book variables significantly increases the goodness-of-fit of the model. Regarding the impact of changes in the market depth and the cumulated pending volume on order and cancellation aggressiveness, we find clear support for the "crowding-out" concept of Parlour (1998). These results are in line with previous empirical studies such as Griffiths et al. (2000), Coppejans and Domowitz (2002), Pascual and Veredas (2004) and Ranaldo (2004). In this context, we also observe relationships which support the notion that the limit order book has information value, i.e. that traders infer from the book with respect to future price movements. However, Parlour's model has no explanatory power for the reaction of order aggressiveness after the occurrence of mid-quote movements in the recent past. In this context, liquidity effects seem to prevail. Furthermore, implications of the Foucault (1999) model regarding the impact of volatility and the size of the bid-ask spread on traders' order aggressiveness are only partly confirmed. Whereas we find a significant crowding out of market order aggressiveness towards limit order aggressiveness after a widening of the spread (as theoretically predicted), we do not observe corresponding effects in response to changes in the volatility. These peculiarities are not suggested by existing equilibrium models.

Nevertheless, our findings suggest that a separate modelling of the single processes in a multivariate setting is a valuable strategy providing a clear-cut picture of how the particular processes are individually affected by the state of the order book. Obviously, limit orders cannot necessarily be treated as less aggressive versions of market orders since they respond in a different way to certain order book variables. In this sense we confirm the results of Coppejans and Domowitz (2002).

It is also demonstrated that the order book effects remain remarkably stable irrespective of whether order book dynamics are taken into account or not. While this finding illustrates the robustness of the results, it also implies that the economic relations hold conditionally on the history of the individual processes as well as unconditionally.

6 Conclusions

We analyze the impact of order book information of traders' order aggressiveness in the electronic trading on the Australian Stock Exchange. The novel feature of the paper is to analyze this issue using a multivariate dynamic intensity framework. Therefore, order aggressiveness in market trading, limit order trading as well as in order cancellations on both sides of the market is modelled on the basis of a six-dimensional version of the autoregressive conditional intensity (ACI) model proposed by Russell (1999). The multivariate intensity function gives the instantaneous order arrival probability per time in each instant and for each order process. Therefore, it has a natural interpretation as a (continuous-time) measure for traders' degree of aggressiveness in the individual dimensions. In this sense, our setting merges approaches where order aggressiveness is modelled in terms of a categorized variable on the basis of the order classification scheme proposed by Biais et al. (1995) (see, for instance, Griffiths et al. 2000, or Ranaldo 2004), and, those approaches which model the intensity of aggressiveness using univariate (ACD-type) dynamic duration models (see e.g. Coppejans and Domowitz 2002, or Pascual and Veredas 2004). A novel feature of this study is to determine order aggressiveness not only based on the type of the order and the corresponding position in the book but also by the posted volume. Hence, we explicitly focus on market and limit orders with volumes which are significantly above the average. Correspondingly, we also classify cancellations by modelling only those with high orders. This strategy allows us to concentrate on the economically most relevant orders and to reduce the impact of noise.

The usefulness of the individual modelling of the single order processes in a multivariate setting is confirmed by the finding that the intensities of market trading, limit order trading, and cancellations have different responses in their dependence on order book variables. This result questions the application of (too simplified) order classification schemes and supports the use of sequential classifications by distinctly distinguishing between market orders, limit orders and cancellations as implemented by Pascual and Veredas (2004).

Our results show that order book information has significant explanatory power in explaining traders' degree of aggressiveness. In particular we find that the inclusion of variables capturing the current state of the order book as well as recent changes in the book improves the model's goodness-of-fit considerably. Analyzing the influence of fundamental market characteristics such as the depth, queued volume, the bid-ask spread, recent movements in the order flow and in the price as well as the recent price volatility during the last trading minutes, we broadly confirm economic theory. Particularly with respect to market depth, clear evidence is provided for "crowding out effects" (cf. Parlour 1998). Depth on one particular side induces a crowding out of aggressive market and limit order trading on that side towards the other side of the market. In addition to crowding out mechanisms we also find evidence for liquidity and volatility effects which are not in line with existing theoretical equilibrium models. These results indicate that traders' order aggressiveness is not only driven by expected execution probabilities but also by price information revealed by the book as well as liquidity considerations.

Our results provide clear evidence that the timing of aggressive market orders, limit orders as well as cancellations is influenced by the state of the order book which is consistent with the findings of Coppejans and Domowitz (2002), but in contrast to those of Pascual and Veredas (2004). A possible explanation for these conflicting results is that Pascual and Veredas (2004) apply a discrete-time duration model which does not allow for time-varying covariates. However, particularly for the processes of the infrequent highly aggressive orders, it seems to be essential to account for changes of the order book during a spell.[16]

Clear evidence for the existence of multivariate dynamic structures in the order arrival processes is found. We observe significant spill-over effects between the both sides of the market and—in a weaker form—between market trading and limit order trading. The fact that these interdependencies are primarily (significantly) positive suggests that order book dynamics are driven by general market activity which simultaneously influences all individual processes rather than by economic "crowding out" arguments which would imply negative spill-over effects. These findings support the notion that the arrival rates of aggressive orders are basically driven by two pieces of information: (1) the state of the market as revealed by the open limit order book and which directs traders' order submission strategy, and (2) general market activity which simultaneously influences the individual arrival rates.[17] Our findings show that order book information plays the dominant role in explaining order aggressiveness. In particular, we observe that in terms of its explanatory power, a model which excludes all dynamics but includes order book covariates significantly outperforms a completely dynamic model that does not account for the state of the market. Nevertheless, the dynamic variables are absolutely necessary in order to obtain a well-specified model. These findings provide support for advocates of greater transparency in electronic trading and indicate that real benefits to traders may result from complete disclosure of the order book.

Acknowledgement Special thanks are due to James McCulloch whose assistance in preparing the data has made this research project feasible.

References

Al-Suhaibani M, Kryzanowski L (2000) An exploratory analysis of the order book, and order flow and execution on the Saudi stock market. J Bank Financ 24:1323–1357

Bauwens L, Giot P (2000) The logarithmic ACD model: an application to the Bid/Ask Quote Process of two NYSE stocks. Ann Econ Stat 60:117–149

Bauwens L, Hautsch N (2003) Dynamic Latent Factor Models for Intensity Processes, Discussion Paper 2003/103, CORE, Université Catholique de Louvain

Bauwens L, Veredas D (2004) The stochastic conditional duration model: a latent factor model for the analysis of financial durations. J Econom 119:381–412

Biais B, Hillion P, Spatt C (1995) An empirical analysis of the limit order book and the order flow in the Paris bourse. J Financ 50:1655–1689

Bisière C, Kamionka T (2000) Timing of orders, orders aggressiveness and the order book at the Paris bourse. Ann Econ Stat 60:43–72

Bowsher CG (2002) Modelling Security Markets in Continuous Time: Intensity based, Multivariate Point Process Models, Discussion Paper 2002-W22, Nuffield College, Oxford

[16] In our setting, an updating of the information set occurs whenever a new point of the pooled process arrives. A further extension would be to account for *any* changes of the order book. However this would considerably increase the computational burden in our multivariate setting.

[17] This result supports the idea of Bauwens and Hautsch (2003) to model the underlying market activity in terms of a latent autoregressive component which simultaneously affects all individual intensity processes.

Brémaud P (1981) Point processes and queues, martingale dynamics. Springer, Berlin Heidelberg New York

Cao C, Hansch O, Wang X (2003) The Informational Content of an Open Limit Order Book, Discussion paper, Pennsylvania State University

Cohen KJ, Maier SF, Schwartz RA, Whitcomb DK (1981) Transaction costs, order placement strategy, and existence of the bid-ask spread. J Polit Econ 89(2):287–305

Coppejans M, Domowitz I (2002) An Empirical Analysis of Trades, Orders, and Cancellations in a Limit Order Market, Discussion paper, Duke University

Dufour A, Engle RF (2000) The ACD Model: Predictability of the Time between Consecutive Trades, Discussion paper, ISMA Centre, University of Reading

Engle RF, Russell JR (1998) Autoregressive conditional duration: a new model for irregularly spaced transaction data. Econometrica 66:1127–1162

Fernandes M, Grammig J (2001) A Family of Autoregressive Conditional Duration Models, Discussion Paper 2001/31, CORE, Université Catholique de Louvain

Foucault T (1999) Order flow composition and trading costs in a dynamic limit order market. J Financ Mark 2:99–134

Glosten LR (1994) Is the electronic open limit order book inevitable. J Financ 49:1127–1161

Grammig J, Maurer K-O (2000) Non-monotonic hazard functions and the autoregressive conditional duration model. Econometrics J 3:16–38

Griffiths MD, Smith BF, Turnbull DAS, White RW (2000) The costs and determinants of order aggressiveness. J Financ Econ 56:65–88

Hall AD, Hautsch N (2004) A continuous-time measurement of the buy-sell pressure in a limit order book market, Discussion Paper 04-07, Institute of Economics, University of Copenhagen

Handa P, Schwartz RA (1996) Limit order trading. J Financ 51:1835–1861

Handa P, Schwartz RA, Tiwari A (2003) Quote setting and price formation in an order driven market. J Financ Mark 6:461–489

Harris L (1998) Optimal dynamic order submission strategies in some stylized trading problems. Financ Mark Inst Instrum 7:1–75

Harris L, Hasbrouck J (1996) Market vs. limit orders: The SuperDOT evidence on order submission strategy. J Financ Quant Anal 31(2):213–231

Hautsch N (2004) Modelling irregularly spaced financial data—theory and practice of dynamic duration models, vol. 539 of Lecture Notes in Economics and Mathematical Systems. Springer, Berlin Heidelberg New York

Hollifield B, Miller RA, Sandås P, Slive J (2002) Liquidity supply and demand in limit order markets, Discussion paper, Centre for Economic Policy Research, London

Lo I, Sapp SG (2003) Order submission: the choice between limit and market orders, Discussion paper, University of Waikato, University of Western Ontario

Lunde A (2000) A generalized gamma autoregressive conditional duration model, Discussion paper, Aarlborg University

Parlour CA (1998) Price dynamics and limit order markets. Rev Financ Stud 11:789–816

Pascual R, Veredas D (2004) What Pieces of Limit Order Book Information are Informative? Discussion Paper 2004/33, CORE, Université Catholique de Louvain

Ranaldo A (2004) Order aggressiveness in limit order book markets. J Financ Mark 7:53–74

Russell JR (1999) Econometric modeling of multivariate irregularly-spaced high-frequency data, Discussion paper, University of Chicago

Seppi DJ (1997) Liquidity provision with limit orders and strategic specialist. Rev Financ Stud 1 (1):103–150

Zhang MY, Russell JR, Tsay RS (2001) A nonlinear autoregressive conditional duration model with applications to financial transaction data. J Econom 104:179–207

Roman Liesenfeld · Ingmar Nolte · Winfried Pohlmeier

Modelling financial transaction price movements: a dynamic integer count data model

Abstract In this paper we develop a dynamic model for integer counts to capture fundamental properties of financial prices at the transaction level. Our model relies on an autoregressive multinomial component for the direction of the price change and a dynamic count data component for the size of the price changes. Since the model is capable of capturing a wide range of discrete price movements it is particularly suited for financial markets where the trading intensity is moderate or low. We present the model at work by applying it to transaction data of two shares traded at the NYSE traded over a period of one trading month. We show that the model is well suited to test some theoretical implications of the market microstructure theory on the relationship between price movements and other marks of the trading process. Based on density forecast methods modified for the case of discrete random variables we show that our model is capable to explain large parts of the observed distribution of price changes at the transaction level.

An earlier version of this paper has been presented at seminars in Helsinki, Munich, Louvain and the ESEM 2003 in Stockholm.

R. Liesenfeld
Christian-Albrechts-Universität, Kiel, Germany

I. Nolte
University of Konstanz, CoFE, Konstanz, Germany

W. Pohlmeier
University of Konstanz, CoFE, ZEW, Konstanz, Germany

W. Pohlmeier (✉)
Department of Economics, University of Konstanz,
Box D124, 78457 Konstanz, Germany
E-mail: winfried.pohlmeier@uni-konstanz.de

Keywords Financial transaction prices · Autoregressive conditional multinomial model · GLARMA · Count data · Market microstructure effects

JEL Classification C22 · C25 · G10

1 Introduction

Financial transaction data, often called ultra high frequency data, is marked by two main features: the irregularity of time intervals and the discreteness of price changes. Based on the seminal work by Russell and Engle (1998) and Engle (2000), a large body of studies has been centered around the further development of autoregressive conditional duration (ACD) models in order to characterize the transaction intensities. This paper is concerned with appropriately modelling the discreteness of the price process at the transaction level within a count data framework. While quantal response approaches seem to be more suitable in modelling price changes on the transaction level if the outcome space consists only of a few possible outcomes, the approach presented here is particularly designed for shares where the possible outcome space for the price changes is a larger range of integer values. This holds for most of the assets traded on European asset markets. But our approach is also attractive for the analysis of transaction price movements at more liquid markets such as the NYSE. With the decimalization on the 29th January 2001, the minimum tick size at the NYSE was reduced from 1/16-th of a US-Dollar to one cent for stocks selling at prices greater than or equal to US $1. This leads to larger price jumps in ticks (the smallest possible price change) and a larger range of observable discrete trade-by-trade price jumps. For example, in the month before decimalization for the *IBM* stocks, 95% of all price changes at the tick level were in a range of ±3 ticks, while this range changed to ±10 ticks in the month thereafter.

Since transaction price changes are quoted as multiples of a smallest divisor, the use of continuous distributions to characterize price changes is far from being appropriate in particular for markets with high transaction intensities. Accordingly, Hausman et al. (1992) proposed an ordered probit model with conditional heteroscedasticity to analyze stock price movements at the NYSE. The same approach is used by Bollerslev and Melvin (1994) to model the bid-ask spread at FX-markets. Contrary to the older rounding approaches by Ball (1988), Cho and Frees (1988) and Harris (1990), conditioning information can be incorporated in the ordered response models quite easily. A drawback of the ordered probit approach is that the parameters result from a threshold crossing latent variable model, where the underlying continuous latent dependent variable has to be given some more or less arbitrary economic interpretation (e.g., latent price pressure). Moreover, since the parameters are only identified up to a factor of proportionality, the estimates of the moments of the latent price variable are only identifiable using additional identifying restrictions.

An alternative to the ordered response models is the autoregressive conditional multinomial (ACM) model proposed by Russel and Engle (2002). Similar to the ordered response models, this approach also rests on the assumption that the distribution of observed transaction price changes is discrete with a finite number of outcomes. A drawback of the ACM model is the necessity that all potential outcomes have to occur in the sample period to guarantee the identification and estimation of the true dimension of the multinomial process. This creates a serious limitation if the ACM is used for forecasting purposes. In the multinomial approaches, as well as in the ordered response models, the number of parameters increases with the outcome space. As long as one is not willing to categorize the outcomes at the expense of a loss of information, both approaches are more suited for the empirical analysis of financial markets which are characterized by a limited number of discrete price changes.

In the following we propose a model that does not suffer from the drawbacks of the discrete response models sketched above. We propose a dynamic model which is based on a probability density function for an integer count variable, and which can be interpreted as a count data hurdle model. Our integer count hurdle (ICH) model is closely related to the components approach by Rydberg and Shephard (2003), who suggest decomposing the process of transaction price changes into three distinct processes: a binary process indicating whether a price change occurs from one transaction to the next, a binary process indicating the direction of the price change conditional on a price change having taken place, and a count process for the size of the price change conditional on the direction of the price change. By combining the above mentioned two binary processes into one trinomial ACM model (no price change or price movement downwards or upwards), and using a count process for the size of the price change based on a dynamic count data specification, our approach is more parsimonious than the one proposed by Rydberg and Shephard. The distribution of price changes used is that of a count data hurdle model extended for the domain of negative integer counts. For both components of the price process, the dynamics are modelled using a generalized ARMA specification.

Our model exhibits a number of desirable features and can be extended in many respects. The decomposition allows for a detailed analysis of the price direction process and volatility as well as the analysis of tail behavior. Inclusion of contemporaneous marks of the transaction price process as conditioning information (e.g. transaction time and volume), can generate insights into the validity of various hypotheses of market microstructure theory. In our empirical application of the ICH model, we will analyze the distribution of price changes conditional on transaction time and volume. Our model can also serve as a building block for the joint process of transaction price and transaction times. In this sense, our approach is more flexible than the competing risks ACD model by Bauwens and Giot (2003), which focuses on the direction of the price process whereby neglecting information on the size of the price changes.

For our empirical application of the ICH model, we use transaction data of the stocks of the Halliburton Company (*HAL*) and Jack in the Box Inc. (*JBX*) traded at the NYSE. Our sample period includes 35021 (*HAL*) and 4566 (*JBX*) transactions observed from 1st to 30th March 2001.[1] The two stocks are chosen for reasons of representativeness. *HAL* is a stock with medium market capitalization (about US

[1] The data used stems from the NYSE Trade and Quote database. We have removed all trades outside the regular trading hours and each day's first trade, to circumvent contamination due to the opening call auction at the NYSE. Besides, all trades are treated as split transactions, if they exhibited exactly the same timestamp. In this case we have aggregated their volume to one transaction and we have assigned the last price in the sequence to the aggregated transaction.

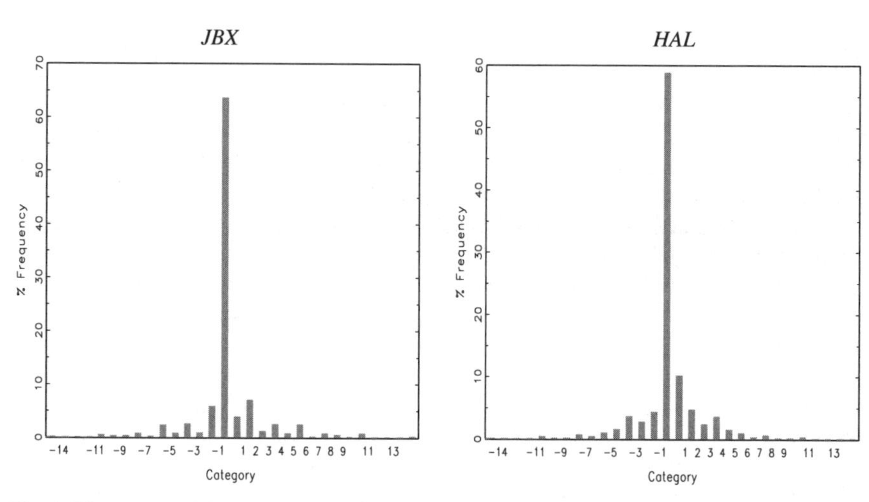

Fig. 1 Histograms of the transaction price changes for Jack in the Box Inc. (*JBX*) and Halliburton Company (*HAL*). The smallest possible price change is 0.01 US-Dollar

$17.5 billion) and considerable trade intensity, while *JBX* may be seen as representative for a share with lower market capitalization (about US$2.5 billion) and less trade intensity.

Figure 1 depicts the histograms for the transaction price changes for *JBX* and *HAL*. Rather typical for transaction data is the large fraction of zero price changes (around 60%). The remaining observations are proportioned between positive (around 25%) and negative price changes (around 15%). With a significantly higher frequency of positive one and two tick price jumps in comparison to the negative one and two tick price changes, the distributions for both stocks turn out to be somewhat skewed. Finally, we can observe price jumps of more than ±5 ticks for 11% (*JBX*) and 6% (*HAL*) of the transactions, which supports our view that both modelling transaction returns as a continuous random variable and quantal response representation, are too crude to pick up the true nature of the dependent variable, and neglect valuable information about the true data generating process. For JBX (HAL) the mean price change is given by 0.009 (−0.009) ticks, with corresponding standard deviation of 3.367 (2.743) ticks. Figures 2 and 3 display the autocorrelation functions of the price changes and the squared price changes for both stocks. Considering the price changes, the few positive first autocorrelation coefficients for JBX can be related to feedback trading and the first negative first order autocorrelation coefficient for HAL can be related to the bid-ask bounce. The bid-ask bounce refers to the fact that the transaction prices often bounce back and forth between the ask and bid prices creating a negative autocorrelation in the transaction price changes.[2] Figure 3 shows that the second moments for both stocks are subject to positive serial dependence, which represents an expression of volatility clustering in the transaction price changes.

[2] See, for example, Campbell et al. (1997), Chap. 3.2.

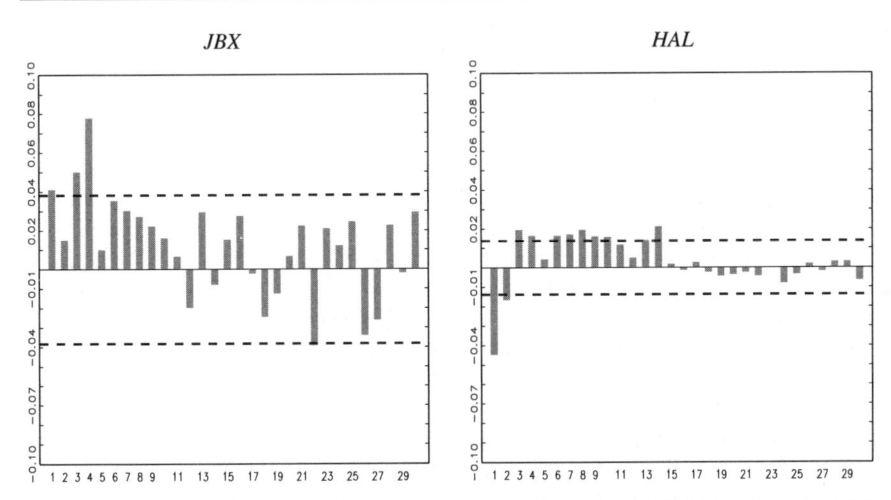

Fig. 2 Autocorrelation function of the transaction price changes for Jack in the Box Inc. (*JBX*) and Halliburton Company (*HAL*). The *dashed lines* mark off the approximate 99% confidence interval $\pm 2.58/\sqrt{n}$

The paper is organized as follows. In Section 2 the ICH model is introduced in its basic form as a time series model, where we analyze the two components of the price process separately. In Section 3 we augment the model by introducing transaction volume and transaction time, which play a crucial role in the literature on market microstructure. We use the ICH model to check some popular hypotheses of market microstructure theory. In Section 4 we take a closer look on the overall price process of discrete price changes by density forecast methods which we extend to the case of discrete distributions. Section 5 concludes and gives an outlook on possible extensions.

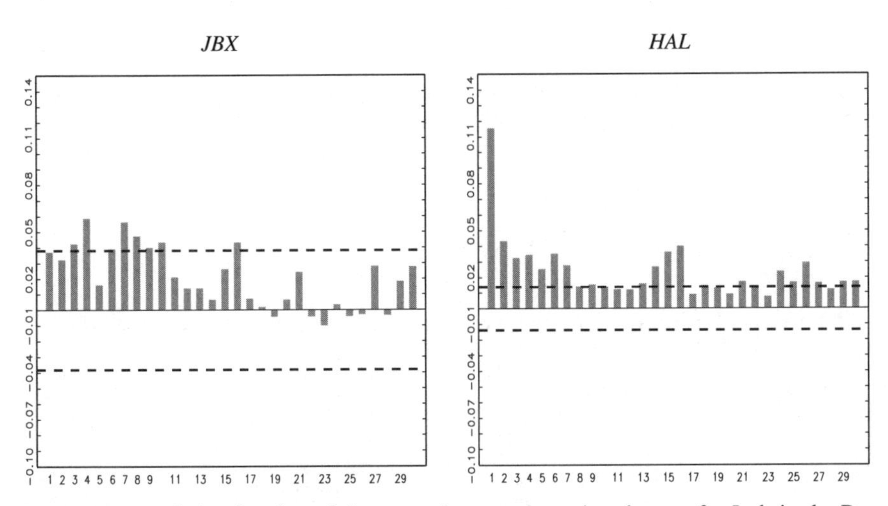

Fig. 3 Autocorrelation function of the squared transaction price changes for Jack in the Box Inc. (*JBX*) and Halliburton Company (*HAL*). The *dashed lines* mark off the approximate 99% confidence interval $\pm 2.58/\sqrt{n}$

2 The hurdle approach to integer counts

Consider a sequence of transaction prices $\{P(t_i), i: 1 \rightarrow n\}$ observed at times $\{t_i$, i: $1 \rightarrow n\}$. Let $\{Y_i, i: 1 \rightarrow n\}$ be a sequence of price changes, where $Y_i = P(t_i) - P(t_{i-1})$ is an integer multiple of a fixed divisor (tick), then $Y_i \in \mathbb{Z}$. Our interest lies in modelling the conditional distribution of the discrete price changes $Y_i|\mathcal{F}_{i-1}$, where $\mathcal{F}_{i-1}$ denotes the information set available at the time transaction i takes place. For this, we generalize the hurdle approach proposed by Mullahy (1986) and Pohlmeier and Ulrich (1995) for the Poisson and the negative binomial (Negbin) distribution, respectively, to the domain of negative counts. The basic idea of this approach is to decompose the overall process of transaction price changes into three components. The first component determines the direction of the process (positive price change, negative price change, or no price change) and will be specified as a dynamic multinomial response model. Given the direction of the price change, count data processes determine the size of positive and negative price changes, representing the second and third component of our model. This yields the following structure for the p.d.f. of $Y_i|\mathcal{F}_{i-1}$:

$$\Pr[Y_i = y_i|\mathcal{F}_{i-1}] = \begin{cases} \Pr[Y_i < 0|\mathcal{F}_{i-1}] \Pr[Y_i = y_i|Y_i < 0, \mathcal{F}_{i-1}] & \text{if} \quad y_i < 0 \\ \Pr[Y_i = 0|\mathcal{F}_{i-1}] & \text{if} \quad y_i = 0 \\ \Pr[Y_i > 0|\mathcal{F}_{i-1}] \Pr[Y_i = y_i|Y_i > 0, \mathcal{F}_{i-1}] & \text{if} \quad y_i > 0. \end{cases} \tag{2.1}$$

The process driving the direction of the price changes is represented by $\Pr[Y_i < 0|\mathcal{F}_{i-1}]$, $\Pr[Y_i = 0|\mathcal{F}_{i-1}]$ and $\Pr[Y_i > 0|\mathcal{F}_{i-1}]$, while the two processes for the size of the price changes conditional on the price direction, are defined by $\Pr[Y_i = y_i|Y_i < 0, \mathcal{F}_{i-1}]$ and $\Pr[Y_i = y_i|Y_i > 0, \mathcal{F}_{i-1}]$. Note that $\Pr[Y_i = y_i| Y_i > 0, \mathcal{F}_{i-1}]$ is a process defined over the set of strictly positive integers and $\Pr[Y_i = y_i|Y_i < 0, \mathcal{F}_{i-1}]$ is the corresponding p.d.f. for strictly negative counts. This decomposition allows us to model the stochastic behavior of the transaction price changes successively.

We follow Mullahy's (1986) idea by modelling the size of positive price changes as a truncated-at-zero count process.[3] Let $f^+(\cdot)$ be the p.d.f. of a standard count data distribution, then the p.d.f. for the size of positive price changes conditional on the fact that the prices are positive is a truncated-at-zero count data distribution:

$$\Pr[Y_i = y_i|Y_i > 0, \mathcal{F}_{i-1}] = h^+(y_i|\mathcal{F}_{i-1}) = \frac{f^+(y_i|\mathcal{F}_{i-1})}{1 - f^+(0|\mathcal{F}_{i-1})}. \tag{2.2}$$

The process for the size of negative price jumps is treated in the same way:

$$\Pr[Y_i = y_i|Y_i < 0, \mathcal{F}_{i-1}] = h^-(y_i|\mathcal{F}_{i-1}) = \frac{f^-(-y_i|\mathcal{F}_{i-1})}{1 - f^-(0|\mathcal{F}_{i-1})}, \tag{2.3}$$

[3] Alternatively, one could specify the p.d.f. of the transformed count $Y_i - 1$ conditional on $Y_i > 0$ using a standard count data approach. This approach was adopted by Rydberg and Shephard (2003) in their decomposition model.

where $f^{-}(\cdot)$ denotes the p.d.f. of a standard count data model. Combining the single components leads to the following p.d.f. for the transaction price changes:

$$\Pr[Y_i = y_i|\mathcal{F}_{i-1}] = \left[\Pr[Y_i < 0|\mathcal{F}_{i-1}]h^{-}(y_i|\mathcal{F}_{i-1})\right]^{\delta_i^-} \left[\Pr[Y_i = 0|\mathcal{F}_{i-1}]\right]^{\delta_i^0} \times \left[\Pr[Y_i > 0|\mathcal{F}_{i-1}]h^{+}(y_i|\mathcal{F}_{i-1})\right]^{\delta_i^+}, \tag{2.4}$$

where $\delta_i^- = \mathbb{1}_{\{Y_i<0\}}$, $\delta_i^0 = \mathbb{1}_{\{Y_i=0\}}$ and $\delta_i^+ = \mathbb{1}_{\{Y_i>0\}}$ are binary variables indicating positive, negative, or no price change for transaction i.

A more parsimonious distribution results if one assumes that $h^{-}(\cdot)$ and $h^{+}(\cdot)$ arise from the same parametric family of probability density functions. Based on this assumption, the stochastic behavior of positive and negative price movements can be summarized in a conditional p.d.f. for the absolute price changes $S_i \equiv |Y_i|$ conditional on the price direction:

$$\Pr[S_i = s_i|S_i > 0, D_i, \mathcal{F}_{i-1}] = h(s_i|D_i, \mathcal{F}_{i-1}) \quad \text{with} \quad D_i = \begin{cases} -1 & \text{if} \quad Y_i < 0, \\ 0 & \text{if} \quad Y_i = 0, \\ 1 & \text{if} \quad Y_i > 0, \end{cases} \tag{2.5}$$

where $h(\cdot)$ is the p.d.f. of a truncated-at-zero count data model. For the parsimonious specification, the p.d.f. for a transaction price change is:

$$\Pr[Y_i = y_i|\mathcal{F}_{i-1}] = \Pr[Y_i < 0|\mathcal{F}_{i-1}]^{\delta_i^-} \Pr[Y_i = 0|\mathcal{F}_{i-1}]^{\delta_i^0} \Pr[Y_i > 0|\mathcal{F}_{i-1}]^{\delta_i^+} \times [h(|y_i||D_i, \mathcal{F}_{i-1})]^{(1-\delta_i^0)}. \tag{2.6}$$

In this case, the resulting sample log-likelihood function of the ICH-model consists of two additive components:

$$L = \sum_{i=1}^{n} \ln \Pr[Y_i = y_i|\mathcal{F}_{i-1}] = \sum_{i=1}^{n} L_{1,i} + \sum_{i=1}^{n} L_{2,i}, \tag{2.7}$$

where:

$$L_{1,i} = \delta_i^- \ln \Pr[Y_i < 0|\mathcal{F}_{i-1}] + \delta_i^0 \ln \Pr[Y_i = 0|\mathcal{F}_{i-1}] + \delta_i^+ \ln \Pr[Y_i > 0|\mathcal{F}_{i-1}] \tag{2.8}$$

$$L_{2,i} = \left(1 - \delta_i^0\right) \ln h(|y_i||D_i, \mathcal{F}_{i-1}). \tag{2.9}$$

The component $\sum L_{1,i}$ is the log-likelihood of the multinomial process determining the direction of prices, while $\sum L_{2,i}$ is the log-likelihood of the truncated-at-zero count process for the absolute size of the price change. If there are no parametric restrictions across the two likelihoods, we can maximize the complete likelihood (2.7) by separately maximizing its components (2.8) and (2.9). This reduces the computational burden considerably. In the following, we now specify the parametric form of the p.d.f. for the price direction and the absolute size of the price changes.

2.1 Dynamics of the price direction

The parametric model for the direction of the transaction price change D_i=j, (j=−1, 0, 1) is taken from the class of logistic ACM (autoregressive conditional multinomial) models suggested by Russel and Engle (2002). In order to relate the probability $\pi_{ji} = \Pr\left[D_i = j|\mathcal{F}_{i-1}\right]$ for the occurrence of price direction j to subsets of $\mathcal{F}_{i-1}$ (and further explanatory variables), we use a logistic link function. This leads to a multinomial logit model of the form:

$$\pi_{ji} = \frac{\exp\{\Lambda_{ji}\}}{\sum_{j=-1}^{1}\exp\{\Lambda_{ji}\}}, \quad j = -1, 0, 1, \tag{2.10}$$

where Λ_{ji} represents a function of some subset of $\mathcal{F}_{i-1}$ to be specified below. As a normalizing constraint, we use Λ_{0i}=0, $\forall i$.

Due to the observed dynamic behavior of the transaction price changes associated with the bid-ask bounce or the volatility clustering, one can expect that the process of the price direction also exhibits serial dependence. In order to shed light on this serial dependence, which has to be taken into account when modelling the conditional distribution of the price direction, we define the following state vector

$$x_i = (x_{-1i}, x_{1i})' = \begin{cases} (1,0)' & \text{if} \quad Y_i < 0 \\ (0,0)' & \text{if} \quad Y_i = 0 \\ (0,1)' & \text{if} \quad Y_i > 0, \end{cases} \tag{2.11}$$

and consider its corresponding sample autocorrelation matrix. For a lag length ℓ, this matrix is given by:

$$\Upsilon(\ell) = D^{-1}\Gamma(\ell)D^{-1}, \quad \ell = 1, 2, \ldots, \tag{2.12}$$

with

$$\Gamma(\ell) = \frac{1}{n-\ell-1}\sum_{i=\ell+1}^{n}(x_i - \bar{x})(x_{i-\ell} - \bar{x})'.$$

D denotes a diagonal matrix containing the standard deviations of x_{-1i} and x_{1i}.

Figure 4 below depicts the cross-correlation function of up to 30 lagged transactions. The significant, but not very large, first order cross-correlations provide empirical support for the existence of a bid-ask bounce: The probability of a price reduction is significantly (*HAL*) positively correlated with the price increase in the previous period (upper right panel), and also significant (both stocks), a price increase is more likely if a negative price change is observed for the previous transaction (lower left panel). The cross-correlation effects turn out to be asymmetric, in the sense that the correlation of a negative price change with a previous positive one is smaller than the effect vice versa. For *HAL* we observe negative first

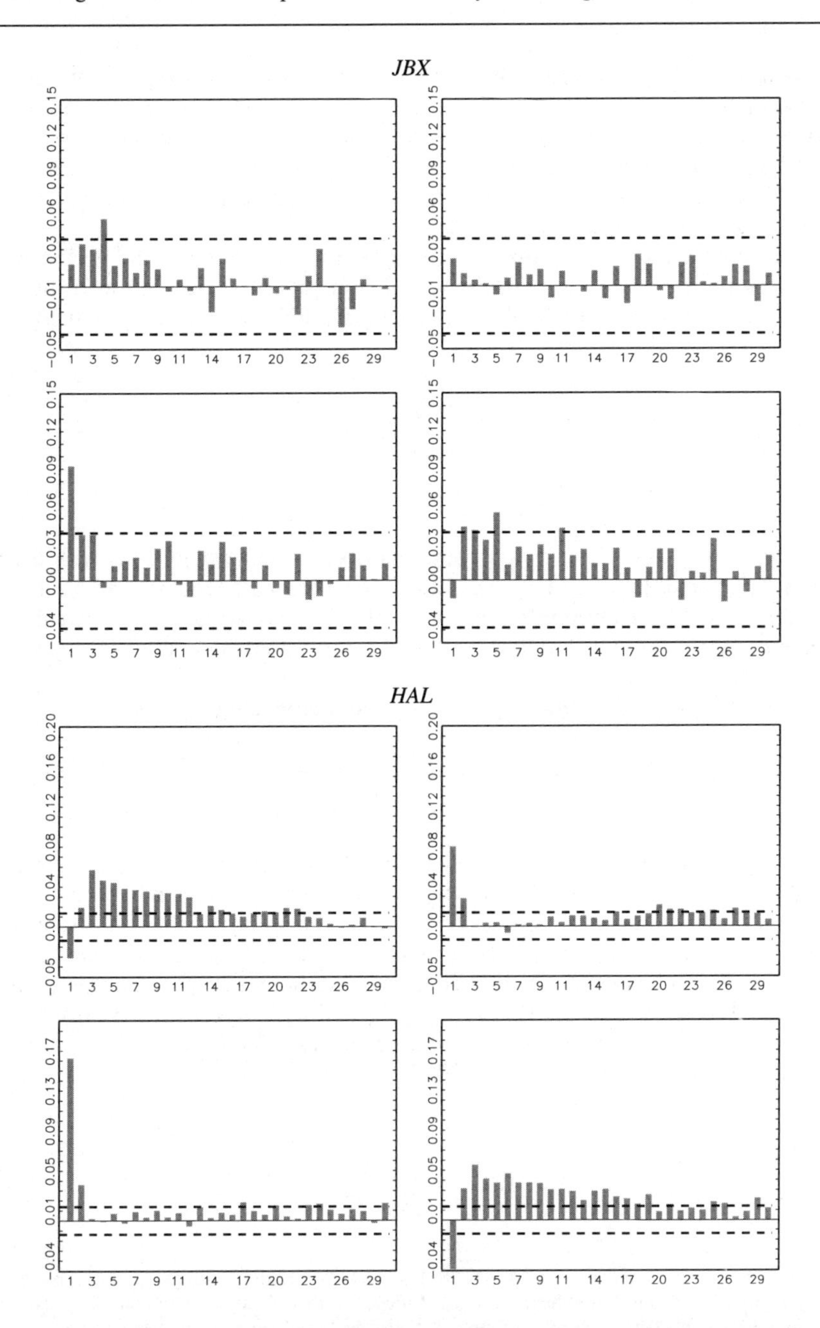

Fig. 4 Multivariate autocorrelation function for the price change directions for Jack in the Box Inc. (*JBX*) and Halliburton Company (*HAL*). Upper *left panel*: $\text{corr}(x_{-1i}, x_{-1i-\ell})$; *upper right panel*: $\text{corr}(x_{-1i}, x_{1i-\ell})$; *lower left panel*: $\text{corr}(x_{-1i-\ell}, x_{1i})$ and *lower right panel*: $\text{corr}(x_{1i}, x_{1i-\ell})$. The *dashed lines* mark the approximate 99% confidence interval $\pm 2.58/\sqrt{n}$

order serial correlation for price changes in the same directions (upper left and lower right panel), which underpins the existence of a bid-ask bounce. For *JBX* an analogue pattern is not observable. Moreover, for *HAL* the positive autocorrelations at longer lags indicate that the bounce effect will be compensated in later periods. Finally, note that the negative serial correlation caused by the bid-ask bounce is a short-run phenomenon.

In order to capture the dynamics of the price direction variable, the vector of log–odds ratios $\Lambda_i = (\Lambda_{-1i}, \Lambda_{1i})' = (\ln[\pi_{-1i}/\pi_{0i}], \ln[\pi_{1i}/\pi_{0i}])'$ is specified as a multivariate ARMA process. The final form including possible explanatory variables is:

$$\begin{aligned} \Lambda_i &= \sum_{l=0}^{m} G_l Z_{i-l}^D + \alpha_i \\ \alpha_i &= \mu + \sum_{l=1}^{p} C_l \alpha_{i-1} + \sum_{l=1}^{q} A_l \xi_{i-l} \end{aligned} \tag{2.13}$$

with $\{C_l,\ l{:}\ 1 \rightarrow p\}$ and $\{A_l,\ l{:}\ 1 \rightarrow q\}$ being matrices of dimension (2×2) with the elements $\{c_{hk}^{(l)}\}$ and $\{a_{hk}^{(l)}\}$ and $\mu = (\mu_1, \mu_2)'$. The vector Z_i^D contains additional explanatory variables capturing other marks of the trading process (market microstructure variables) with $\{G_l,\ l{:}\ 0 \rightarrow m\}$ as the corresponding coefficient matrix and typical element $\{g_{hk}^{(l)}\}$.

The vector of log–odds ratios is driven by the martingale differences:

$$\xi_i = (\xi_{-1i}, \xi_{1i})', \quad \text{with} \quad \xi_{ji} = \frac{x_{ji} - \pi_{ji}}{\sqrt{\pi_{ji}(1 - \pi_{ji})}}, \quad j = -1, 1, \tag{2.14}$$

which is the standardized state vector x_i. In this ACM-ARMA(p,q) specification, the conditional distribution of the direction of price changes depends on lagged conditional distributions of the process and the lagged values of the standardized state vector.[4] The process is stationary if all values of z that satisfy $|I - C_1 z - C_2 z^2 - \cdots - C_p z^p| = 0$ lie outside the unit circle. Furthermore note, that the existence of a bid-ask bounce would imply that $a_{12}^{(1)} > a_{11}^{(1)}$ and $a_{21}^{(1)} > a_{22}^{(1)}$, which means that the probability of an immediate reversal of the price direction is higher than that of an unchanged price direction.[5] The log likelihood of the logistic ACM model, the

[4] According to the classification by Cox (1981), our ACM model belongs to the class of observationally driven models where time dependence arises from a recursion on lagged endogenous variables. Alternatively, our model could be based on a parameter driven specification, in which the log–odds ratios Λ_i are determined by a dynamic latent process. However, the estimation and the diagnostics of the latter approach results in a substantially higher computational burden than for the ACM model. On the other hand, models driven by latent processes are usually more parsimonious than comparable dynamic models based on lagged dependent variables. A comparison of the two alternatives should be the subject of future research.

[5] See Russel and Engle (2002) for a more detailed discussion of the stochastic properties of the ACM-ARMA(p,q) model.

first component of the likelihood of the overall model, takes on the familiar form presented below:

$$L_1 = \sum_{i=1}^{n} \left[\delta_i^- \ln \pi_{-1i} + \delta_i^0 \ln \pi_{0i} + \delta_i^+ \ln \pi_{1i} \right]. \tag{2.15}$$

Concerning the coefficient matrix A_l, our empirical analysis will be based on two alternative specifications: an unrestricted one, and one including symmetry restrictions as suggested by Russel and Engle (2002). In particular, we impose the symmetry restriction $a_{12}^{(1)}=a_{21}^{(1)}$. This implies that the marginal effect of a negative price change on the conditional probability of a future positive price change is of the same size as the marginal effect of a positive change on the probability of a future negative change. Moreover, we impose the symmetry restriction $a_{11}^{(1)}=a_{22}^{(1)}$ which guarantees that the impact of a negative change on the probability of a future negative change is the same as the corresponding effect for positive price changes. The symmetry of impacts on the conditional price direction probabilities will also be imposed for all lagged values of the probabilities and normalized state variables. Following Russel and Engle (2002), we set in the model with symmetry restrictions $c_2^{(l)}=0, \forall l$, which implies that shocks in the log–odds ratios vanish at an exponential rate determined by the diagonal element $c_1^{(l)}$. This simplifies the ARMA specification (2.13) for the symmetric model to:

$$\mu = \begin{pmatrix} \mu_1 \\ \mu_2 \end{pmatrix}, \quad C_l = \begin{pmatrix} c_1^{(l)} & 0 \\ 0 & c_1^{(l)} \end{pmatrix}, \quad A_l = \begin{pmatrix} a_1^{(l)} & a_2^{(l)} \\ a_2^{(l)} & a_1^{(l)} \end{pmatrix}. \tag{2.16}$$

Although the reasoning behind these restrictions seems appropriate due to the explorative evidence of the state variable x_i, the validity of these restrictions can, of course, be easily tested by standard ML based tests.

2.2 Empirical results for the ACM model

In search of the best specification, we use the Schwarz information criterion (SIC) to determine the order of the ARMA process. For the selected specification, its standardized residuals will be subject to diagnostic checks. For the estimates of the conditional expectations and the variances $\hat{E}[x_i|\mathcal{F}_{i-1}] = \hat{\pi}_i$ and $\hat{V}[x_i|\mathcal{F}_{i-1}] = \text{diag}(\hat{\pi}_i) - \hat{\pi}_i\hat{\pi}_i'$, respectively, the standardized residuals can be computed as:

$$v_i = (v_{-1i}, v_{1i})' = \hat{V}\left[x_i|\mathcal{F}_{i-1}\right]^{-1/2}\left[x_i - \hat{E}\left[x_i|\mathcal{F}_{i-1}\right]\right], \tag{2.17}$$

where $\hat{V}[x_i|\mathcal{F}_{i-1}]^{-1/2}$ is the inverse of the Cholesky factor of the conditional variance. For a correctly specified model, the standardized residuals evaluated at the true parameter values should be serially uncorrelated in the first two moments with the following unconditional moments: $E[v_i]=0$ and $E[v_i v_i']=I$. The null

Table 1 ML estimates of the logistic ACM-ARMA model for Jack in the Box Inc. (*JBX*) and Halliburton Company (*HAL*)*

Parameter	*JBX*				HAL			
	Symmetric		Non-symmetric		Symmetric		Non-symmetric	
	Estimate	Std. dev.	Estimate	Std. dev.	Estimate	Std. dev.	Estimate	Std. dev.
μ_1	−0.0998	0.0263	−0.1331	0.0382	−0.0355	0.0035	−0.0368	0.0036
μ_2	−0.0998	0.0263	−0.1204	0.0345	−0.0355	0.0035	−0.0349	0.0035
$c_1^{(1)}$	0.9116	0.0230	0.8881	0.0316	1.1412	0.0294	1.1309	0.0287
$c_1^{(2)}$					−0.1895	0.0292	−0.1797	0.0283
$a_{11}^{(1)}$	0.1164	0.0186	0.1367	0.0309	0.0568	0.0109	0.0934	0.0147
$a_{12}^{(1)}$	0.0955	0.0180	0.0512	0.0219	0.2510	0.0099	0.1458	0.0135
$a_{21}^{(1)}$	0.0955	0.0180	0.1606	0.0335	0.2510	0.0099	0.3539	0.0135
$a_{22}^{(1)}$	0.1164	0.0186	0.1124	0.0196	0.0568	0.0109	0.0253	0.0144
$a_{11}^{(2)}$					0.0473	0.0116	0.0140	0.0152
$a_{12}^{(2)}$					−0.1853	0.0100	−0.0897	0.0136
$a_{21}^{(2)}$					−0.1853	0.0100	−0.2755	0.0136
$a_{22}^{(2)}$					0.0473	0.0116	0.0773	0.0148
log-likelihood	−0.928726		−0.926585		−1.005572		−1.003333	
SIC	0.932423		0.933055		1.006618		1.005126	
$Q(30)$	132.9 (0.149)		121.1 (0.331)		302.4 (0.000)		163.6 (0.001)	
$Q(50)$	215.1 (0.179)		204.5 (0.306)		401.7 (0.000)		261.4 (0.001)	
Resid. mean	(−0.019, 0.029)		(0.000, 0.001)		(−0.004, 0.012)		(−0.000, −0.000)	
Resid. variance	$\begin{pmatrix} 0.902 & 0.024 \\ 0.024 & 1.107 \end{pmatrix}$		$\begin{pmatrix} 0.934 & 0.023 \\ 0.023 & 1.069 \end{pmatrix}$		$\begin{pmatrix} 0.887 & 0.033 \\ 0.033 & 1.132 \end{pmatrix}$		$\begin{pmatrix} 0.890 & 0.037 \\ 0.037 & 1.132 \end{pmatrix}$	
LR-test on symmetry	19.514 (0.0002)				156.79 (0.0000)			

*Dependent variable is the direction of the price changes, D_i, $n = 4{,}566$ (*JBX*), $n = 35{,}021$ (*HAL*), p-values in parenthesis

hypothesis of absence of serial and cross-correlation in v_i can be tested by the multivariate version of the Portmanteau statistic:[6]

$$Q(L) = n \sum_{\ell=1}^{L} \mathrm{tr}\left[\Gamma_v(\ell)'\Gamma_v(0)^{-1}\Gamma_v(\ell)\Gamma_v(0)^{-1}\right], \tag{2.18}$$

where $\Gamma_v(\ell) = \sum_{i=\ell+1}^{n} v_i v'_{i-\ell} / (n-\ell-1)$. Under the null hypothesis, $Q(L)$ is asymptotically χ^2-distributed with degrees of freedom equal to the difference between four times L and the number of parameters to be estimated.

The ML-estimation results for the pure time series specifications (obtained using the optimization procedure of Berndt et al. (1974) (BHHH)) and the results of the corresponding diagnostic checks are summarized in Table 1. The Schwarz criterion suggests to select an ARMA(1,1)-specification for *JBX* and an ARMA (2,2)-specification for *HAL* regardless whether symmetry on the coefficient matrices is imposed or not. All coefficient estimates are at least significant at the 5% level. The LR-test for symmetry of the coefficient matrices rejects the null of symmetric responses for both stocks.

Our estimates reveal some differences in the dynamics of the price direction variable for the two stocks. For the *JBX* stock the estimate of 0.912 (0.888 for the non-symmetric specification) for the coefficient $c_1^{(1)}$ indicates a high degree of persistence in the price direction variable. Independent of the direction of the price movement, the probability of a price change is comparatively high if the probability of a price change for the previous transaction was high. Because the probability of a non-zero price change can be interpreted as a specific measure of price volatility, this finding reflects a clustering of volatility. Since under the estimated non-symmetric specification $a_{12}^{(1)} < a_{11}^{(1)}$ and $a_{22}^{(1)} < a_{21}^{(1)}$ we can conclude that there is no clear evidence for a bid-ask bounce. Hence, our estimates confirm the explorative findings on the cross-correlations of the price direction variable (Fig. 4) and the simple autocorrelation function of the transaction prices in Fig. 2. The results for *HAL* are more along the lines of previous empirical findings on the dynamics of the transaction price process. Since $c_1^{(1)} + c_1^{(2)}$ is positive and close to unity, a high degree of persistence can be found as well. Moreover, there is now evidence for a negative first-order serial cross-correlation for the price change indicators with $a_{12}^{(1)} > a_{11}^{(1)}$ and $a_{22}^{(1)} < a_{21}^{(1)}$, indicating the existence of a bid-ask bounce.

Our diagnostic checks show that most of the dynamics of the price direction variable is captured for the *JBX* data, but not for the *HAL*. For *JBX* the generalized Portmanteau statistic $Q(30)$ with a value of 132.9 (121.1) does not reject the null hypothesis of no cross-correlations at the 10% significance level for the symmetric and the nonsymmetric specification. For the raw data x_i, the corresponding value of the $Q(30)$ statistic was found to be 303.9. For *HAL* the generalized Portmanteau statistics indicate that the standardized residuals are still cross-correlated, although the model reduces the value of the $Q(30)$ statistic quite substantially from value of 3590.0 for the raw data to 163.6 (non-symmetric specification).

Figure 5 depicts the cross-correlation functions of the standardized residuals v_{-1i} and v_{1i}. For *JBX* all but one correlations lie within the 99% confidence band

[6] See, for example, Lütkepohl (1993).

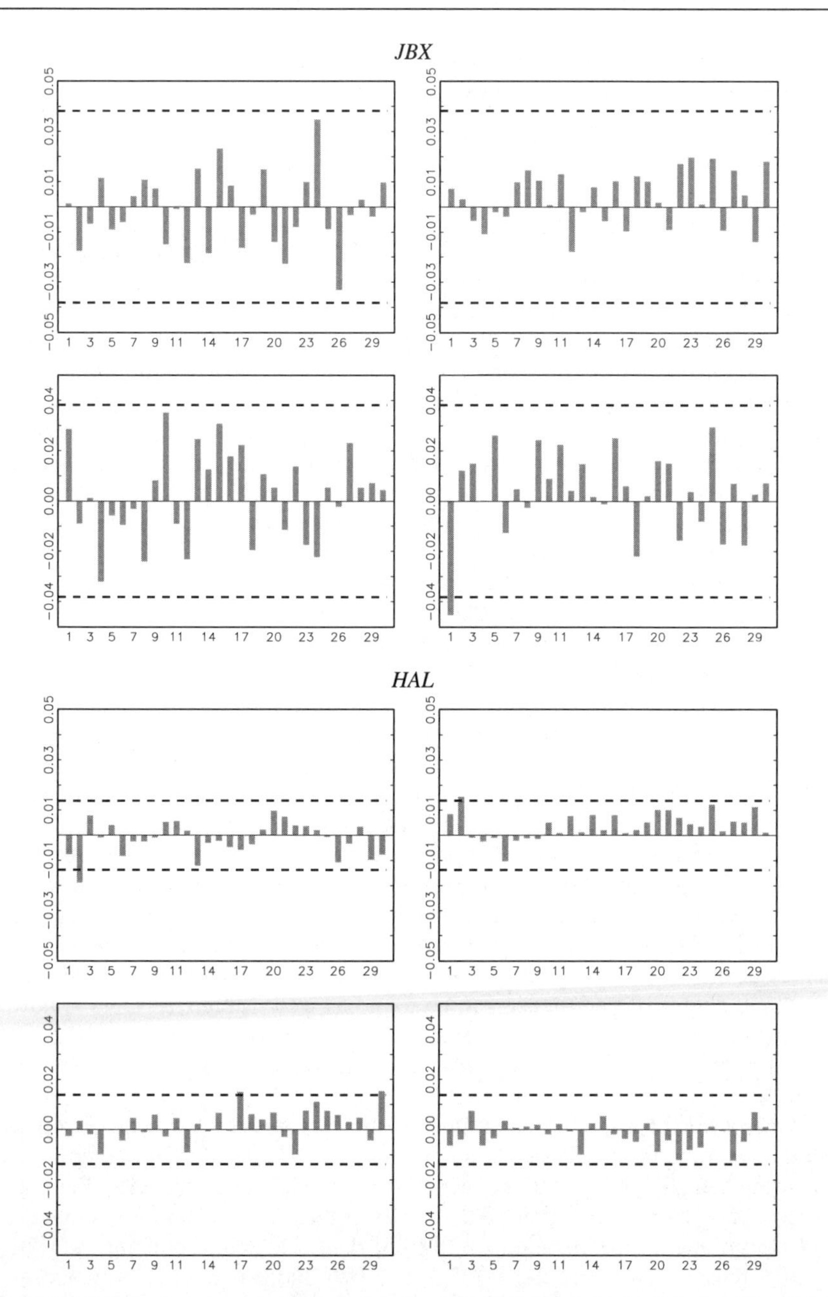

Fig. 5 Multivariate autocorrelation function for the residuals of the logistic ACM model for Jack in the Box Inc. (*JBX*) and Halliburton Company (*HAL*) up to 30 lagged transactions. The correlations are the following: *upper left panel*: $\text{corr}(v_{-1i}, v_{-1i-\ell})$, *upper right panel*: $\text{corr}(v_{-1i}, v_{1i-\ell})$, *lower left panel*: $\text{corr}(v_{-1i-\ell}, v_{1i})$ and *lower right panel*: $\text{corr}(v_{1i}, v_{1i-\ell})$. The *dashed lines* mark off the approximate 99% confidence interval $\pm 2.58/\sqrt{n}$

and for *HAL* all but two correlations lie inside the 99% band. The means of standardized residuals reported in Table 1 are close to zero, which should be expected from a well specified model. However, the estimated variance–covariance matrix of the standardized residuals deviate slightly from the identity matrix. This may hint to a distributional misspecification or a misspecification of log–odds ratios Λ_i, which is not fully compatible with the variation in the observed variation of price change direction.

2.3 Dynamics of the size of price changes

In order to analyze the size of the non-zero price changes, we use a GLARMA (generalized linear autoregressive moving average) model based on a truncated-at-zero Negative Binomial (Negbin) distribution. The choice of a Negbin in favor of a Poisson distribution is motivated by the fact, that the unconditional distributions of the non-zero price changes show over-dispersion for both stocks. For JBX (HAL) the dispersion coefficient[7] is given by 3.770 (2.911). Moreover, note, that an at-zero-truncated Poisson distribution would allow only for under-dispersion.

Similar to the ACM model, the dynamic structure of this count data model rests on a recursion on lagged observable variables. A comprehensive description of this class of models can, for instance, be found in Davis et al. (2003). Note that the time scale for absolute price changes (defined by transactions associated with non-zero price changes) is different from the one of the ACM model for the direction of the price changes, which is defined on the ticktime scale. Let u be a random variable following a Negbin distribution with the p.d.f.[8]

$$f(u) = \frac{\Gamma(\kappa + u)}{\Gamma(\kappa)\Gamma(u+1)} \left(\frac{\kappa}{\kappa + \omega}\right)^{\kappa} \left(\frac{\omega}{\omega + \kappa}\right)^{u}, \quad u = 0, 1, 2, \ldots, \tag{2.19}$$

with $E(u) = \omega > 0$ and $\mathrm{Var}(u) = \omega + \omega^2/\kappa$. The overdispersion of the Negbin distribution depends on parameter $\kappa > 0$. As $\kappa \to \infty$, the Negbin collapses to a Poisson distribution. The corresponding truncated-at-zero Negbin distribution is obtained as $h(u) = f(u)/[1 - f(0)]$, $(u = 1, 2, 3, \ldots)$, with $f(0) = [\kappa/(\kappa + \omega)]^{\kappa}$. This flexible class of distributions will be used to model the size of non-zero price changes conditional on filtration $\mathcal{F}_{i-1}$ and price direction D_i. Thus, for $S_i \mid S_i > 0, D_i, \mathcal{F}_{i-1}$ we assume the following p.d.f.:

$$h(s_i | D_i, \mathcal{F}_{i-1}) = \frac{\Gamma(\kappa + s_i)}{\Gamma(\kappa)\Gamma(s_i + 1)} \left(\left[\frac{\kappa + \omega_i}{\kappa}\right]^{\kappa} - 1\right)^{-1} \left(\frac{\omega_i}{\omega_i + \kappa}\right)^{s_i}, \quad s_i = 1, 2, \ldots, \tag{2.20}$$

with the conditional moments:

$$E[S_i | S_i > 0, D_i, \mathcal{F}_{i-1}] = \mu_{S_i} = \frac{\omega_i}{1 - \vartheta_i} \tag{2.21}$$

[7] Computed as variance over mean.
[8] See, for example, Cameron and Trivedi (1998) (Ch. 4.2.2.).

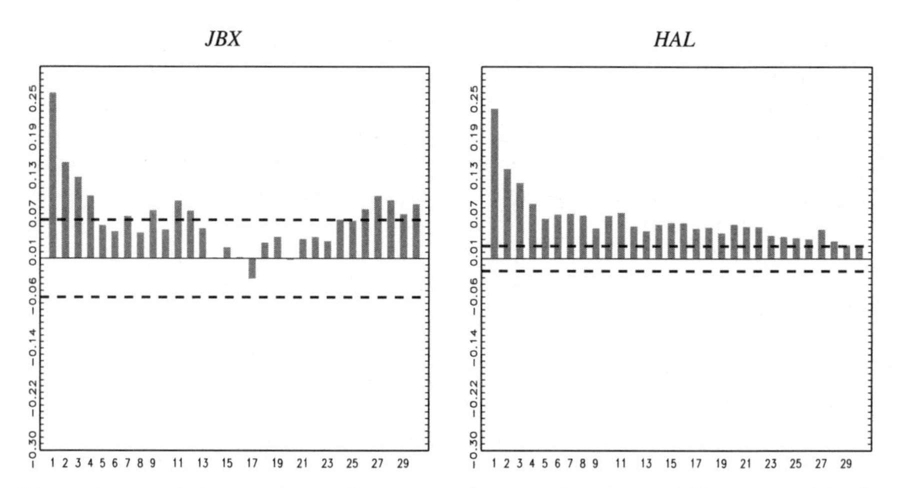

Fig. 6 Autocorrelation function of the non-zero absolute price changes $S_i|S_i>0$ for Jack in the Box Inc. (*JBX*) and Halliburton Company (*HAL*). The *dashed lines* mark the approximate 99% confidence interval $\pm 2.58/\sqrt{\tilde{n}}$, where ñ is the number of non-zero price changes

$$V[S_i|S_i > 0, D_i, \mathcal{F}_{i-1}] = \sigma^2_{S_i} = \frac{\omega_i}{1-\vartheta_i} - \frac{\omega_i^2}{(1-\vartheta_i)^2}\left(\vartheta_i - \frac{1-\vartheta_i}{\kappa}\right), \tag{2.22}$$

where $\vartheta_i = \left[\kappa/(\kappa+\omega_i)\right]^{\kappa}$. In this specification, both mean and variance are monotonic increasing functions of the variable ω_i that is assumed to capture the variation of the conditional distribution depending on D_i and $\mathcal{F}_{i-1}$.

As noted above, volatility clustering of asset returns is a well-known property, which also occurs at high frequencies. This is confirmed by the autocorrelation function of the nonzero absolute price changes shown in Fig. 6, which reveals a significant autocorrelation of this specific volatility measure. For the less frequently traded stock *JBX*, the correlations die out quicker than for *HAL* where significant but small correlations can be observed even after more than 25 trades.

In order to take into account the dynamics of S_i we follow Rydberg and Shephard (2002) and impose a GLARMA structure on $\ln \omega_i$ as follows:[9]

$$\begin{aligned} \ln \omega_i &= \beta_d' \check{D}_i + \sum_{l=0}^{m} B_l Z^S_{i-l} + \lambda_i \\ \text{with} \quad \lambda_i &= \gamma_0 + S(\nu, \tau, K) + \sum_{l=1}^{p} \gamma_l \lambda_{i-l} + \sum_{l=1}^{q} \delta_l \varepsilon_{i-1} + \sum_{l=1}^{r} \zeta_l |\varepsilon_{i-l}|. \end{aligned} \tag{2.23}$$

Since it is a well-known stylized fact for high frequent financial data that there exists intraday seasonality in price volatility, we introduce the seasonal component $S(\nu, \tau, K) = \nu_0\tau + \sum_{k=1}^{K} \nu_{2k-1} \sin(2\pi(2k-1)\tau) + \nu_{2k}\cos(2\pi(2k)\tau)$. This Fourier flexible form is to capture intraday seasonality in the absolute price changes, where τ is

[9] Similar to the alternative specification discussed in the context of the ACM model one could also specify a dynamic latent process for ω_i. See Zeger (1988) and Jung and Liesenfeld (2001) for examples.

the intraday trading time standardized on [0, 1] and ν is a $2K$+1 dimensional parameter vector.

In Eq. (2.23) the standardized absolute price change, $\varepsilon_i = (S_i - \mu_{S_i})/\sigma_{S_i}$, drives the extended ARMA process in λ_i. Similar to the dynamics structure in the EGARCH model of Nelson (1991) we allow λ_i to depend on the innovation term, and additionally on its absolute value to be able to capture a wide range of possible news impact functions. The vector $\check{D}_i = (D_i, D_{i-1}, \ldots D_{i-l})'$ contains information about the contemporaneous and lagged price directions with a corresponding coefficient vector $\beta_d = (\beta_{d0}, \beta_{d1}, \ldots, \beta_{dl})'$. The inclusion of $D_i \in \{-1, 1\}$ is to capture potential differences in the behavior of the absolute price changes depending on the direction of the price process. Thus, a negative β_{d0} implies that with negative price changes ($D_i = -1$) large absolute price changes are more likely to occur than with positive price changes ($D_i = 1$). If one interprets the absolute price changes as an alternative volatility measure, this asymmetry reflects a kind of leverage effect, where downward price movements imply a higher volatility than upward movements.[10] The inclusion of lagged terms D_{i-l} l=1, 2,... allows for a dynamic variant of the leverage effect. Z_i^S denote further explanatory variables, which will be discussed below.

2.4 Empirical results for the GLARMA model

For estimation of parameters of the GLARMA(p,q) model, given by Eqs. (2.20) to (2.23), we maximize the log-likelihood $L_2 = \sum_{i=1}^{n} \left(1 - \delta_i^0\right) \ln h(s_i|D_i, \mathcal{F}_{i-1})$ using the BHHH algorithm. We use the Schwarz information criterion to determine the optimal order of p and q. The diagnostic checks are based on the standardized residuals

$$e_i = \frac{S_i - \hat{\mu}_{Si}}{\hat{\sigma}_{si}}. \tag{2.24}$$

For a correctly specified model, the residuals evaluated at the true parameter values should be uncorrelated in the first two moments with $E[e_i] = 0$ and $E[e_i^2] = 1$.

The ML-estimates of the pure time series component of the GLARMA model for two specifications allowing for a plain ($\zeta_l = 0$) and a more flexible news impact curve ($\zeta_l \neq 0$) are given in Table 2. Since the coefficients of the seasonal component $S(\nu, \tau, K)$ are jointly significant for all specifications estimated, we refrain from reporting the results for the models without any seasonal component in the absolute price change variable. The diurnal seasonalities of the absolut price changes for *JBX* and *HAL* are depicted in Fig. 7. We observe a standard intraday volatility (or market activity) seasonality pattern: A high volatility when trading starts, which declines until lunch-time (after around 9,000 s, 12:00 EST), a slight recovery in the early afternoon, and another decrease when the European stock markets close (after around 16,200 s, 14:00 EST). As for the price direction variable for the *JBX* data, a parsimonious GLARMA(1,1)-specification turns out to be best specification, while

[10] See, for instance, Nelson (1991).

Table 2 ML estimates of the GLARMA model for Jack in the Box Inc. (*JBX*) and Halliburton Company (*HAL*)*

Parameter	*JBX*				*HAL*			
	Plain news impact		Flexible news impact		Plain news impact		Flexible news impact	
	Estimate	Std. dev.	Estimate	Std. dev.	Estimate	Std. dev.	Estimate	Std. dev.
γ_0	0.2079	0.0545	−0.0133	0.0426	0.0806	0.0505	−0.0119	0.0039
γ_1	0.8753	0.0356	0.9275	0.0292	0.3821	0.5177	1.5058	0.0717
γ_2					0.3914	0.3761	−0.5219	0.0682
δ_1	0.1643	0.0312	0.1234	0.0288	0.2240	0.0181	0.2053	0.0183
δ_2					0.0262	0.1411	−0.1724	0.0164
ζ_1			0.2008	0.0389			0.1602	0.0262
ζ_2							−0.1307	0.0257
ν_0	−0.0705	0.0507	−0.0879	0.0475	−0.2036	0.1452	−0.0143	0.0041
ν_1	0.0078	0.0080	0.0045	0.0076	0.0097	0.0097	0.0008	0.0005
ν_2	0.0169	0.0085	0.0165	0.0080	0.0115	0.0093	0.0012	0.0006
ν_3	−0.0184	0.0183	−0.0251	0.0173	−0.0440	0.0330	−0.0027	0.0011
ν_4	−0.0095	0.0119	−0.0154	0.0115	−0.0176	0.0147	−0.0011	0.0007
$\kappa^{-\frac{1}{2}}$	1.5755	0.1255	1.5078	0.1190	2.5513	0.1907	2.4963	0.1405
Log-likelihood	−0.841361		−0.837540		−0.845520		−0.843532	
SIC	0.849680		0.846783		0.847164		0.845474	
$Q(30)$	124.8 (0.000)		57.6 (0.001)		187.8 (0.000)		43.5 (0.009)	
$Q(50)$	160.7 (0.000)		84.5 (0.001)		215.4 (0.000)		61.5 (0.042)	
Resid. mean	0.001		0.006		0.000		0.005	
Resid. variance	0.974		0.953		1.167		1.152	

*Dependent variable is the absolute value of the size of a non-zero price change, $S_i \mid S_i > 0$, $\tilde{n}$=1,809 (*JBX*), $\tilde{n}$=17,329 (*HAL*), *p*-values in parenthesis

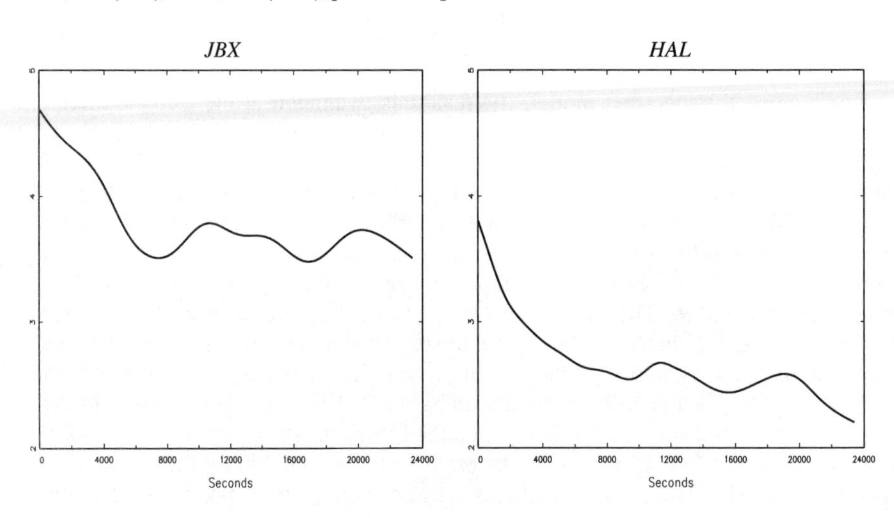

Fig. 7 Estimated diurnal seasonality function of the non-zero absolute price changes $S_i \mid S_i > 0$ for Jack in the Box Inc. (*JBX*) and Halliburton Company (*HAL*). The *x*-axes represents the 23,400 trading-seconds in the NYSE-trading period (from 9:30 to 16:00 EST)

for *HAL* a richer GLARMA(2,2) yields the best goodness of fit in terms of the Schwarz criterion.

For all estimates, the dispersion parameter $\kappa^{-1/2}$ is significantly different from zero so that we have to reject the null of a truncated-at-zero Poisson distribution in favor of a Negbin distribution. The parameters for the ARMA components are significant as well. For *JBX* the estimates for the roots of the AR components are 0.8753 (plain model) and 0.9275 (flexible model), so that stationarity is guaranteed. For *HAL* the implicit estimates for the two roots[11] of the AR components are 0.8452 and −0.4631 (plain model) and 0.9649 and 0.5409 (flexible model), so that stationarity is guaranteed as well. Since for both stocks roots are close to one, the GLARMA model predicts a strong persistence in the non-zero absolute price changes.

For both data sets the specification allowing for more flexible news impacts are superior to the ones with standard news impact curves. The coefficients on $|\varepsilon_{i-1}|$ are significantly positive, but also the fit of the observed dynamics improves over the specification ignoring the possibility of the additional news impact component. For *HAL* the effect is considerably lowered by a negative effect on $|\varepsilon_{i-2}|$.

The value of Ljung-Box Q-statistics for the standardized residuals e_i reported in Table 2 indicate that the GLARMA(1,1)-dynamics chosen for the *JBX* absolute price changes does not remove autocorrelation in the residuals completely. However, based on the $Q(50)$-statistic for the flexible GLARMA(2,2)-specification for *HAL* the null of no serial correlation in the residuals cannot be rejected at the 1% level.

3 Transaction price dynamics and market microstructure

One of the fundamental questions of the market microstructure theory of financial markets is concerned with the determinants of the price process and the specific role of the institutional set-up.[12] Generally, the goal is to figure out how new price relevant information affects the price process. Approaches based on the rational expectation hypothesis typically assume some kind of heterogeneity among the market participants with respect to their level of information. The corresponding transaction process generally leads to successive revelations of price information. This leads to empirically testable hypotheses about the joint process of transaction price changes and other marks of the trading process, such as transaction intensities and transaction volume.

Provided that short-selling is infeasible, Diamond and Verrecchia (1987) infer that longer times between transactions can be taken as a signal for the existence of bad news implying negative price reactions. The absence of a short selling mechanism prevents market participants from profiting by exploiting the negative information through corresponding transactions. Therefore, one can expect that low transaction rates (longer times between transactions) are associated with higher volatility in the transaction price process and vice versa.

---

[11] Computed as $z_{1,2} = \frac{\gamma_1}{2} \pm \frac{1}{2}\sqrt{\gamma_1^2 + 4_{\gamma 2}}$.

[12] See O'Hara (1995) for a comprehensive survey on the theoretical literature on market microstructure.

Easley and O'Hara (1992) provide an alternative explanation of the relationship between transaction intensities and transaction price changes. In their model, higher transaction rates occur when a larger share of informed traders is active, which is anticipated by less informed traders. Consequently, the price reacts more sensitively when the market is marked by high transaction intensities than at times when the transaction intensity is low. Hence, contrary to Diamond and Verrechia, Easley and O'Hara predict a negative relationship between transaction times and volatility.

Similar predictions about the link between price dynamics and transaction volume result from the model proposed by Easley and O'Hara (1987). In their model, informed market participants try to trade comparatively large volumes per transaction in order to profit from their current informational advantage. It is assumed that this advantage exists only temporarily. The occurrence of those large transactions are seen by uninformed traders as evidence for new information. Hence, one can expect that the price reacts to larger orders more sensitively than to smaller ones. In general, price volatility should be larger when larger trading volumes are observed.

A further theoretical explanation for the positive association between trading volume and volatility goes back to the mixture of distribution model of Clark (1973) and Tauchen and Pitts (1983). In a standard set-up of the model, the positive association results from a joint dependence on the news arrival rate.[13]

A suitable framework for quantifying the relationship between transaction price changes and other marks of the trading process, such as trading volumes and transaction rates, is their joint distribution. Let Z_i be the vector representing the marks of the trading process with the joint p.d.f. $\Pr\left[Y_i = y_i, Z_i | \mathcal{F}_{i-1}^{(y,z)}\right]$ for transaction price changes and the marks conditional on partial filtration on y and z.

Without any loss of generality, the joint p.d.f. can be decomposed into the p.d.f. of the price changes conditional on the marks and the marginal density of the marks $f\left(Z_i | \mathcal{F}_{i-1}^{(y,z)}\right)$:

$$\Pr\left[Y_i = y_i, Z_i | \mathcal{F}_{i-1}^{(y,z)}\right] = \Pr\left[Y_i = y_i | Z_i, \mathcal{F}_{i-1}^{(y,z)}\right] f\left(Z_i | \mathcal{F}_{i-1}^{(y,z)}\right), \tag{3.1}$$

where the p.d.f. of the ICH model can be used as the basis for specifying the conditional p.d.f. of the price changes. In the following we correspondingly extend the ICH model by introducing the transaction rate and trading volume as conditioning information.

Let T_i be the time between transaction $i-1$ and i (measured in seconds) and V_i the transaction volume (measured as the number of shares) we enrich the ACM model by introducing in the ARMA specification of the log–odds ratios (2.13):

$$Z_i^D = (\ln V_i, \ln T_i)', \tag{3.2}$$

without imposing any symmetry restriction on the coefficient matrix G_l. Imposing such a restriction would mean, for example, that an increase in the transaction intensity T_i has the same impact on the probability of a positive price response as on a negative one. This, however, contradicts the implications of the theoretical

[13] See Andersen (1996) and Liesenfeld (1998, 2001) for extensions of the mixture models.

hypothesis of Diamond and Verrecchia (1987) who predict a negative correlation between price changes and the time between transactions. Hence, one would expect asymmetric effects on the probabilities of a certain price reaction.[14]

In a similar way, the transaction rate and the transaction volume are introduced into the model as conditioning information for the size of the price changes (2.23):

$$\ln \omega_i = \beta_{d0} D_i + \beta_{d1} D_{i-1} + \beta_{v0} V_i + \beta_{v1} V_{i-1} + \beta_{t0} T_i + \beta_{t1} T_{i-1} + \lambda_i. \qquad (3.3)$$

Note that the conditional distribution of the price change $\Pr\left[Y_i = y_i | Z_i, \mathcal{F}_{i-1}^{(y,z)}\right]$ resulting from the specifications (3.2) and (3.3) does not explicitly rest on a structural theoretical model for the joint process of price changes, volume, and transaction rates, which would treat each of these variables as an endogenous quantity. Equations (3.2) and (3.3), rather, reflect ad-hoc assumptions with respect to the distribution of the price changes conditional on volume and transaction rate (as it could result from a joint distribution of these variables). Correspondingly, the estimated relations cannot be interpreted as structural economic relations. Nevertheless, the augmented ICH model can serve as an instrument for capturing and quantifying the relationship between important marks of the trading process. This allows us to shed light on the empirical relevance of the theoretical implications sketched above.

Tables 3 and 4 contain the estimation results for the augmented ICH model with the transaction rate and trading volume as additional covariates. For the two submodels, we have chosen the same order of the process that was found to be optimal for the pure time series specification. In the price direction model (Table 3) both log trading volume V_i and the log time between transactions T_i have a significantly positive impact on the log–odds ratios (i.e., the probability that the transaction price changes increases with the size of the transaction volume and the time between transactions). Since the probability of a nonzero price change can be interpreted as a specific measure of price volatility, it implies that low transaction rates go along with higher price volatility. This provides empirical support for the implications of the model proposed by Diamond and Verrecchia (1987), where no transactions indicate bad news, which contradicts the theoretical implications of Easley and O'Hara (1992) where no transactions indicate lack of news in the market. Our finding that high transaction volumes are positively correlated with volatility is consistent with the implication of the model proposed by Easley and O'Hara (1987), where large volumes correspond to the existence of additional news in the market. The effect of volume on the probability of a price change is partly compensated for by the subsequent transaction. The effect of the transaction time on the probability of a price change is asymmetric in the sense that the major reaction for negative price change occurs immediately, while parts of the reaction on log–odds ratio for a positive price change occurs also with the subsequent transaction. This interesting reaction pattern holds for *JBX* and *HAL*.

Finally, the inclusion of the microstructure variables greatly improves the value of the Schwarz criterion, but worsens the dynamic properties of the model as indicated by the Q-statistics. Our empirical results for the direction of the price changes are in accordance with those put forth by Rydberg and Shephard (2003) for

[14] The LR-test clearly rejects the null hypothesis of symmetric price reactions.

Table 3 ML estimates of the logistic ACM-ARMA model with microstructure variables and non-symmetric response coefficients*

Par.	*JBX*		*HAL*	
	Estimate	Std. dev.	Estimate	Std. dev.
μ_1	−0.5443	0.2146	−0.0684	0.0079
μ_2	−0.6294	0.2470	−0.0806	0.0087
$c_1^{(1)}$	0.8319	0.0628	1.1283	0.0277
$c_1^{(2)}$			−0.1738	0.0275
$a_{11}^{(1)}$	0.1494	0.0366	0.1157	0.0148
$a_{12}^{(1)}$	0.0561	0.0269	0.1648	0.0138
$a_{21}^{(1)}$	0.1798	0.0442	0.3795	0.0137
$a_{22}^{(1)}$	0.1075	0.0233	0.0431	0.0146
$a_{11}^{(2)}$			−0.0047	0.0153
$a_{12}^{(2)}$			−0.1053	0.0139
$a_{21}^{(2)}$			−0.2967	0.0139
$a_{22}^{(2)}$			0.0653	0.0150
$g_{v1}^{(0)}$	0.1969	0.0316	0.1073	0.0111
$g_{v2}^{(0)}$	0.2259	0.0312	0.1960	0.0112
$g_{t1}^{(0)}$	0.3499	0.0251	0.2599	0.0132
$g_{t2}^{(0)}$	0.3289	0.0253	0.1546	0.0133
$g_{v1}^{(1)}$	−0.0792	0.0327	−0.0761	0.0112
$g_{v2}^{(1)}$	−0.0262	0.0316	−0.0949	0.0114
$g_{t1}^{(1)}$	−0.0001	0.0238	−0.0106	0.0132
$g_{t2}^{(1)}$	0.0546	0.0234	0.0298	0.0132
Log-lik.	−0.885792		−0.990106	
SIC	0.897808		0.993094	
$Q(30)$	125.6 (0.106)		169.9 (0.000)	
$Q(50)$	211.1 (0.109)		269.3 (0.001)	
Res. mean	(−0.003, −0.001)		(0.001, −0.002)	
Res. var.	$\begin{pmatrix} 0.898 & 0.035 \\ 0.035 & 1.092 \end{pmatrix}$		$\begin{pmatrix} 0.886 & 0.044 \\ 0.044 & 1.141 \end{pmatrix}$	

*Dependent variable is the direction of the price changes, D_i, p-values in brackets

IBM transaction prices. In particular, they also find a positive impact of transaction volume and time between transactions on the activity of transaction prices.

Table 4 reports the estimation results of the augmented GLARMA model for the absolute (non-zero) price changes. We find no evidence for a strong leverage effect when accounting for lagged effects. For both shares the contemporaneous effect of D_i on the volatility measure S_i is negative, supporting the hypothesis of a leverage effect. But this effect is completely over-compensated for in *JBX* and nearly compensated for in *HAL* by the positive effect of D_{i-1}. This result stands in contrast to the findings by Rydberg and Shephard (2003) who find a leverage effect for transaction prices of the IBM share traded at the NYSE. Again, volume and transaction rate have a positive impact on the size of the price changes. Since the size of the price changes as well as the probability of a non-zero price change are volatility measures, our previous conclusions based upon the ACM component

Table 4 ML estimates of the GLARMA model with microstructure variables and leverage effect*

Par.	*JBX*		*HAL*	
	Estimate	Std. dev.	Estimate	Std. dev.
γ_0	−0.1565	0.0439	−0.0279	0.0069
γ_1	0.9321	0.0219	1.6822	0.0484
γ_2			−0.6891	0.0468
δ_1	0.1234	0.0183	0.1630	0.0112
δ_2			−0.1445	0.0102
ζ_1	0.1852	0.0282	0.1494	0.0158
ζ_2			−0.1304	0.0153
ν_0	−0.0491	0.0422	−0.0045	0.0016
ν_1	0.0077	0.0071	0.0002	0.0003
ν_2	0.0156	0.0074	0.0001	0.0003
ν_3	−0.0140	0.0151	−0.0007	0.0005
ν_4	−0.0095	0.0102	−0.0001	0.0003
$\kappa^{-\frac{1}{2}}$	1.2476	0.0808	1.3026	0.0349
β_{d0}	−0.0699	0.0288	−0.0960	0.0109
β_{d1}	0.0748	0.0390	0.0709	0.0157
β_{v0}	0.2227	0.0208	0.3212	0.0087
β_{v1}	0.0561	0.0228	0.0809	0.0097
β_{t0}	0.0427	0.0216	0.2723	0.0133
β_{t1}	0.0214	0.0178	0.0297	0.0109
Log-lik.	−0.824800		−0.811089	
SIC	0.839589		0.813928	
$Q(30)$	55.0 (0.000)		49.7 (0.000)	
$Q(50)$	76.6 (0.001)		76.2 (0.000)	
Res. mean	0.004		0.002	
Res. var.	0.981		1.028	

*Dependent variable is the absolute value of the size of a non-zero price change, $S_i|S_i>0$, p-values in brackets

regarding the empirical confirmation of various implications from market microstructure theory are confirmed.

4 Diagnostics based on the predicted price change distribution

So far we have analyzed the individual components of the ICH model – the ACM-ARMA part for the price direction and the GLARMA part for the price change size given the price direction – separately. However, the ICH model is a specification for the overall conditional distribution of the transaction price changes. Hence, in this section, we check to what extend the merged components of the ICH model are capable to capture the features of the observed price change distribution. In particular, based on the estimates for the model components with microstructure variables (see Tables 3 and 4), we analyze the goodness-of-fit of the ICH model.

For this purpose, we use appropriate residuals obtained for the complete ICH model.

First we consider the standardized residuals of the ICH model computed as

$$w_i = \frac{y_i - \widehat{E}[Y_i|\mathcal{F}_{i-1}, Z_i]}{\widehat{V}[Y_i|\mathcal{F}_{i-1}, Z_i]^{\frac{1}{2}}},$$

where the estimated conditional mean is

$$\widehat{E}[Y_i|\mathcal{F}_{i-1}, Z_i] = \sum_{j\in\mathbb{Z}} j \cdot \widehat{\Pr}[Y_i = j|F_{i-1}, Z_i].$$

$\widehat{\Pr}[\cdot]$ represents the estimated counterpart of the conditional probability given in Eq. (2.6) (augmented by the additional conditioning variable Z_i). This probability is calculated according to Eqs. (2.10), (2.13), (2.20), and (2.23). The estimated conditional variance $\widehat{V}[\cdot]$ is obtained analogously. If the ICH model is correctly specified, the standardized residuals evaluated at the true parameter values should be uncorrelated in the first two moments with mean zero and unit variance.

Figure 8 shows the autocorrelation function of the standardized residuals for both stocks and Fig. 9 depicts the corresponding autocorrelation functions of the squared standardized residuals. Comparing these autocorrelation functions with those of the raw price changes (see Figs. 2 and 3), we observe that most of the serial dependence in the first and second moments is captured by the ICH model. Moreover, a comparison of the Ljung-Box Q-statistic for the residuals and squared residuals with that for the price changes and squared price changes (see Table 5) confirms that a large part of the dynamics can be explained by the ICH model.

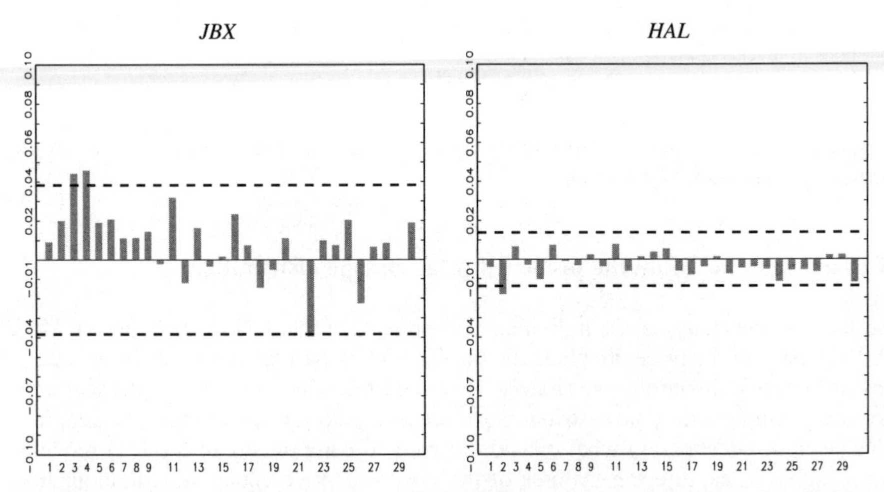

Fig. 8 Autocorrelation functions of the residuals of the entire ICH models for Jack in the Box Inc. (*JBX*) and Halliburton Company (*HAL*). The *dashed lines* mark off the approximate 99% confidence interval $\pm 2.58/\sqrt{n}$

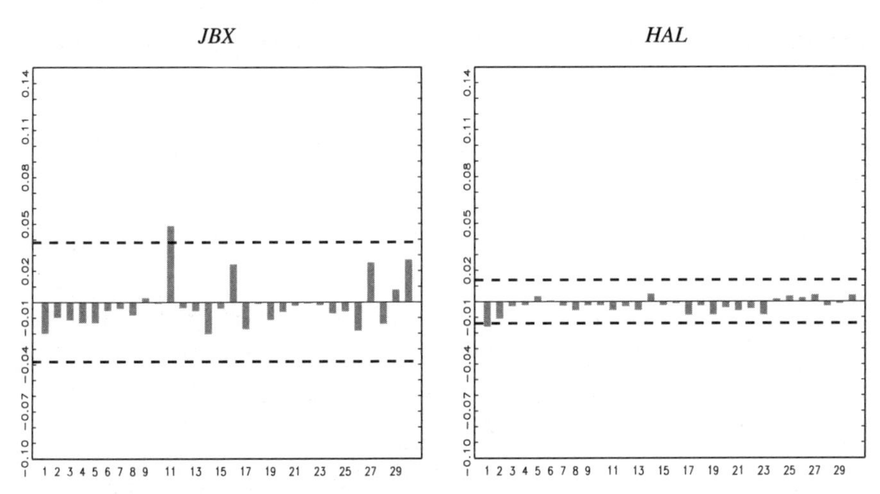

Fig. 9 Autocorrelation functions of the squared residuals of the entire ICH models for Jack in the Box Inc. (*JBX*) and Halliburton Company (*HAL*). The *dashed lines* mark off the approximate 99% confidence interval $\pm 2.58/\sqrt{n}$

However, the null of absence of serial correlation in w_i and w_i^2 has to be rejected at all standard significance levels.

To assess the ability of the ICH model to characterize the density of the price change process, we extend the density forecast test by Diebold et al. (1998) originally developed for continuous densities processes to the case of discrete ones. For discrete data generating processes the key assumption that the cumulative density function of the true data generating process has to be invertible is violated. Our modification of the density forecast test rests on the idea of a continuization by adding random noise to the discrete random variable under consideration such that

Table 5 Properties of the raw series and the ICH model residuals for Jack in the Box Inc. (*JBX*) and Halliburton Company (*HAL*)

	JBX	HAL	JBX	HAL
	Price changes y_i		Squared price changes y_i^2	
$Q(30)$ (p-value)	107.8 (0.000)	186.4 (0.000)	100.8 (0.000)	996.2 (0.000)
$Q(50)$ (p-value)	134.9 (0.000)	212.4 (0.000)	141.2 (0.000)	1105.2 (0.000)
Mean	0.009	−0.009		
Variance	11.34	7.526		
	Residuals w_i		Squared residuals w_i	
$Q(30)$ (p-value)	50.9 (0.000)	44.9 (n.a.*)	31.3 (0.000)	29.2 (n.a.*)
$Q(50)$ (p-value)	78.5 (0.000)	59.1 (0.000)	58.5 (0.000)	49.2 (0.000)
Mean	0.006	0.008		
Variance	1.042	1.019		

*Number of parameters are larger than the number of included lags

the invertibility condition holds. The proposed modified test procedure works as follows:

i) Construct u_i^u as the probability that Y_i is less than the actually observed price change y_i:

$$u_i^u \equiv \widehat{\Pr}[Y_i \le y_i | \mathcal{F}_{i-1}, Z_i] = \sum_{j=-\infty}^{y_i} \widehat{\Pr}[Y_i = j | \mathcal{F}_{i-1}, Z_i] \tag{4.1}$$

for each i: $1 \rightarrow n$, and construct u_i^l as the probability that Y_i is less than the actually observed price change minus one:

$$u_i^l \equiv \widehat{\Pr}[Y_i \le y_i - 1 | \mathcal{F}_{i-1}, Z_i] = \sum_{j=-\infty}^{y_i - 1} \widehat{\Pr}[Y_i = j | \mathcal{F}_{i-1}, Z_i]. \tag{4.2}$$

ii) Then use this sequence of probabilities to generate a sequence of artificial random numbers from the conditional uniform distributions on the intervals $[u_i^l, u_i^u]$, i.e.,

$$u_i \sim \mathcal{U}\left(u_i^l, u_i^u\right), \quad i : 1 \rightarrow n. \tag{4.3}$$

If the model is correctly specified, the u_i's drawn under the true parameter values (to calculate u_i^l and u_i^u) are *i.i.d.* following a uniform distribution on the interval [0,1] (see the Appendix). Moreover, one can map the u_i's into a standard normal distribution using $u_i^* \equiv \Phi^{-1}(u_i)$. Under the correct model, the normalized u_i^*'s are *i.i.d.* standard normal distributed.

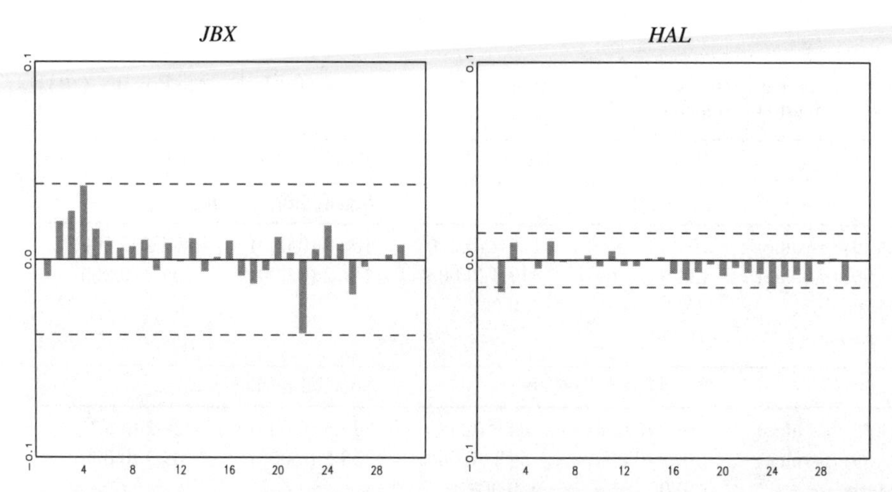

Fig. 10 Autocorrelation function of the density forecast variable u_i^* for Jack in the Box Inc. (*JBX*) and Halliburton Company (*HAL*). The *dashed lines* mark off the approximate 99% confidence interval $\pm 2.58/\sqrt{n}$

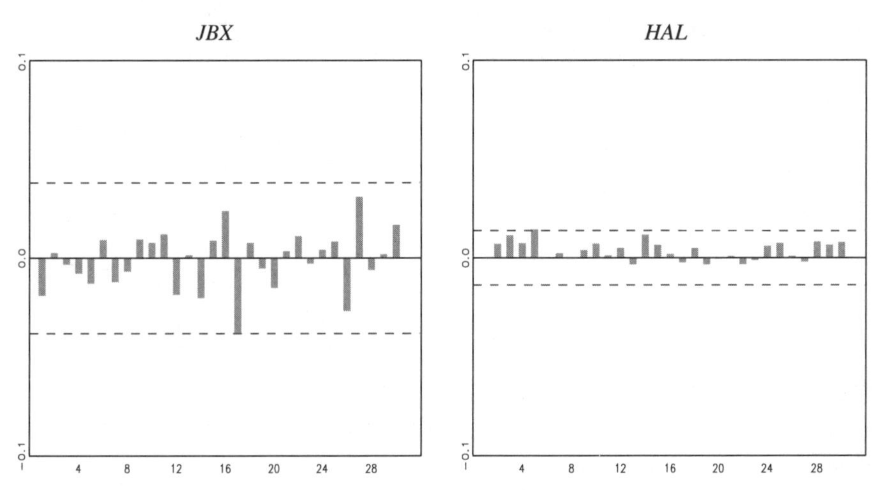

Fig. 11 Autocorrelation function of the squared density forecast variable u_i^* for Jack in the Box Inc. (*JBX*) and Halliburton Company (*HAL*). The *dashed lines* mark off the approximate 99% confidence interval $\pm 2.58/\sqrt{n}$

Figures 10 and 11 plot the autocorrelation function for a sequence of normalized residuals u_i^* and of squared normalized residuals $(u_i^*)^2$. The plots indicate that for both stocks there is nearly no significant autocorrelation left in the first and second moments. Table 6 represents summary statistics of the normalized residuals. They are computed as the corresponding sample means based on 1,000 repeated draws of the trajectory $\{u_i^*, i: 1 \rightarrow N\}$. For JBX the Jarque-Bera statistic indicates that we cannot reject the null of a normal distribution for u_i*, whereas for HAL we have to reject the null at 1% significance level.

The quantile–quantile (QQ) plot of a sequence of u_i^*'s against the standard normal distribution displayed in Fig. 12 reveals that the ICH model approximates the distributional properties of the transaction price changes for both stocks fairly well. However, the deviation from normality in the tails of the distribution of the u_i^*'s indicates slight difficulties characterizing extreme price changes appropriately.

Table 6 Properties of the density forecast variable u_i^* for Jack in the Box Inc. (*JBX*) and Halliburton Company (*HAL*)

	JBX	HAL
Mean	−0.003 [0.0000]	−0.001 [0.0000]
Variance	1.000 [0.0000]	1.002 [0.0000]
Skewness	0.027 [0.0000]	0.029 [0.0000]
Kurtosis	2.989 [0.0000]	3.048 [0.0000]
Jarque-Bera	0.793 [0.0006]	9.910 [0.0019]
p-value	(0.6727)	(0.007)

The values of the statistics are sample means based upon 1,000 repeated draws of $\{u_i^*, i: 1 \rightarrow N\}$. Values in brackets report the sampling standard error, due to repeated sampling of $\{u_i^*, i: 1 \rightarrow N\}$ in the density forecast procedure

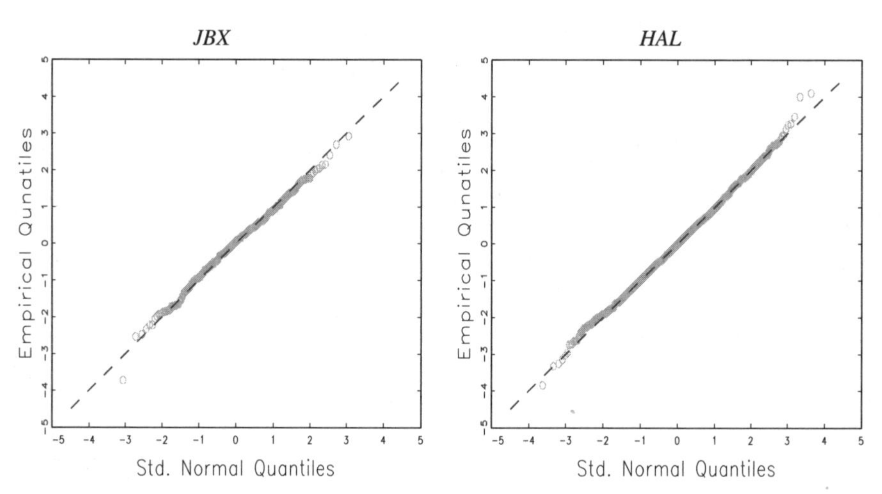

Fig. 12 QQ-Plot against std. normal quantiles of the density forecast variable u_i^* for Jack in the Box Inc. (*JBX*) and Halliburton Company (*HAL*)

5 Conclusions

In this paper, we introduce a new approach to analyze transaction price movements in financial markets. It relies on a hurdle count data approach that has been extended to include negative counts. The parsimonious form of our model consists of two processes: a process for the price direction, and one for the size of the price movement. Our approach is particularly suited for financial markets where the outcome space of the transaction returns is countable. Since the decimalization at many stock exchanges, the model is applicable for a wide range of stocks.

We show the approach at work by analyzing the transaction price dynamics of the frequently traded stocks of Jack in the Box Inc. and the stocks of the Halliburton Company with considerably higher market capitalization and trade intensity. The pure time series approach can easily be extended to test various implications of market microstructure theory. We show that our approach does fairly well in modelling the overall observed distribution of the transaction price changes. Using a density forecast test designed for discrete distributions, we show that for our two samples the vast majority of small and moderate price changes can be well-explained while there is evidence for some misspecification with respect to the tail behavior.

In order to assess the potential of our approach, the model has to be subjected to intensive checks of its forecasting properties. Comparative studies with respect to such properties of various approaches and applications to other financial assets and to exchanges with different trading platforms should provide more insights into the potential applicability of our approach. Alternatively, the quality of the model could be assessed by using it as the basis for a trading strategy. Finally, the ICH model should be embedded into a joint model of the transaction price movements and trading times.

Acknowledgements For helpful comments and suggestions we like to thank Bernd Fitzenberger, Nikolaus Hautsch, Neil Shephard, Gerd Ronning, Timo Teräsvirta and Pravin Trivedi. The work of the second and third co-author is supported by the Friedrich Thyssen Foundation and the European Community's Human Potential Programme under contract HPRN-CT-2002-00232, Microstructure of Financial Markets in Europe (MICFINMA), respectively.

Appendix

This appendix shows that under a correctly specified model for Y_i, the u_i's drawn from the uniform distributions (4.3) follow a uniform distribution on the interval [0, 1].[15] Consider a discrete random variable Y with support $\Delta \subseteq \mathbb{Z}$, and let u be a continuous random variable with the following conditional uniform distribution

$$u \sim \mathcal{U}\left(u_y^l, u_y^u\right), \tag{A.1}$$

where the boundaries are $u_y^l = \Pr(Y \le y-1)$, $u_y^u = \Pr(Y \le y)$ (for ease of notation we ignore the index i for the variables u and Y). Then, the c.d.f. of the unconditional distribution of u is

$$\Pr(u \le c) = \sum_{y \in \Delta} \Pr(u \le c | Y = y) \Pr(Y = y), \quad c \in [0, 1], \tag{A.2}$$

with

$$\Pr(u \le c | Y = y) = \frac{c - u_y^l}{u_y^u - u_y^l} \mathcal{I}_{[u_y^l, u_y^u)}(c) + \mathcal{I}_{[u_y^u, 1]}(c) \tag{A.3}$$

$$\Pr(Y = y) = u_y^u - u_y^l, \tag{A.4}$$

where $\mathcal{I}_A(z)$ is an indicator function which is 1 if $z \in A$ and zero for $z \notin A$. Inserting Eqs. (A.3) and (A.4) into Eq. (A.2), we obtain

$$\Pr(u \le c) = \sum_{y \in \Delta} \left\{ \left(c - u_y^l\right) \mathcal{I}_{[u_y^l, u_y^u)}(c) + \left(u_y^u - u_y^l\right) \mathcal{I}_{[u_y^u, 1]}(c) \right\}. \tag{A.5}$$

Assuming that $c \in [u_j^l, u_j^u]$, $j \in \Delta$, we find

$$\begin{aligned} \Pr(u \le c) &= c\mathcal{I}_{[0,1]}(c) - \Pr(Y \le j-1) + \ldots \\ &\quad + \Pr(Y \le j-3) - \Pr(Y \le j-4) \\ &\quad + \Pr(Y \le j-2) - \Pr(Y \le j-3) \\ &\quad + \Pr(Y \le j-1) - \Pr(Y \le j-2) \\ &= c\mathcal{I}_{[0,1]}(c), \end{aligned} \tag{A.6}$$

which represents the c.d.f. of a uniform distribution on the interval [0, 1].

---

[15] This technique of continuization is widely used to describe the properties of the p.d.f. of discrete random variables, see e.g. Stevens (1950) and Denuit and Lambert (2005).

References

Andersen TG (1996) Return volatility and trading volume: an information flow interpretation of stochastic volatility. J Financ 51:169–204, 169–231

Ball C (1988) Estimation bias induced by discrete security prices. J Financ 43:841–865

Bauwens L, Giot P (2003) Asymmetric ACD models: introducing price information in ACD models with a two-state transition model. Empir Econ 28:709–731

Berndt EK, Hall BH, Hall RE, Hausman JA (1974) Estimation and inference in nonlinear structural models. Ann Econ Soc Meas 3/4:653–665

Bollerslev T, Melvin M (1994) Bid-ask spreads and volatility in the foreign exchange market – an empirical analysis. J Int Econ 36:355–372

Cameron AC, Trivedi PK (1998) Regression analysis of count data. Cambridge University Press, Cambridge

Campbell JY, Lo AW, MacKinlay AC (1997) The econometrics of financial markets. Princeton University Press

Cho DC, Frees EW (1988) Estimating the volatility of discrete stock prices. J Financ 43:451–466

Clark PK (1973) A subordinated stochastic process model with finite variance for speculative prices. Econometrica 41:135–155

Cox D (1981) Statistical analysis of time series: some recent developments. Scand J Statist 8:93–115

Davis R, Dunsmuir W, Streett S (2003) Observation driven models for poisson counts. Biometrika 90:777–790

Denuit M, Lambert P (2005) Constraints on concordance measures in bivariate discrete data. J Multivar Anal 93:40–57

Diamond DW, Verrecchia RE (1987) Constraints on short-selling and asset price adjustment to private information. J Financ Econ 18:277–311

Diebold FX, Gunther TA, Tay AS (1998) Evaluating density forecasts, with applications to financial risk management. Int Econ Rev 39:863–883

Easley D, O'Hara M (1987) Price, trade size, and information in securities markets. J Financ Econ 19:69–90

Easley D, O'Hara M (1992) Time and the process of security price adjustment. J Financ 47:577–607

Engle R (2000) The econometrics of ultra-high-frequency data. Econometrica 68(1):1–22

Harris L (1990) Estimation of stock variances and serial covariances from discrete observations. J Financ Quant Anal 25:291–306

Hausman JA, Lo AW, MacKinlay AC (1992) An ordered probit analysis of transaction stock prices. J Financ Econ 31:319–379

Jung R, Liesenfeld R (2001) Estimating time series models for count data using efficient importance sampling. Allg Stat Arch 85:387–407

Liesenfeld R (1998) Dynamic bivariate mixture models: modelling the behavior of prices and trading volume. J Bus Econ Stat 16:101–109

Liesenfeld R (2001) A generalized bivariate mixture model for stock price volatility and trading volume. J Econ 104:141–178

Lütkepohl H (1993) Introduction to multiple time series analysis. Springer, Berlin Heidelberg New York

Mullahy J (1986) Specification and testing of some modified count data models. J Econ 33:341–365

Nelson D (1991) Conditional heteroskedasticity in asset returns: a new approach. J Econ 43:227–251

O'Hara M (1995) Market microstructure theory. Blackwell Publishers, Oxford

Pohlmeier W, Ulrich V (1995) An econometric model of the two-part decision process in the demand for health. J Hum Resour 30:339–361

Russel J, Engle R (2002) Econometric analysis of discrete-valued, irregularly-spaced financial transactions data using a new autoregressive conditional multinomial model. University of California, San Diego (revised version of Discussion Paper 98–10)

Russell JR, Engle RF (1998) Econometric analysis of discrete-valued, irregularly-spaced financial transactions data using a new autoregressive conditional multinomial model, presented at Second International Conference on High Frequency Data in Finance, Zurich, Switzerland

Rydberg T, Shephard N (2002) Dynamics of trade-by-trade price movements: decomposition and models. Discussion paper, Nuffiled College, Oxford University, will be published in Journal of Financial Econometrics

Rydberg T, Shephard N (2003) Dynamics of trade-by-trade price movements: decomposition and models. J Financ Econ 1:2–25

Stevens W (1950) Fiducial limits of the parameter of a discontinuous distribution. Biometrika 37:117–129

Tauchen GE, Pitts M (1983) The price variability-volume relationship on speculative markets. Econometrica 51:485–505

Zeger S (1988) A regression model for time series of counts. Biometrika 75:621–629

Walid Ben Omrane · Hervé Van Oppens

The performance analysis of chart patterns: Monte Carlo simulation and evidence from the euro/dollar foreign exchange market

Abstract We investigate the existence of chart patterns in the euro/dollar intra-daily foreign exchange market. We use two identification methods of the different chart patterns: one built on 5-min close prices only, and one based on both 5-min low and high prices. We look for twelve types of chart patterns and we study the detected patterns through two criteria: predictability and profitability. We run a Monte Carlo simulation to compute the statistical significance of the obtained results. We find an apparent existence of some chart patterns in the currency market. More than one half of detected charts present a significant predictability. Nevertheless, only two chart patterns imply a significant profitability which is however too small to cover the transaction costs. The second extrema detection method provides higher but riskier profits than the first one.

Keywords Foreign exchange market · Chart patterns · High frequency data · Technical analysis

JEL Classification C13 · C14 · F31

1 Introduction

Technical analysis is the oldest method for analyzing market behavior. It is defined by Murphy (1999) as the study of market action, primarily through the use of charts, for the purpose of forecasting future price trends. The term 'market action' includes three main sources of information available to the technician: price, volume and open interest. Béchu and Bertrand (1999) distinguish three categories of technical analysis. Traditional analysis is entirely based on the study of charts and the location of technical patterns like the Head and Shoulders pattern. Modern

---

W. B. Omrane (✉) · H. V. Oppens
IAG Business School, Université Catholique de Louvain, 1 place des Doyens,
1348 Louvain-la-Neuve, Belgium
E-mail: benomrane@fin.ucl.ac.be

analysis is composed of more quantitative methods like moving averages, oscillators, etc. The third category, qualified as philosophical, has the ambition to explain more than the overall market behavior. One of the most famous examples is the Elliot wave theory (for more details see Prost and Prechter (1985)) which assumes that every price movement can be decomposed into eight phases or waves: five impulse waves and three corrective ones.

In this paper, we focus on the traditional approach of technical analysis and particularly on chart patterns. These patterns have been studied, among others, by Levy (1971); Osler (1998); Dempster and Jones (1998a); Chang and Osler (1999), and Lo et al. (2000) who have mainly focused on the profitability of trading rules related to chart patterns and also on the informational content that could generate such patterns. All these investigations conclude to the lack of profitability of technical patterns. However, Lo et al. (2000) find that these patterns present an informational content that affect stock returns.

We investigate twelve chart patterns in the euro/dollar foreign exchange market. Currency markets seem especially appropriate for testing technical signals because of their very high liquidity, low bid-ask spread, and round-the-clock decentralized trading (Chang and Osler (1999)). As our empirical evidence is built upon high frequency data, we rather focus on the speed of convergence to market efficiency than on the hypothesis of market efficiency per se. As argued by Chordia et al. (2002), information takes a minimum of time to be incorporated into prices so that markets may be at the same time inefficient over a short-time (e.g. 5-min) interval and efficient over a longer (e.g. daily) interval.

To test the existence of twelve chart patterns in the euro/dollar foreign exchange market, we use two identification methods (M1, M2) for detecting local extrema. The first method (M1), also used in the literature, considers only prices at the end of each time interval (they are called close prices). The second method (M2), which is new compared to those used in the literature, takes into account both the highest and the lowest price in each interval of time corresponding to a detected pattern.

The detected extrema are analyzed through twelve recognition pattern algorithms, each of them corresponding to a defined chart pattern. Our purpose is to analyze the predictability and profitability of each type of chart pattern. In addition, we intend to test the usefulness of our contribution regarding the extrema detection method M2. Although Osler (1998) and Chang and Osler (1999) briefly mention these prices, most of the previous studies focused on chart patterns do not give much interest to high and low prices. This is in sharp contrast to the majority of practitioners (in particular dealers and traders) who use high and low prices in their technical strategies through bar charts and candlesticks. In addition, Fiess and MacDonald (2002) show that high, low and close prices carry useful information for forecasting the volatility as well as the level of future exchange rates. Consequently, in our framework, we investigate also the sensitivity of the chart patterns to the extrema detection methods M1 and M2. To evaluate the statistical significance of our results, we run a Monte Carlo simulation. We simulate a geometric Brownian motion to construct artificial series. Each of them has the same length, mean, variance and starting value as the original observations.

Our results show the apparent existence of some chart patterns in the euro/dollar intra-daily foreign exchange rate. More than one half of the detected patterns, according to M1 and M2, seem to have a significant predictive success. Nevertheless, only two patterns from our sample of twelve present a significant profitability which is however too small to cover the transaction costs. We show, moreover, that the extrema detection method M2 provides higher but riskier profits than those provided by M1. These findings are in accordance with those found by Levy (1971); Osler (1998); Dempster and Jones (1998b); Chang and Osler (1999).

The paper is organized as follows. In Section 2, we summarize the most recent empirical studies which have focused on technical analysis, particularly on chart patterns. Section 3 is dedicated to the methodology adopted for both the extrema detection methods M1 and M2, and to the pattern recognition algorithms. The section also includes details about the two criteria used for the analysis of the observed technical patterns: predictability and profitability. In Section 4, we analyze and describe the data. Empirical results are exposed in Section 5. We conclude in Section 6.

2 Technical analysis

Technical analysis is widely used in practice by several dealers, also called technical analysts or chartists. According to Cheung and Wong (1999), 25% to 30% of the foreign exchange dealers base most of their trade on technical trading signals. More broadly, Allen and Taylor (1992) show, through questionnaire evidence, that technical analysis is used either as the primary or the secondary information source by more than 90% of the foreign exchange dealers trading in London. Furthermore, 60% judge charts to be at least as important as fundamentals. Most of them consider also chartism and fundamental analysis to be largely complementary. Menkhoff (1998) shows in addition that more than half of foreign exchange market participants in Germany give more importance to the information coming from non-fundamental analysis, i.e. technical analysis and order flows. Moreover, Lui and Mole (1998) show that technical analysis is the most used method for short term horizon on the foreign exchange market in Hong Kong.

Despite its broad use by practitioners, academics have historically neglected technical analysis, mainly because it contrasts with the most fundamental hypothesis in finance, namely market efficiency. Indeed, the weak form of the market efficiency hypothesis implies that all information available in past prices must be reflected in the current price. Then, according to this hypothesis, technical analysis, which is entirely based on past prices (Murphy (1999)), cannot predict future price behavior.

Recently, several studies have focused on technical analysis. Brock et al. (1992) support the use of two of the simplest and most popular trading rules: moving average and trading range break (support and resistance levels). They show that these trading rules help to predict return variations in the Dow Jones index. These simple trading rules were studied, amongst others, by Dooley and Shafer (1984); Sweeney (1986); Levich and Thomas (1993); Neely (1997) and LeBaron (1999) in the context of the foreign exchange rate dynamics. Moreover, Andrada-Felix et al. (1995); Ready (1997) and Detry (2001) investigate the use of these rules in stock markets. Still with the moving average trading rules, Gençay and Stengos (1997); Gençay (1998) and Gençay (1999) examine the predictability of stock market and foreign exchange market returns by using past buy and sell signals, and they find an evidence of nonlinear predictability of such returns.

In addition to these simple trading rules, technical analysis abounds of methods in order to predict future price trends. These methods have also been considered in empirical research. Jensen (1970) tests empirically the 'relative strength' trading rule.[1] The estimated profit provided by this trading rule is not significantly bigger than the one obtained by the 'Buy and Hold' strategy.[2] Osler (2000) finds that the support and resistance technique provides a predictive success. Other studies make use of genetic programs to develop trading rules likely to realize significant profits (e.g., Neely et al. (1997); Dempster and Jones (1998a) and Neely and Weller (1999)). Furthermore, Blume et al. (1994) demonstrate that sequences of volume can be informative. This would explain the widespread use by practitioners of technical analysis based upon volumes.

The different studies mentioned above have mainly focused on linear price relations. However, other researchers have oriented their investigations to non-linear price relations. Technical patterns, also called chart patterns, are considered as non-linear patterns. Both Murphy (1999) and Béchu and Bertrand (1999), argue that these kinds of patterns present a predictive success which allows traders to acquire profit by developing specific trading rules. In most studies, technical patterns are analyzed through their profitability. Levy (1971) focuses on the predictive property of the patterns based on a sequence of five price extrema and conclude, after taking into account the transaction costs, to the unprofitability of such configurations. Osler (1998) analyzes the most famous chart pattern, the head and shoulders pattern.[3] She underlines that agents who adopt this kind of technical pattern in their strategy must be qualified as noise traders because they generate important order flow and their trading is unprofitable. Dempster and Jones (1998b) and Chang and Osler (1999) obtain the same conclusion regarding the non profitability of the trading rules related to chart patterns. In contrast, Lo et al. (2000) show that the informational content of chart patterns affects significantly future stock returns.

Some studies go beyond the scope of testing the performance of trading models. For example, Gençay et al. (2002, 2003) employ a widely used commercial real-time trading model as a diagnostic tool to evaluate the statistical properties of foreign exchange rates. They consider that the trading model on real data out-performs some sophisticated statistical models implying that these latter are not relevant for capturing the data generating process. They add that in financial markets, the data generating process is a complex network of layers where each layer corresponds to a particular frequency.

In our paper we choose to deal with high frequency data, believing that all our results are sensitive to the time scale. The results carried out from 1 h or 30 min time scale are certainly different from those triggered by 5 min frequency. However, the goal of our study is to analyze the performance of some chart pattern at a specific time scale without generalizing our results to other frequencies. Our

[1] Once computing the ratio $P_t/\bar{P}_t$ where $\bar{P}_t$ corresponds to the mean of prices preceding the moment t, the relative strength trading rule consists in buying the asset if the ratio is bigger than a particular value and selling it when the ratio reaches a specific threshold.

[2] This strategy consists in buying the asset at the beginning of a certain period and keeping it until the end.

[3] This chart pattern is defined in Section 3.2.

choice of high frequency data, as emphasized by Gençay et al. (2003), is motivated by two main reasons. First, any position recommended by our strategy (defined below in subsection 3.3.2) have to be closed quickly within a short period following the chart completion. The stop-loss objectives need to be satisfied and the high frequency data provides an appropriate platform for this requirement. Second, the trading positions and strategies, can only be replicated with a high statistical degree of accuracy by using high frequency data in a real time trading model.

However in practice, technical analysts often combine high and low time scales in order to monitor their positions in the short (5-min to 1 h) and long run (one day to one month).

3 Methodology

The methodology adopted in this paper consists in identifying regularities in the time series of currency prices by extracting nonlinear patterns from noisy data. We take into consideration significant price movements which contribute to the formation of a specific chart pattern and we ignore random fluctuations considered as noise. We do this by adopting a smoothing technique in order to average out the noise. The smoothing technique allows to identify significant price movements which are only characterized by sequences of extrema.

In the first subsection we present two methods used to identify local extrema. Then, we explain the pattern recognition algorithm which is based on the quantitative definition of chart patterns. In the third subsection, we present the two criteria chosen for the analysis of the detected charts: predictability and profitability. The last subsection is dedicated to the way we compute the statistical significance of our results. It is achieved by running a Monte Carlo simulation.

3.1 Identification of local extrema

Each chart pattern can be characterized by a sequence of local extrema, that is by a sequence of alternate maxima and minima. Two methods are used to detect local extrema. The first method, largely used in the literature, is based on close prices, i.e. prices which take place at the end of each time interval. The second method, which is one of the contribution of this paper to the literature, is built on the highest and the lowest prices in the same time intervals. We examine the usefulness of using high and low prices in the identification process of chart patterns. Taking these prices into account is more in line with practice as dealers use bars or candlestick charts to build their technical trading rules.[4] Moreover, Fiess and MacDonald (2002) show that high and low prices carry useful information about the level of future exchange rates.

---

[4] Béchu and Bertrand (1999) stipulate that line charts are imprecise because they do not display all the information available as they are only based upon close prices of each time interval. In contrast, bar charts and candlesticks involve the high, low, open and close prices of each time interval.

The extrema detection method based on closed prices (M1) works as follows. The first step consists in smoothing the price curve to eliminate the noise in prices and locate the different extrema on the smoothed curve. To smooth the estimated curve we use the Nadaraya–Watson kernel estimator.[5] We then determine different extrema by finding the moments at which the kernel first derivative changes its sign. We therefore guarantee the alternation between maxima and minima. This smoothing technique has been also used by Lo et al. (2000). Other methods to detect extrema have been adopted by Levy (1971); Osler (1998); Dempster and Jones (1998b) and Chang and Osler (1999). The second step involves orthogonal projections of the smoothed extrema on the original price curve. In other words, we deduce corresponding extrema on the original curve through orthogonal projection.

The second method (M2) is based on high and low prices. Local maxima must be determined on the high price curve and local minima on the low one. We smooth both curves and we select the corresponding extrema when there is a change of the sign for the kernel first derivative function. In such a case, alternation between extrema is not automatically obtained. Thus, we start by projecting the first extremum on the corresponding original price curve. If this extremum is a maximum (minimum), we project it into the high price curve (low price curve) and then we alternate between a projection of a minimum (maximum) on the low price curve (high price curve) and a projection of a maximum (minimum) on the high price curve (low price curve).

To detect local extrema we use a rolling window which goes through all the time periods with an increment of a single time interval. For each window, we apply both extrema detection methods and the pattern recognition algorithms in order to test if the detected sequence of extrema corresponds to one of our twelve chart pattern definitions (see the following section). The advantage of a rolling window is to concentrate on patterns that sequentially develop in the same window and therefore to cancel the risk of look-ahead bias. This implies that the future evolution of the price curve is not yet known at the time of detection of technical patterns. A technical pattern is thus recorded only if all extrema have been detected in windows of identical time duration. Furthermore, we add a filter rule to keep only one record of each detected chart pattern. We present in Appendix B a detailed description of the two extrema detection methods.

3.2 Chart patterns quantitative definitions

By looking at specialized books on technical analysis like Murphy (1999) and Béchu and Bertrand (1999), which provide graphical descriptions of technical patterns, we build twelve quantitative definitions corresponding to the most famous chart patterns. Only the Head and Shoulders definition is presented in this Section. This pattern (HS) is defined from a particular sequence of extrema detected by the method presented in Appendix B. The other pattern definitions are presented in Appendix C. The eleven remaining chart patterns are the following: Inverse Head and Shoulders (IHS), Double Top (DT), Double Bottom (DB), Triple Top (TT), Triple Bottom (TB), Rectangle Top (RT), Rectangle Bottom

[5] Details about this estimation are given in Appendix A.

(RB), Broadening Top (BT), Broadening Bottom (BB), Triangle Top (TRIT) and Triangle Bottom (TRIB).

From a series of price P_t, we denote by E_i $(i = 1, \ldots, I)$ the local extremum i from a sequence composed of I extrema and t_{E_i} the moment when it occurs. The slope, $p(E_i, E_j)$, of the line passing through E_i and E_j and the y-coordinate at t_k of a point of this line, $V_{t_k}(E_i, E_j)$, are defined as follows:

$$p(E_i, E_j) = \frac{E_j - E_i}{t_{E_j} - t_{E_i}} \tag{1}$$

$$V_{t_k}(E_i, E_j) = E_i + (t_k - t_{E_i}) \times p(E_i, E_j). \tag{2}$$

Figure 1 presents the theoretical Head and Shoulders chart pattern while Fig. 2 illustrates the observed pattern after implementing both extrema detection methods. The theoretical figure serves mainly to help in the comprehension of the following definition:

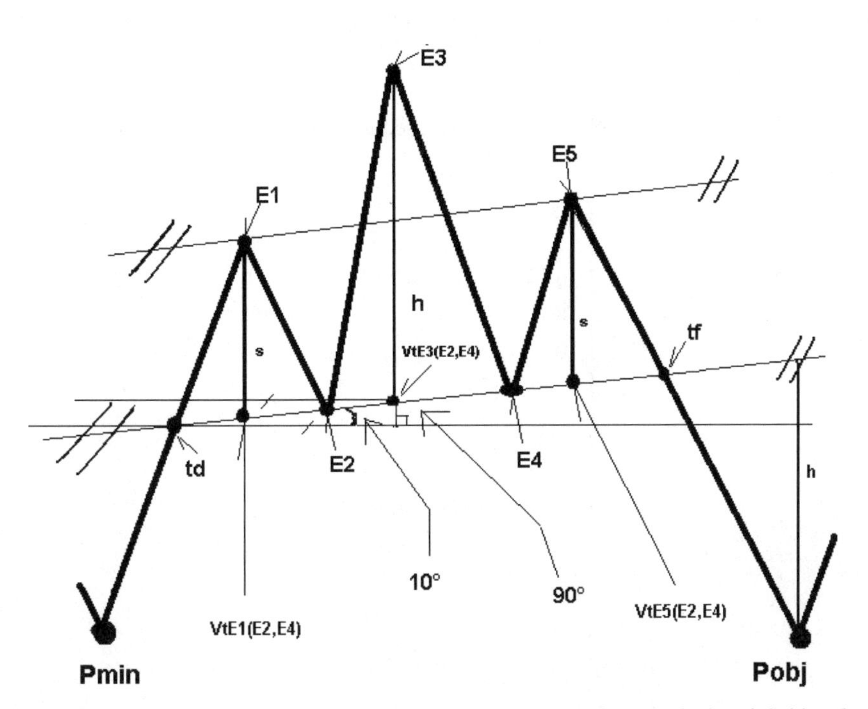

Fig. 1 The head and shoulders: theoretical chart pattern (*HS*). The quantitative definition for such chart is presented in Section 3.2

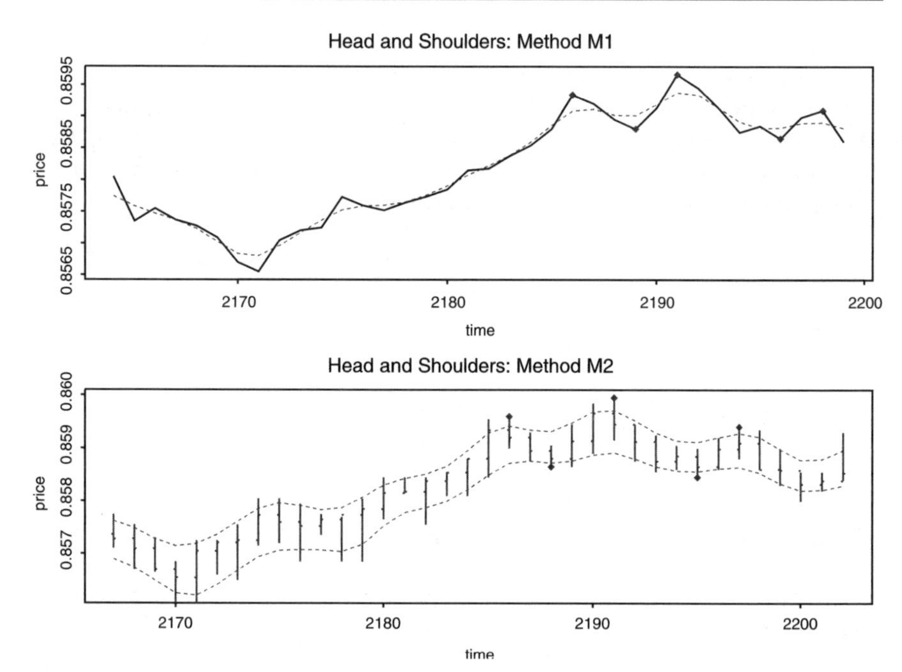

Fig. 2 The head and shoulders: observed chart pattern. This figure shows an observation window in which the Head and Shoulders chart pattern is detected through both M1 and M2 methods (detailed in Appendix B). The *dashed lines* in both graphs illustrates the smoothed price *curves* and the *solid line*, for the first graph, presents the original price *curve*. The second graph shows the original price series through *bar charts*. Each of them involves the maximum, the minimum, the open and the close price for each 5-min time interval

The *HS* chart pattern is characterized by a sequence of five extrema E_i $(i=1,..,5)$ such that:

$$hs \equiv \begin{cases} E_1 > E_2 & (a) \\ E_3 > E_1, E_3 > E_5 & (b) \\ |p(E_1,E_5)| \leq tg(10) & (c) \\ |p(E_2,E_4)| \leq tg(10) & (d) \\ 0.9 \leq \frac{E_1 - V_{t_{E_1}}(E_2,E4)}{E_5 - V_{t_{E_5}}(E_2,E4)} \leq 1.1 & (e) \\ 1.1 \leq \frac{h}{s} \leq 2.5 & (f) \\ \frac{1}{2} \leq \frac{t_{E_2} - t_d}{t_f - t_{E_4}} \leq 2 & (g) \\ \frac{1}{2} \leq \frac{t_{E_4} - t_{E_2}}{m} \leq 2 & (h) \\ (P_{t_d} - P_{t_{\min}}) \geq \frac{2}{3} \times h & (i) \end{cases}$$

where

- h is the height of the head: $h = E_3 - V_{t_{E_3}}(E_2, E_4)$
- s is the average height of the two shoulders:

$$s = \frac{\left(E_1 - V_{t_{E_1}}(E_2, E_4)\right) + \left(E_5 - V_{t_{E_5}}(E_2, E_4)\right)}{2}$$

- t_d is the starting time for the pattern: $t_d = \max_t (P_t \leq V_t(E_2, E_4), t < t_{E_1})$
- t_f is the ending time for the pattern: $t_f = \min_t (P_t \leq V_t(E_2, E_4), t > t_{E_5})$
- $t_{d-(f-d)} = t_d - (t_f - t_d)$
- $t_{f+(f-d)} = t_f + (t_f - t_d)$
- m is the average time that the shoulders take for their total completion: $m = \frac{(t_{E_2} - t_d) + (t_f - t_{E_4})}{2}$
- $P_{t_{\min}}$ is the smallest price observed in the time interval $[t_{d-(f-d)}, t_d]$: $P_{t_{\min}} = \min (P_t) \big| t_{d-(f-d)} \leq t \leq t_d$

If a sequence of five extrema satisfies the above conditions, they build up a Head and Shoulders chart pattern. Theoretically, at the completion of this chart pattern, the price must go down for at least the height of the head, h. Furthermore, the objective price detected by the chart pattern has to be reached within the time interval $[t_f, t_{f+(f-d)}]$. In other words, the price has to reach at least $P(obj)$ such that:

$$P(obj) = P_{t_f} - h. \tag{3}$$

3.3 The performance measures

Detected chart patterns are analyzed in terms of predictability and profitability. In other words, we study the capability of each chart pattern to predict the future price trend just after the chart completion and the profit that a dealer could realize when he applies a trading rule.

3.3.1 Predictability

Following its completion, the chart pattern can be used to forecast the future price trend. More precisely, it predicts the price objective which has to be reached. We denote by h the predicted price variation, and by t_f and t_d, respectively, the time at the end and at the beginning of the chart pattern. If the pattern predicts a downward trend, the price objective is given by Eq. (3). This price objective has to be reached within the time interval $[t_f, t_{f+(f-d)}]$. In such cases, we can measure the actual price reached in this time interval by computing P_a such that:

$$P_a = \min \left\{ P_t \middle| t_f \leq t \leq t_{f+(f-d)} \right\}. \tag{4}$$

The value of the observed trend is then:

$$trend = P_{t_f} - P_a. \tag{5}$$

The predictability criterion is defined as follows:

$$pred = \frac{trend}{h}. \tag{6}$$

We distinguish three possible cases:

- $0 \leq pred < 1$: the price does not reach its predicted objective. It goes in the predicted direction but only for *pred* of the forecasted objective.
- $pred = 1$: the price reaches exactly its objective.
- $pred > 1$: the price exceeds its objective by (*pred*-1).

Consequently if $pred \geq 1$, the chart pattern can be said to predict successfully the future price trend.

3.3.2 Profitability

If a chart pattern presents a predictive success, is it sufficient to get a profit? To answer this question, we investigate the profitability that technical patterns could imply. When the price evolves in the direction predicted by the chart pattern, a trader who takes a position at a precise time could realize a profit. Nevertheless, if the price evolves in the opposite direction, the position taken at the same time would involve a loss. A profit or a loss is the result of the implementation of a trading rule chosen by a chartist trader at a given time according to the completion of the chart pattern.

We propose the following strategy: the trader opens a position at the end of the pattern (at the moment of its completion) and closes it according to the future price direction. We distinguish two cases for the future trend:

- If the price evolves in the predicted direction, the trader closes his position when the price reaches 50% of the predicted price variation, h.
- If the price evolves in the opposite direction, the trader closes his position after a loss corresponding in absolute value to 20% of the forecasted price variation.

However, if at the end of the interval $[t_f, t_{f+(f-d)}]$, the trader position is not yet closed, this latter is automatically closed at $t_{f+(f-d)}$. In both cases, the trader can be considered as risk averse. Indeed, he limits his eventual profit and accept only small losses.

Once the predictability and the profitability criteria of each pattern are computed, we compare the results for the two extrema detection methods M1 and M2. We adopt a test of difference of means in order to infer the statistical significance of such comparisons. It consists in computing the statistic, t, as follows:

$$t = \frac{m_{M1} - m_{M2}}{\sqrt{\left(\frac{s_{M1}^2}{n_{M1}} + \frac{s_{M2}^2}{n_{M2}}\right)}}, \tag{7}$$

where m_{M_i} and $s_{M_i}^2$ are respectively the estimated mean and variance of the outputs (i.e. the number of detected charts, the predictability or the profitability criteria) obtained when method M_i (i=1,2) is adopted. The t-statistic follows a Student distribution with $n_{M1}+n_{M2}-1$ degrees of freedom, where n_{M1} and n_{M2} are respectively the number of observations resulting from the methods M1 and M2.

The last step for the profitability analysis consists in taking into consideration the risk incurred by the strategy. This latter is measured by the variability of the achieved profits. We compute three measures in order to gauge our strategy performance. We start by computing the ratio of mean profit to its standard deviation. However, according to Dacorogna et al. (2001a,b), this kind of Sharpe ratio is numerically unstable, exhibits a lot of deficiencies, and does not take into consideration dealers risk-aversion. For a robust performance evaluation we adopt two other performance measures, proposed by these authors, which are directly related to the utility of a strategy to a risk-averse dealer. Both of them are based on the maximization of the expected utility of a dealer. The first measure, called X_{eff}, considers a constant risk aversion, while the second one, named R_{eff}, supposes an asymmetric risk aversion (a higher risk-aversion when there is a loss). These performance measures adjust the mean profit from a kind of risk premium.[6] We have adapted these two performance measures to our context: for instance, we have computed X_{eff} and R_{eff} using the profit levels instead of returns realized after the completion of the chart pattern.

3.4 Monte Carlo simulation

In order to assess the statistical significance of the obtained results, we run a Monte Carlo simulation. We create 200 artificial exchange rate series[7] and we implement both extrema detection methods and the pattern recognition algorithms. These series follow a geometric Brownian motion process and are characterized by the same length, mean, variance and starting value as the original observations.[8]

Nevertheless, there is an important difference between the artificial series and the original one: the simulated series are built in such a way that any detected pattern is meaningless, whereas in the original exchange rate series, this may or may not be true. The existence of technical patterns in the original series could be generated by trader behaviors which induce a particular pattern in the prices. We test the null hypothesis of the absence of chart patterns in the observed series. This hypothesis involves also the absence of both predictability and profitability. On the other hand, if a chart pattern really exists in the observed series, then the number of chart detections has to be significantly larger than those obtained when we deal with artificial observations. Consequently, the probability of accepting the null hypothesis is computed by the percentage of simulated series for which the results obtained on the simulated series are greater than those obtained on the observed one.

4 Data description

The euro/dollar FOREX market is a market maker based trading system, where three types of market participants interact around the clock (i.e. in successive time zones): dealers, brokers and customers from which the primary order flow

[6] The details about the computation of X_{eff} and R_{eff} are given in Dacorogna et al. (2001a,b).
[7] We limit our simulation to 200 series because the recognition pattern algorithm needs a lot of computer time.
[8] The same methodology was adopted by Chang and Osler (1999); Osler (1998); Gençay (1998); Gençay et al. (2002), and Gençay et al. (2003).

originates. The most active trading centers are New York, London, Frankfurt, Sydney, Tokyo and Hong Kong. A complete description of the FOREX market is given by Lyons (2001).

To compute the mid prices used for the estimation of the models reported in Appendix A, we bought from Olsen and Associates a database made up of 'tick-by-tick' euro/dollar quotes for the period ranging from May 15 to November 14, 2001 (i.e. 26 weeks and three days). This database includes 3,420,315 observations. As in most empirical studies on FOREX data, these euro/dollar quotes are market makers' quotes and not transaction quotes (which are not widely available).[9] More specifically, the database contains the date, the time-of-day time stamped to the second in Greenwich mean time (GMT), the dealer bid and ask quotes, the identification codes for the country, city and market maker bank, and a return code indicating the filter status. According to Dacorogna et al. (1993), when trading activity is intense, some quotes are not entered into the electronic system. If traders are too busy or the system is running at full capacity, quotations displayed in the electronic system may lag prices by a few seconds to one or more minutes. We retained only the quotes that have a filter code value greater than 0.85.[10]

From the tick data, we computed mid quote prices, where the mid quote is the average of the bid and ask prices. As we use 5-min time intervals, we have a daily grid of 288 points. Because of scarce trading activity during the week-end, we exclude all mid prices computed between Friday 21 h 05 min and Sunday 24 h. The mean of the mid-quotes is equal to 0.8853, the minimum and maximum are 0.8349 and 0.9329.

5 Empirical results

Table 1 presents the number of detected chart patterns for the extrema identification methods (M1 and M2). The results show the apparent existence of some chart patterns in the euro/dollar foreign exchange series. Using the first detection method (M1), we have more detected charts in the original price series than in the simulated one for six chart patterns (out of twelve), at the 5% significance level. When we implement the M2 method, we detect significantly only four chart patterns, which are also significantly detected by the method M1: DT, DB, RT and RB. By looking at the last column which represents the total number of detections, we can see that we have more detected chart patterns when only close prices are used (M1). These results confirm the idea that the presence of such chart patterns does not occur by chance, at least for some chart patterns, but it is due, amongst others, to a determined behavior of the chartist dealers.

[9] Danielsson and Payne (2002) show that the statistical properties of 5-min dollar/DM quotes are similar to those of transaction quotes.

[10] Olsen and Associates recently changed the structure of their HF database. While they provided a 0/1 filter indicator some time ago (for example in the 1993 database), they now provide a continuous indicator that lies between 0 (worst quote quality) and 1 (best quote quality). While a value larger than 0.5 is already deemed acceptable by Olsen and Associates, we choose a 0.85 threshold to have high quality data. We remove however almost no data records (Olsen and Associates already supplied us with data which features a filter value larger than 0.5), as most filter values are very close to 1.

Table 1 Detected chart patterns

Meth	HS	IHS	DT	DB	TT	TB	RT	RB	BT	BB	TRIT	TRIB	∑
M1	78	4	7**	12**	5	12	107**	89**	57*	135	38	73*	617
	(0.42)	(0.86)	(0.00)	(0.00)	(0.96)	(0.91)	(0.00)	(0.00)	(0.02)	(0.09)	(0.18)	(0.02)	(0.12)
	40%	25%	57%**	33%**	60%	42%	58%	73%	72%	67%	76%	74%	63%
	(0.92)	(0.84)	(0.00)	(0.00)	(0.51)	(0.90)	(0.79)	(0.17)	(0.39)	(1.00)	(0.38)	(0.72)	(0.66)
M2	28	14	35**	44**	16	20	24**	33**	26	57	15	23	335
	(1.00)	(0.45)	(0.00)	(0.00)	(0.36)	(0.50)	(0.00)	(0.00)	(0.92)	(1.00)	(1.00)	(1.00)	(0.50)
	21%	21%	29%**	43%**	19%	35%	46%	45%	69%**	49%	53%	61%	42%
	(0.44)	(0.64)	(0.00)	(0.00)	(0.69)	(0.34)	(0.19)	(0.24)	(0.00)	(0.10)	(0.35)	(0.60)	(0.23)

Entries are the number of detected chart patterns and the percentage of chart patterns that reached their price objective, according to the extrema detection methods M1 and M2 (described in Appendix B). The *p*-values, computed through a Monte-Carlo simulation, and given in parenthesis represent the percentage of times the results on the simulated series are greater than the one of the original price series. The last column presents results for the whole sample, whatever is the chart pattern

** and * indicate, respectively, significance at 1% and 5%

The percentage of successful chart patterns (i.e. charts for which the price objective has been met) is given by the third and the seventh rows in Table 1. For example, 40% of Head and Shoulders (HS) detected by M1 succeed to meet their objective, but this result is not significant since for 92% of the simulated series we obtain more successful HS. For M1, only two charts, DT and DB present a significant successful percentage.[11] For M2, in addition to DT and DB, the chart pattern BT presents a significant percentage of success.

Nevertheless, this measure of predictive power, i.e. the percentage of charts that succeed to meet their price objective, is too drastic. It does not allow to capture to what extent the price objective is not met or to what extent the price objective is outclassed. That is why we quantified the predictability through the ratio *pred*.

Table 2 presents the average predictability *pred* for all detected chart patterns which succeed or fail to meet their objectives. For example, in the case of M1, HS has an average predictive power of 1,12. This average ratio is not significant at 5% since for 79% of the artificial series, we obtain a higher average ratio. However, the table shows that whatever the extrema detection methods implemented, more than one half of the whole chart patterns sample presents a predictability success statistically significant. At the 5% significance level, predictability varies from 0.86 to 9.45. The triangle chart patterns (TRIT and TRIB) offer the best predictability.

These results are consistent with those obtained in Table 1 in which M1 exhibits more predictability. This observation is even more striking in Table 2. The last column shows that M1 provide on average, a predicted value more than twice larger than M2. This is confirmed by positive significant signs for the difference of means test presented in the last line of the Table 2. Comparatively, Table 1 shows a percentage of 63% of successful chart patterns using M1 against 42% provided by M2.

Table 3 gives the maximum profitability that can be achieved by the use of chart patterns. It is computed in basis points (i.e.: 1/10,000) and provided for each of the twelve chart patterns. It corresponds to the implementation of the trading rule related to each chart pattern whatever its success level. The maximum profit is

[11] Both chart patterns DT and DB have not been detected in any artificial series, whatever the extrema detection method implemented.

Table 2 Predictability of the chart patterns

Meth	HS	IHS	DT	DB	TT	TB	RT	RB	BT	BB	TRIT	TRIB	μ
M1	1.12	0.88	1.93**	0.86**	1.72	1.33	2.56	3.52**	4.38**	4.00	9.35*	9.45**	4.15
	(0.79)	(0.70)	(0.00)	(0.00)	(0.41)	(0.80)	(0.10)	(0.00)	(0.00)	(0.06)	(0.03)	(0.01)	(0.15)
M2	0.70	0.87	0.88**	1.19**	0.74	1.16	1.05	1.46*	2.52**	1.68**	2.58	3.42	1.50
	(0.14)	(0.14)	(0.00)	(0.00)	(0.42)	(0.08)	(0.17)	(0.02)	(0.00)	(0.00)	(0.43)	(0.31)	(0.10)
M1–M2	+**	+	+	–	+	+	+**	+**	+**	+**	+**	+**	+**

This table shows the predictability of different chart patterns according to the extrema detection methods M1 and M2 (described in Appendix B). The predictability criterion is detailed in Section 3.3.1. The last column shows the weighted average predictability for the whole sample of charts. The *p*-values, computed through a Monte-Carlo simulation, are given in parenthesis. The last line of the table reports the sign of the difference between both method's outputs and its statistical significance according to the difference of means test
** and * indicate respectively significance at 1% and 5%

equal to the difference, in absolute value, between the price at the end of the chart and the minimum/maximum[12] of the prices occurring after the chart pattern ($|P_{t_f} - P_a|$). To compute these profits, we suppose that dealers are able to buy or to sell the currency at the mid price. The computed profits vary between three and 52 basis points, but are significant for only three chart patterns: DT, DB and BT.

However, this profit cannot be realized surely by the chartists because they cannot precisely guess if the price is at the end of its right trend or not. That is why they adopt a strategy for their intervention according to their risk aversion. Table 4 presents the results for the strategy described in Section 3.3.2. Profits are computed through the average of the whole detected chart patterns which succeed or fail to meet their objectives. This profit is statistically significant for only two charts, DT and DB whatever the detection method implemented. However, this profit more or less equal to one basis point for three cases out of four, seems too small to cover the transaction costs. Indeed, the transaction costs are often estimated as the observed bid-ask spread which varies on average, in the euro/dollar currency market, between three to five basis points (Chang and Osler (1999)). Consequently, even by choosing a particular risk averse trading rule, strategies using chart patterns seem unprofitable.

Furthermore, the difference of means test shows that M2 is more profitable than M1. For the majority of charts, profitability computed by adopting M2 is sig-

Table 3 Maximum profitability of the chart patterns

Meth	HS	IHS	DT	DB	TT	TB	RT	RB	BT	BB	TRIT	TRIB	μ
M1	8	14	9**	3**	6	9	11	13	16	14	52	37	17
	(1.00)	(0.71)	(0.00)	(0.00)	(0.98)	(0.92)	(0.99)	(0.76)	(0.42)	(0.99)	(0.29)	(0.84)	(0.81)
M2	10	16	10**	12**	8	15	7	12	22*	16	28	51	16
	(0.99)	(0.51)	(0.00)	(0.00)	(0.90)	(0.49)	(0.98)	(0.84)	(0.02)	(0.72)	(0.89)	(0.64)	(0.53)

This table shows the maximum computed profit, according to the extrema detection methods M1 and M2 (described in Appendix B), expressed in basis points. The *p*-values, computed through a Monte-Carlo simulation, are given in parenthesis. The last column shows the weighted average maximum profitability for the whole charts
** and * indicate respectively significance at 1% and 5%

[12] We adopt the minimum if the price evolves, after the completion of the chart, into downward trend and we adopt the maximum when there is an upward trend.

Table 4 Profitability of the trading strategy

Meth	HS	IHS	DT	DB	TT	TB	RT	RB	BT	BB	TRIT	TRIB	μ
Profit													
M1	1.42	2.80	1.10**	0.60**	−0.13	1.54	1.10	1.52	1.25	1.39	2.77	2.13	1.50
	(1.00)	(0.70)	(0.00)	(0.00)	(0.99)	(0.92)	(1.00)	(0.98)	(1.00)	(1.00)	(0.82)	(1.00)	(0.95)
M2	0.07	0.05	1.10**	3.10**	0.99	3.23	0.70	1.87	4.15	2.20	2.44	5.18	2.16
	(0.99)	(0.94)	(0.00)	(0.00)	(0.85)	(0.63)	(0.95)	(0.84)	(0.27)	(0.99)	(0.96)	(0.85)	(0.64)
M1–M2	+*	+*	−	−**	−	−*	+	−	−**	−*	+	−**	−**
Sharpe													
M1	0.44	0.48	0.54**	0.37**	−0.05	0.50	0.61	0.98	0.78	0.79	0.74	0.76	0.71
	(0.99)	(0.65)	(0.00)	(0.00)	(1.00)	(0.91)	(0.98)	(0.57)	(0.92)	(1.00)	(0.94)	(0.99)	(0.88)
M2	0.01	0.01	0.26**	0.73**	0.22	0.64	0.19	0.51	1.30**	0.53	0.42	0.78	0.50
	(1.00)	(0.95)	(0.00)	(0.00)	(0.78)	(0.33)	(0.91)	(0.57)	(0.01)	(0.92)	(0.89)	(0.83)	(0.55)
X_{eff}													
M1	0.92	1.56	0.89**	0.48**	−0.43	1.13	0.93	1.40	1.12	1.24	2.26	1.78	1.26
	(0.99)	(0.64)	(0.00)	(0.00)	(0.99)	(0.91)	(0.99)	(0.96)	(1.00)	(1.00)	(0.84)	(0.99)	(0.95)
M2	−1.06	−1.93	0.29**	2.20**	0.13	2.01	0.10	1.22	3.64*	1.37	0.88	3.19	1.18
	(0.94)	(0.84)	(0.00)	(0.00)	(0.70)	(0.35)	(0.87)	(0.58)	(0.04)	(0.93)	(0.81)	(0.75)	(0.54)
R_{eff}													
M1	0.99	1.90	0.94**	0.50**	−0.65	1.23	0.96	1.45	1.16	1.29	2.44	1.89	1.31
	(1.00)	(0.46)	(0.00)	(0.00)	(0.99)	(0.91)	(0.99)	(0.97)	(1.00)	(1.00)	(0.82)	(0.99)	(0.95)
M2	−1.32	−2.59	0.32**	2.40**	0.13	2.24	0.03	1.33	3.83	1.49	0.80	3.74	1.25
	(0.94)	(0.81)	(0.00)	(0.00)	(0.55)	(0.29)	(0.80)	(0.54)	(0.06)	(0.95)	(0.80)	(0.59)	(0.51)

This table includes the average profits, expressed in basis points, realized after adopting the strategy detailed in Section 3.3.2, according to the extrema detection methods M1 and M2 (described in Appendix B). M1–M2 indicates the computed difference results between the two methods. It shows the sign of this difference and its statistical significance through the difference of means test. The Sharpe ratio measure the profit adjusted for risk. However, X_{eff} and R_{eff} are also a measure of profit adjusted for risk but they take into account respectively symmetric and asymmetric dealers risk aversion (more details for the computation of these two measures are provided in Dacorogna et al. 2001a). The p-values, computed through a Monte-Carlo simulation, are given in parenthesis. The last column shows the weighted average profitability for the whole charts

** and * indicate respectively significance at 1% and 5%

nificantly larger than the one provided by M1. We observe in Table 4 five significant negative signs versus two positive. This observation is confirmed by the significant negative sign for the weighted average profitability for all chart sample, presented in the last column.

This finding is quite important since at the light of the predictability results, we might conclude that only close prices matter. However, when the profitability is taken into consideration, the use of high and low prices seems to have an importance which is more in accordance with what is observed in practice (dealers use bar charts and only profit matters).

Nevertheless, if we consider the profit adjusted for the inherent risk, the same two mean profits of one basis point obtained for DT have different risk levels. Taking into account the risk level by computing the three different performance measures; Sharpe ratio, X_{eff} , and R_{eff} , we obtain a smaller value for M2. This means that the second method M2 generates riskier profits than M1. Moreover, X_{eff} and R_{eff} performance measures carry out the same outputs as the Sharpe risk-adjusted profits which implies the robustness of our results in terms of performance evaluation.

6 Conclusion

Using 5-min euro/dollar mid-quotes for the May 15 through November 14, 2001 time period, we shed light on the predictability and profitability of some chart patterns. We compare results according to two extrema detection methods. The first method (M1), traditionally used in the literature, considers only prices which occur at the end of each time interval (they are called close prices). The second method (M2) takes into account both the highest and the lowest price of each interval of time. To evaluate the statistical significance of the results, we run a Monte Carlo simulation.

We conclude on the apparent existence of some technical patterns in the euro/dollar intra-daily foreign exchange rate. More than one half of the detected patterns, according to M1 and M2, seem to have some significant predictive success. Nevertheless, only two out of twelve patterns present significant profitability, which is however too small to cover the transaction costs. We also show that the extrema detection method using high and low prices provides higher but riskier profits than those provided by the M1 method.

To summarize, chart patterns seems to really exist in the euro/dollar foreign exchange market at the 5 min level. They also show some power to predict future price trends. However, trading rules based upon them seem unprofitable.

Acknowledgements While remaining responsible for any error in this paper, we thank Luc Bauwens, Winfried Pohlmeier and David Veredas (the editors of the special issue) and two anonymous referees for helpful comments which improved the results in the paper. We thank also Eric Debodt, Nihat Aktas, Hervé Alexandre, Patrick Roger and participants at the 20th AFFI International Conference (France) and SIFF 2003 seminar (France). This paper presents the results of research supported in part by the European Community Human Potential Programme under contract HPRN-CT-2002-00232, Microstructure of Financial Markets in Europe.

Appendices

A. Price curve estimation

Before adopting the Nadaraya–Watson kernel estimator, we tested the cubic splines and polynomial approximations but we conclude empirically that the appropriate smoothing method is the kernel. Because the two first methods carry out too smoothed results and they are not flexible as the kernel method.

From the complete series of the price, P_t ($t=1,\ldots,T$), we take a window k of l regularly spaced time intervals,[13] such that:

$$P_{j,k} \subset \{P_t | \, k \leq t \leq k+l-1\}, \tag{8}$$

$j=1,\ldots,l$ and $k=1,\ldots,T-l+1$. For each window k, we consider the following relation:

$$P_{j,k} = m\left(X_{P_{j,k}}\right) + \epsilon_{P_{j,k}}, \tag{9}$$

where $\epsilon_{P_{j,k}}$ is a white noise and $m\left(X_{P_{j,k}}\right)$ is an arbitrarily fixed but unknown non linear function of a state variable $X_{P_{j,k}}$. Like Lo et al. (2000) to construct a smooth

[13] We fix l at 36 observations.

function in order to approximate the time series of prices $P_{j,k}$, we set the state variable equal to time, $X_{P_{j,k}} = t$. For any arbitrary x, a smoothing estimator of $m(x)$ may be expressed as:

$$\hat{m}(x) = \frac{1}{l}\sum_{j=1}^{l} \omega_j(x)P_{j,k}, \tag{10}$$

where the weight $\omega_j(x)$ is large for the prices $P_{j,k}$ with $X_{P_{j,k}}$ near x and small for those with $X_{P_{j,k}}$ far from x. For the kernel regression estimator, the weight function $\omega_j(x)$ is built from a probability density function $K(x)$, also called a kernel:

$$K(x) \geq 0, \int_{-\infty}^{+\infty} K(u)du = 1. \tag{11}$$

By rescaling the kernel with respect to a parameter $h>0$, we can change its spread:

$$K_h(u) \equiv \frac{1}{h}K(u/h), \int_{-\infty}^{+\infty} K_h(u)du = 1 \tag{12}$$

and define the weight function to be used in the weighted average (10) as:

$$\omega_{j,h} \equiv K_h\left(x - X_{P_{j,k}}\right)/g_h(x) \tag{13}$$

$$g_h(x) \equiv \frac{1}{l}\sum_{j=l}^{l} K_h\left(x - X_{P_{j,k}}\right). \tag{14}$$

Substituting (14) into (10) yields the Nadaraya–Watson kernel estimator $\hat{m}_h(x)$ of $m(x)$:

$$\hat{m}_h(x) = \frac{1}{l}\sum_{j=l}^{l} \omega_{j,h}(x)P_{j,k} = \frac{\sum_{j=1}^{l} K_h\left(x - X_{P_{j,k}}\right)P_{j,k}}{\sum_{j=1}^{l} K_h\left(x - X_{P_{j,k}}\right)}. \tag{15}$$

If h is very small, the averaging will be done with respect to a rather small neighborhood around each of the $X_{P_{j,k}}$'s. If h is very large, the averaging will be over larger neighborhoods of the $X_{P_{j,k}}$'s. Therefore, controlling the degree of averaging amounts to adjusting the smoothing parameter h, also known as the bandwidth. Choosing the appropriate bandwidth is an important aspect of any local-averaging technique. In our case we select a Gaussian kernel with a bandwidth, $h_{opt,j}$, computed by Silverman (1986):

$$K_h(x) = \frac{1}{h\sqrt{2\pi}}e^{-\frac{x^2}{2h^2}} \tag{16}$$

$$h_{opt,k} = \left(\frac{4}{3}\right)^{1/5} \sigma_k l^{-1/5}, \tag{17}$$

where σ_k is the standard deviations for the observations that occur within the window *k*. However, the optimal bandwidth for Silverman (1986) involves a fitted function which is too smooth. In other words this optimal bandwidth places too much weight on prices far away from any given time *t*, inducing too much averaging and discarding valuable information in local price movements. Like Lo et al. (2000), through trial and error, we found that an acceptable solution to this problem is to use a bandwidth equal to 20% of $h_{opt,k}$:

$$h^* = 0.2 \times h_{opt,k}. \tag{18}$$

B. Extrema detection methods

Technical details for both extrema detection methods and projection procedure are presented below:

B.1. M1

M1 is the extrema detection method using the close prices. After smoothing the data by estimating the Nadaraya–Watson kernel function, $\hat{m}_h(X_{P_{j,k}})$, we compute maxima and minima respectively noted by $max_{\hat{m}_h(X_{P_{j,k}})}$ and $min_{\hat{m}_h(X_{P_{j,k}})}$:

$$max_{\hat{m}_h}(X_{P_{j,k}}) = \left\{\hat{m}_h(X_{P_{j,k}}) \middle| S(\hat{m}'_h(X_{P_{j,k}})) = +1, S(\hat{m}'_h(X_{P_{j+1,k}})) = -1\right\}$$
$$min_{\hat{m}_h}(X_{P_{j,k}}) = \left\{\hat{m}_h(X_{P_{j,k}}) \middle| S(\hat{m}'_h(X_{P_{j,k}})) = -1, S(\hat{m}'_h(X_{P_{j+1,k}})) = +1\right\},$$

where $S(X)$ is the sign function, equal to +1 (−1) when the sign of X is positive (negative), and $\hat{m}'_h(X_{P_{j,k}})$ is the first derivative of the kernel function $\hat{m}_h(X_{P_{j,k}})$. By construction we obtain alternate extrema. We denote respectively by $t_M(\hat{m}_h(X_{P_{j,k}}))$ and $t_m(\hat{m}_h(X_{P_{j,k}}))$ the moments correspondent to detected extrema such that:

$$t_M(\hat{m}_h(X_{P_{j,k}})) = \left\{j \,\middle|\, j \in max_{\hat{m}_h(X_{P_{j,k}})}\right\} \tag{19}$$

$$t_m(\hat{m}_h(X_{P_{j,k}})) = \left\{j \,\middle|\, j \in min_{\hat{m}_h(X_{P_{j,k}})}\right\}. \tag{20}$$

After recording the moments of the detected extrema we realize an orthogonal projection of selected extrema, from the smoothing curve, to the original one. We deduce the corresponding extrema to construct the series involving both maxima, $max_{P_{j,k}}$ and minima, $min_{P_{j,k}}$ such that:

$$max_{P_{j,k}} = max\left(P_{t_M(\hat{m}_h(X_{P_{j,k}}))-1,k}, P_{t_M(\hat{m}_h(X_{P_{j,k}})),k}, P_{t_M(\hat{m}_h(X_{P_{j,k}}))+1,k}\right)$$
$$min_{P_{j,k}} = min\left(P_{t_m(\hat{m}_h(X_{P_{j,k}}))-1,k}, P_{t_m(\hat{m}_h(X_{P_{j,k}})),k}, P_{t_m(\hat{m}_h(X_{P_{j,k}}))+1,k}\right).$$

For each window *k* we get alternate maxima and minima. This is assured by the bandwidth *h* which provide at least two time intervals between two consecutive extrema. The final step consists to scan the extrema sequence to identify an eventual chart pattern. If the same sequence of extremum was observed in more than one window, only the first sequence is retained for the recognition study to avoid the duplication of results.

B.2. M2

M2 is the extrema detection method built on high and low prices. According to this method, maxima and minima have to be detected onto separate curves. Maxima on high prices curve and minima on the low one.

Let H_t and L_t ($t=1,\ldots,T$), be respectively the series for the high and the how prices, and k a window containing l regularly spaced time intervals such that:

$$H_{j,k} \subset \{H_t \mid k \leq t \leq k+l-1\} \tag{21}$$

$$L_{j,k} \subset \{L_t \mid k \leq t \leq k+l-1\}, \tag{22}$$

$j=1,\ldots,l$ and $k=1,\ldots,T-l+1$. We smooth these series through the kernel estimator detailed in Appendix A to obtain $\hat{m}_h\left(X_{H_{j,k}}\right)$ and $\hat{m}_h\left(X_{L_{j,k}}\right)$. We detect maxima on the former series and minima on the latter one in order to construct two separate extrema series $max_{\hat{m}_h(X_{H_{j,k}})}$ and $min_{\hat{m}_h(X_{L_{j,k}})}$ such that:

$$\begin{aligned} max_{\hat{m}_h(X_{H_{j,k}})} &= \left\{\hat{m}_h\left(X_{H_{j,k}}\right) \middle| S\left(\hat{m}_h'\left(X_{H_{j,k}}\right)\right) = +1, S\left(\hat{m}_h'\left(X_{H_{j+1,k}}\right)\right) = -1\right\} \\ min_{\hat{m}_h(X_{L_{j,k}})} &= \left\{\hat{m}_h\left(X_{L_{j,k}}\right) \middle| S\left(\hat{m}_h'\left(X_{L_{j,k}}\right)\right) = -1, S\left(\hat{m}_h'\left(X_{L_{j+1,k}}\right)\right) = +1\right\}, \end{aligned}$$

where $S(x)$ is the sign function defined in the previous Section.

We record the moments for such maxima and minima, denoted respectively by $t_M\left(\hat{m}_h\left(X_{H_{j,k}}\right)\right)$ and $t_m\left(\hat{m}_h\left(X_{L_{j,k}}\right)\right)$ and we project them on the original high and how curves to deduce the original extrema series $max_{H_{j,k}}$ and $min_{L_{j,k}}$, such that:

$$\begin{aligned} max_{H_{j,k}} &= max\left(H_{t_M(\hat{m}_h(X_{H_{j,k}}))-1,k}, H_{t_M(\hat{m}_h(X_{H_{j,k}})),k}, H_{t_M(\hat{m}_h(X_{H_{j,k}}))+1,k}\right) \\ min_{L_{j,k}} &= min\left(L_{t_m(\hat{m}_h(X_{L_{j,k}}))-1,k}, L_{t_m(\hat{m}_h(X_{H_{j,k}})),k}, L_{t_m(\hat{m}_h(X_{H_{j,k}}))+1,k}\right). \end{aligned}$$

However, this method does not guarantee alternate occurrences of maxima and minima. It is easy to observe, in the same window k, the occurrence of two consecutive minima on the low series before observing a maximum on high series. To resolve this problem we start by recording the moments for the selected maxima on high curve, $t_M(H_{j,k})$, and minima in low curve, $t_m(L_{j,k})$. Then we select, for window k the first extremum from these two series, $E_{1,k}$, and its relative moment, $t_{E_{1,k}}$, such that:

$$t_{E_{1,k}} = \min_t \left(t_M\left(H_{j,k}\right), t_m\left(H_{j,k}\right)\right) \tag{23}$$

$$E_{1,k} = \left(\left\{max_{H_{j,k}}\right\} \cup \left\{min_{L_{j,k}}\right\} \middle| j = t_{E_{1,k}}\right) . \tag{24}$$

If we meet a particular case such that a minimum and a maximum occur at the same first moment, then we retain arbitrarily the maximum. To build the alternate series, we have to know the type of the last extremum introduced into the series. If it is a maximum (minimum) then the next extremum has to be a minimum (maximum) selected from the low (high) series such that:

$$E_{j,k|E_{(j-1),k}\in\left\{max_{H_{j,k}}\right\}} = \left\{min_{L_{j,k}} \,\middle|\, j = min\left(t_m\left(L_{j,k}\right)\right), t_m\left(L_{j,k}\right) > t_{E_{(j-1),k}}\right\}$$
$$E_{j,k|E_{(j-1),k}\in\left\{\min_{L_{j,k}}\right\}} = \left\{max_{H_{j,k}} \,\middle|\, j = min\left(t_M\left(H_{j,k}\right)\right), t_M\left(H_{j,k}\right) > t_{E_{(j-1),k}}\right\},$$

where $E_{j,k}$ is the extremum detected on original series.

Finally, the obtained series is scanned by the recognition pattern algorithms to identify an eventual chart pattern.

C. Definition of chart patterns

C.1. Inverse head and shoulders (IHS):

IHS is characterized by a sequence of five extrema E_i (i=1, ..., 5) such that:

$$IHS \equiv \begin{cases} E_1 < E_2 \\ E_3 < E_1, E_3 < E_5 \\ |p(E_2, E_4)| \leq tg(10) \\ |p(E_1, E_5)| \leq tg(10) \\ 0.9 \leq \frac{V_{t_{E_1}}(E_2,E_4)-E_1}{V_{t_{E_5}}(E_2,E_4)-E_5} \leq 1.1 \\ 1.1 \leq \frac{h}{s} \leq 2.5 \\ \frac{1}{2} \leq \frac{t_{E_2}-t_d}{t_f-t_{E_4}} \leq 2 \\ \frac{1}{2} \leq \frac{t_{E_4}-t_{E_2}}{m} \leq 2 \\ (P_{t_{max}} - P_{t_d}) \geq \frac{2}{3} \times h \end{cases}$$

where

- h is the height of the head: $h = V_{t_{E_3}}(E_2, E_4) - E_3$
- s is the height average of the two shoulders:
$$s = \frac{\left(V_{t_{E_1}}(E_2,E_4)-E_1\right)+\left(V_{t_{E_5}}(E_2,E_4)-E_5\right)}{2}$$
- $P_{t_{max}}$ is the highest price observed into the time interval $[t_{d-(f-d)}, t_d]$: $P_{t_{max}} = max(P_t) \mid t_{d-(f-d)} \leq t \leq t_d$
- t_d is the starting time for the pattern
- t_f is the ending time for the pattern
- $t_{d-(f-d)}=t_d-(t_f-t_d)$
- $t_{f+(f-d)}=t_f+(t_f-t_d)$
- m is the average time which the shoulders take for their total completion

C.2. Double top (DT)

DT is characterized by a sequence of three extrema E_i ($i=1,\ldots,3$), such that:

$$DT \equiv \begin{cases} E_1 > E_2 \\ \frac{E_1 - E_2}{V_{t_{E_2}}(E_1,E_3) - E_2} = 1 \\ \frac{E_3 - E_2}{V_{t_{E_2}}(E_1,E_3) - E_2} = 1 \\ \frac{1}{2} \leq \frac{t_{E_2} - t_d}{(t_f - t_d)/2} \leq 2 \\ \frac{1}{2} \leq \frac{t_f - t_{E_2}}{(t_f - t_d)/2} \leq 2 \\ (P_{t_d} - P_{t_{min}}) \geq \frac{2}{3} \times (V_{t_{E_2}}(E_1, E_3) - E_2) \end{cases}$$

C.3. Double bottom (DB)

DB is characterized by a sequence of three extrema E_i ($i=1,\ldots,3$), such that:

$$DB \equiv \begin{cases} E_1 < E_2 \\ \frac{E_2 - E_1}{E_2 - V_{t_{E_2}}(E_1,E_3)} = 1 \\ \frac{E_2 - E_3}{E_2 - V_{t_{E_2}}(E_1,E_3)} = 1 \\ \frac{1}{2} \leq \frac{t_{E_2} - t_d}{(t_f - t_d)/2} \leq 2 \\ \frac{1}{2} \leq \frac{t_f - t_{E_2}}{(t_f - t_d)/2} \leq 2 \\ (P_{t_{max}} - P_{t_d}) \geq \frac{2}{3} \times (E_2 - V_{t_{E_2}}(E_1, E_3)) \end{cases}$$

C.4. Triple top (TT)

TT is characterized by a sequence of five extrema E_i ($i=1,\ldots,5$) such that:

$$TT \equiv \begin{cases} E_1 > E_2 \\ |p(E_1, E_5)| \leq tg(10) \\ |p(E_2, E_4)| \leq tg(10) \\ 0.9 \leq \frac{h}{E_1 - V_{t_{E_1}}(E_2,E_4)} \leq 1.1 \\ 0.9 \leq \frac{h}{E_5 - V_{t_{E_5}}(E_2,E_4)} \leq 1.1 \\ \frac{1}{2} \leq \frac{t_{E_2} - t_d}{(t_f - t_d)/3} \leq 2 \\ \frac{1}{2} \leq \frac{t_{E_4} - t_{E_2}}{(t_f - t_d)/3} \leq 2 \\ \frac{1}{2} \leq \frac{t_f - t_{E_4}}{(t_f - t_d)/3} \leq 2 \\ (P_{t_d} - P_{t_{min}}) \geq \frac{2}{3} \times h \end{cases}$$

C.5. Triple bottom (TB)

TB is characterized by a sequence of five extrema E_i $(i=1,\ldots,5)$ such that:

$$TB \equiv \begin{cases} E_1 < E_2 \\ |p(E_2,E_4)| \leq tg(10) \\ |p(E_1,E_5)| \leq tg(10) \\ 0.9 \leq \frac{h}{V_{t_{E_1}}(E_2,E_4)-E_1} \leq 1.1 \\ 0.9 \leq \frac{h}{V_{t_{E_5}}(E_2,E_4)-E_5} \leq 1.1 \\ \frac{1}{2} \leq \frac{t_{E_2}-t_d}{(t_f-t_d)/3} \leq 2 \\ \frac{1}{2} \leq \frac{t_{E_4}-t_{E_2}}{(t_f-t_d)/3} \leq 2 \\ \frac{1}{2} \leq \frac{t_f-t_{E_4}}{(t_f-t_d)/3} \leq 2 \\ (P_{t_{max}} - P_{t_d}) \geq \frac{2}{3} \times h \end{cases}$$

C.6. Rectangle top (RT)

RT is characterized by a sequence of six extrema E_i $(i=1,\ldots,6)$ such that:

$$RT \equiv \begin{cases} E_1 > E_2 \\ |p(E_1,E_5)| \leq 0.001 \\ |p(E_2,E_6)| \leq 0.001 \\ \frac{V_{t_{E_3}}(E_1,E_5)}{E_3} = 1 \\ \frac{E_4}{V_{t_{E_4}}(E_2,E_6)} = 1 \\ (P_{t_d} - P_{t_{min}}) \geq \frac{2}{3} \times h \end{cases}$$

C.7. Rectangle bottom (RB)

RB is characterized by a sequence of six extrema E_i $(i=1,\ldots,6)$ such that:

$$RB \equiv \begin{cases} E_1 < E_2 \\ |p(E_2,E_6)| \leq 0.001 \\ |p(E_1,E_5)| \leq 0.001 \\ \frac{E_3}{V_{t_{E_3}}(E_1,E_5)} = 1 \\ \frac{V_{t_{E_4}}(E_2,E_6)}{E_4} = 1 \\ (P_{t_{max}} - P_{t_d}) \geq \frac{2}{3} \times h \end{cases}$$

C.8. Broadening Top(BT)

BT is characterized by a sequence of five extrema E_i (i=1,...,5) such that:

$$BT \equiv \begin{cases} E_1 > E_2 \\ E_3 > E_1, E_4 < E_2, E_5 > E_3 \\ (P_{t_d} - P_{t_{min}}) \geq \frac{2}{3} \times h \end{cases}$$

C.9. Broadening bottom(BB)

BB is characterized by a sequence of five extrema E_i (i=1,...,5) such that:

$$BB \equiv \begin{cases} E_1 < E_2 \\ E_3 < E_1, E_4 > E_2, E_5 < E_3 \\ (P_{t_{max}} - P_{t_d}) \geq \frac{2}{3} \times h \end{cases}$$

C.10. Triangle Top (TRIT)

TRIT is characterized by a sequence of four extrema E_i ($i = 1,\ldots,4$) such that:

$$TRIT \equiv \begin{cases} E_1 > E_2 \\ p(E_1, E_3) \leq tg(-30) \\ 0.9 \leq \frac{|p(E_1,E_3)|}{p(E_2,E_4)} \leq 1.1 \\ t_f \leq t_{E_1} + 0.75 \times (t_{int} - t_{E_1}) \\ \left(P_{t_{E_1}} - P_{t_{min}}\right) \geq \frac{2}{3} \times h \end{cases}$$

where t_{int} is the moment of support and resistance lines intersection:

$$t_{int} = \min_t \left(V_t(E_1, E_3) \leq V_t(E_2, E_4), t > t_{E_4}\right).$$

C.11. Triangle bottom(TRIB)

TRIB is characterized by a sequence of four extrema E_i ($i = 1,\ldots,4$) such that:

$$TRIB \equiv \begin{cases} E_1 < E_2 \\ p(E_2, E_4) \leq tg(-30) \\ 0.9 \leq \frac{|p(E_2,E_4)|}{p(E_1,E_3)} \leq 1.1 \\ t_f \leq t_{E_1} + 0.75 \times (t_{int} - t_{E_1}) \\ \left(P_{t_{max}} - P_{t_{E_1}}\right) \geq \frac{2}{3} \times h \end{cases}$$

where t_{int} is the moment of support and resistance lines intersection:

$$t_{int} = \min{}_t(V_t(E_2, E_4) \leq V_t(E_1, E_3), t > t_{E_4}).$$

References

Allen H, Taylor MP (1992) The use of technical analysis in the foreign exchange market. J Int Money Financ 11:304–314

Andrada-Felix J, Fernandez-Rodriguez F, Sosvilla-Rivero S (1995) Technical analysis in the Madrid stock exchange, Foundation for Applied Economic Research (FEDEA) Working Paper, No. 99–05

Béchu T, Bertrand E (1999) L'analyse technique: pratiques et méthodes, Economica, gestion

Blume L, Easley D, O'Hara M (1994) Market statistics and technical analysis: the role of volume. J Finance 49(1):153–181

Brock W, Lakonishok J, LeBaron B (1992) Simple technical trading rules and the stochastic properties of stock returns. J Finance 47:1731–1764

Chang PHK, Osler CL (1999) Methodical madness: technical analysis and the irrationality of exchange-rate forecasts. Econ J 109:636–661

Cheung Y-W, Wong CY-P (1999) Foreign exchange traders in Hong Kong, Tokyo and Singapore: a survey study. Adv Pac Basin Financ Mark 5:111–134

Chordia T, Roll R, Subrahmanyam A (2002) Evidence on the speed of Convergence to Market Efficiency, UCLA Working paper

Dacorogna MM, Müller UA, Nagler RJ, Olsen RB, Pictet OV (1993) A geographical model for the daily and weekly seasonal volatility in the foreign exchange market. J Int Money Financ 12:413–438

Dacorogna M, Gençay R, Müller UA, Pictet OV (2001a) Effective return, risk aversion and drawdowns. Physica A 289:229–248

Dacorogna M, Gençay R, Müller UA, Olsen RB, Pictet OV (2001b) An introduction to high frequency finance, Academic

Danielsson J, Payne R (2002) Real trading patterns and prices in the spot foreign exchange markets. J Int Money Financ 21:203–222

Dempster MAH, Jones CM (1998a) Can technical pattern trading be profitably automated? 1. Channel, Centre for Financial Research, Judge Institute of Management Studies, University of Cambridge, working paper

Dempster MAH., Jones CM (1998b) Can technical pattern trading be profitably automated? 2. The head & shoulders, Centre for Financial Research, Judge Institute of Management Studies, University of Cambridge, working paper

Detry PJ (2001) Other evidences of the predictive power of technical analysis: the moving averages rules on European indexes, EFMA 2001 Lugano Meetings

Dooley M, Shafer J (1984) Analysis of short-run exchange rate behavior: March 1973 to 1981. Floating exchange rates and the state of world trade and payments, 43–70

Fiess N, MacDonald R (2002) Towards the fundamentals of technical analysis: analysing the information content of high, low and close prices. Econ Model 19:353–374

Gençay R (1998) The predictability of security returns with simple technical trading rules. J Empir Finance 5:347–359

Gençay R (1999) Linear, non-linear and essential foreign exchange rate prediction with simple technical trading rules. J Int Econ 47:91–107

Gençay R, Stengos T (1997) Technical trading rules and the size of the risk premium in security returns. Stud Nonlinear Dyn Econ 2:23–34

Gençay R, Ballocchi G, Dacorogna M, Olsen R, Pictet OV (2002) Real-time trading models and the statistical properties of foreign exchange rates. Int Econ Rev 43:463–491

Gençay R, Dacorogna M, Olsen R, Pictet OV (2003) Foreign exchange trading models and market behavior. J Econ Dyn Control 27:909–935

Jensen MC (1970) Random walks and technical theories: some additional evidence. J Finance 25 (2):469–482

LeBaron B (1999) Technical trading rule profitability and foreign exchange intervention. J Int Econ 49:125–143

Levich RM, Thomas LR (1993) The significance of the trading-rule profits in the foreign exchange market: a bootstrap approach. J Int Money Financ 12:1705–1765

Levy RA (1971) The predictive significance of five-point patterns. J Bus 41:316–323

Lo AW, Mamaysky H, Wang J (2000) Foundations of technical analysis: computational algorithms, statistical inference, and empirical implementation. J Finance 56:1705–1765

Lui YH, Mole D (1998) The Use of Fundamental and Technical Analyses by Foreign Exchange Dealers: Hong Kong Evidence. J Int Money Financ 17:535–545
Lyons RK (2001) The microstructure approach to exchange rates. MIT
Menkhoff L (1998) The noise trading approach – questionnaire evidence from foreign exchange. J Int Money Financ 17:547–564
Murphy JJ (1999) Technical analysis of the financial markets, New York Institute of Finance
Neely CJ (1999) Technical analysis in the foreign exchange market: a layman's guide, Federal Reserve Bank of Saint Louis Working Paper
Neely CJ, Weller PA (2003) Intraday technical trading in the foreign exchange market. J Int Money Financ 22(2):223–237
Neely CJ, Weller PA, Dittmar R (1997) Is technical analysis in the foreign exchange market profitable? A genetic programming approach. J Financ Quant Anal 32(4):405–426
Osler CL (1998) Identifying noise traders: the head-and-shoulders pattern in U.S. equities, Federal Reserve Bank of New York Working Paper
Osler C (2000) Support for resistance: technical analysis and intraday exchange rates, Federal Reserve Bank of New York Working Paper
Prost AJ, Prechter R (1985) Eliott waves principle, New classics library
Ready MJ (1997) Profits from technical trading rules, Working Paper, University of Wisconsin-Madison
Silverman BW (1986) Density estimation for statistics and data analysis. Chapman and hall
Sweeney RJ (1986) Beating the foreign exchange market. J Finance 41:304–314

Juan M. Rodríguez-Poo · David Veredas · Antoni Espasa

Semiparametric estimation for financial durations

Abstract We propose a semiparametric model for the analysis of time series of durations that show autocorrelation and deterministic patterns. Estimation rests on generalized profile likelihood, which allows for joint estimation of the parametric—an ACD type of model—and nonparametric components, providing consistent and asymptotically normal estimators. It is possible to derive the explicit form for the nonparametric estimator, simplifying estimation to a standard maximum likelihood problem.

Keywords Generalized profile likelihood · ACD model · Seasonality

JEL Classification C14 · C15 · C22 · C32

1 Introduction

Modeling financial durations has been a very active area of research since Engle and Russell (1998) introduced the autoregressive conditional duration (ACD) model. Their analysis is justified from an economic and a statistical point of view. Market microstructure theory shows that the time between events in stock markets conveys information that is used by market participants. On the other hand, financial durations are one-dimensional point processes (with time as space) and the analysis of these processes has a long tradition in statistics.

Since the ACD model, a plethora of modifications and alternatives have been proposed. Bauwens and Giot (2000) introduce the Log-ACD model, an exponential

Juan M. Rodríguez-Poo
Departamento de Economía, niversidad de Cantabria, E-39005 Santander, Belgium

David Veredas
Universite Libre de Bruxelles (ECARES), 50 Ave Jeanne CP114 B1050 Brussels, Belgium

David Veredas
CORE, Université Catholique de Louvain, Louvain, Belgium

Antoni Espasa
Departamento de Estadística y Econometría, Universidad Carlos III de Madrid, E-28903 Getafe, Spain

version of the ACD. Grammig and Mauer (2000) use a Burr distribution in the ACD model. Zhang et al. (2001) introduce a threshold ACD (TACD). Fernandes and Grammig (2001) introduce the augmented ACD model, a very general model that covers almost all the existing ones. Drost and Werker (2004) provide a method for obtaining efficient estimators of the ACD model with no need to specify the distribution. Meitz and Terasvirta (2005) introduce smooth transition ACD models and testing evaluation procedures. Alternative models are the stochastic conditional duration (SCD) model of Bauwens and Veredas (2004) and the stochastic volatility duration (SVD) model of Ghysels et al. (2004), both based on latent variables.

In the application of most of the above studies, durations show a strong intradaily seasonality, i.e., an inverted U shape pattern along the day, as it is shown in Figs. 2 and 3 for price and volume durations of two stocks traded at the NYSE. The study of seasonality for regularly spaced variables, i.e., observed at equidistant periods of time, is well known. It has focused mainly on the intradaily behavior of volatility either on stock markets or on foreign exchange markets—see, among others, Baillie and Bollerslev (1990), Bollerslev and Domowitz (1993), Andersen and Bollerslev (1997 and 1998) and Beltratti and Morana (1999). All these articles used data sampled at different frequencies: hourly, half-hourly, every 15 min or every 5 min.

Durations do not fit into this category as they are themselves the main characteristic of irregularly spaced data. For tick-by-tick data, Engle and Russell (1998) introduce a method for dealing with intradaily seasonality. It consists of decomposing the expected duration into a deterministic part, that depends on the time-of-the-day at which the duration starts, and a stochastic part. The deterministic component accounts for the seasonal effect whereas the stochastic component models the dynamics. If both components are assumed to belong to a parametric family of functions, the two sets of parameters can be jointly estimated by maximum likelihood (ML) techniques. And, under standard regularity conditions, ML estimators of the parameters of interest are consistent and asymptotically normal (see Engle and Russell, 1998; corollary, p. 1135).

In many situations the researcher does not have enough information to fully specify the seasonality functions or seasonality itself is not the main subject of analysis but it has to be taken into account. In these cases, the choice of a particular parametric function can be delicate. An alternative is to approximate the seasonal component through nonparametric functions (mainly Fourier series, spline functions, or other types of smoothers) and then estimate the parameters of the dynamic component using maximum likelihood techniques. Unfortunately, standard maximum likelihood for finite dimensional parameters, in the presence of infinite "incidental" parameters, may yield inconsistency and slow rates of convergence (see, for examples of inconsistency: Kiefer and Wolfowitz (1956); Grenander (1981); and Shen and Wong (1994) and for examples of slow rates of convergence Birgé and Massart (1994)). In order to solve this problem many alternative solutions have been proposed. If the seasonal component is estimated through splines, Fourier series, neural networks, or wavelets, then the method of sieve extremum estimation can be used to make inference on the seasonal term. Chen and Shen (1998) give sufficient conditions for splines and Fourier series under regularly spaced dependent data. If, instead, other methods such as kernels or local polynomials are used, then the sieve method is no longer valid and other nonsieve ML estimation methods are needed. Among others, the so-called generalized profile likelihood approach (see Severini and Wong (1992) and Severini and Staniswalis (1994)).

In this paper we propose a new method to jointly estimate the parametric dynamic and nonparametric seasonal components in an ACD framework. Estimation is based on generalized profile likelihood techniques. To make inference on the parameters of interest, we need to extend some previous results on i.i.d. data, Severini and Wong (1992), to a dependent data setup. Our estimation method presents several advantages against other methods in the literature: (1) It presents closed form seasonal estimators that are very intuitive. The resulting nonparametric estimator of the seasonal component is a simple transformation of the Nadaraya–Watson estimator. (2) The statistical properties of both the parametric and nonparametric estimators are well established. We present the asymptotic distribution of both the nonparametric and the parametric estimator. This enables us to make correct inference in the different components. (3) This methodology provides a data driven method for computing the bandwidth. On the contrary, polynomial spline techniques do not have a method for choosing the number and location of nodes and the proportionality coefficients for the end-point restrictions. (4) Multivariate extensions of the seasonal estimator are straightforward. For example, considering a multivariate exponential distribution, the estimator is easily adapted to the multivariate case. (5) The decomposition presented in the paper can be easily extended to cope with other specifications that are frequently used in the econometric analysis of tick-by-tick data. For example, to capture nonlinear relationships between financial durations and market microstructure variables (see, for instance, Spierdik et al. (2004), and references therein). Likewise, it is also possible to replace the dependent variable by any other tick-by-tick market microstructure variable and make it a function of its own lags, through the parametric component, and any other variable through the nonparametric component. In sum, although focused on durations and its nonlinear dependency with the time-of-the-day, the potential applications of this model are very ample.

As an illustration, we apply our method to price and volume durations of two stocks traded on the NYSE. We show that the model is able to correctly capture the seasonal pattern and it is able to adjust to changes on this pattern.

The structure of the paper is as follows. Section 2 develops a general ML framework for analyzing tick-by-tick data, and proposes the new estimator for seasonality. Its asymptotic properties are also analyzed. Second, it develops the same method but in a generalized linear model (GLM) framework, which allows us to use quasi maximum likelihood (QML). Section 3 is devoted to the empirical application comparing the results with others existing in the literature. We use density forecast to evaluate the out-of-sample goodness of fit. Section 4 concludes. The assumptions and proofs of the main results are relegated to the Appendix.

2 Econometric model and estimators

Let t_i be the time at which the ith event occurs and let $d_i = t_i - t_{i-1}$, where $t_{i-1} < t_i, i = 1, \ldots, n$ be the ith duration between two consecutive events. The sequence of times t_i is measured as accumulated seconds from the starting date of the sample. At the time t_i, the ith event occurs and k characteristics associated to this event are observed. We gather these characteristics in a k-dimensional vector y_i. Thus, the following set of observations is available

$$\{(d_i, y_i)\}_{i=1,\ldots,n}.$$

One possible way to obtain information about the whole process is to assume that the ith duration has a prespecified conditional parametric density $d_i|I_{i-1} \sim p\left(d_i|\bar{d}_{i-1}, \bar{y}_{i-1}; \delta\right)$, where $(\bar{d}_{i-1}, \bar{y}_{i-1})$ is the past information and δ is a set of finite-dimensional parameters. Under these conditions it is possible to estimate the parameter vector δ through maximum likelihood techniques. But sometimes assumptions about the knowledge of the whole conditional density are too strong and the researcher prefers to make assumptions about some of its conditional moments. Let

$$E[d_i|\bar{d}_{i-1}, \bar{y}_{i-1}] = \psi(\bar{d}_{i-1}, \bar{y}_{i-1}; \vartheta_1) \tag{1}$$

be the expectation of the ith duration conditional on the past filtration. The ACD class of models consists of parameterizations of (1) and the assumption that

$$d_i = \psi(\bar{d}_{i-1}, \bar{y}_{i-1}; \vartheta_1)\varepsilon_i,$$

where ε_i is an i.i.d. random variable with density function $p(\varepsilon_i; \xi)$ depending on a set of parameters ξ and mean equal to one. The vector ϑ_1 contains the parameters that measure the dynamics of the scale and ξ contains the shape parameters. The conditional log-likelihood function can be written as

$$L_n(d; \vartheta_1, \xi) = \sum_{i=1}^{n} \log p\left(d_i|\bar{d}_{i-1}, \bar{y}_{i-1}; \vartheta_1, \xi\right). \tag{2}$$

If the conditional density is correctly specified and ϑ_1 and ξ are finite-dimensional parameters, then, under some standard regularity conditions, the maximum likelihood estimators of the parameters of interest are consistent and asymptotically normal (for conditions see Engle and Russell (1998)).

The specification described above is sometimes too simple and/or rigid since the expected duration can vary systematically over time and can be subject to many different time effects. One way to extend the above model is to decompose the conditional mean into different effects. In standard time series literature a stochastic process can be decomposed into a combination of cycle and trend, seasonality and noise. This decomposition, with a long tradition in time series analysis, has already been used in volatility (see, among others, Andersen and Bollerslev (1998)) and duration analysis (e.g. Engle and Russell (1998)). Instead of (1), the following nonlinear decomposition is proposed:

$$E[d_i|\bar{d}_{i-1}, \bar{y}_{i-1}] = \varphi\left(\psi(\bar{d}_{i-1}, \bar{y}_{i-1}; \vartheta_1), \phi(\bar{d}_{i-1}, \bar{y}_{i-1}; \vartheta_2)\right). \tag{3}$$

Durations are modeled as a possibly nonlinear function of two components, $\psi(\cdot; \vartheta_1)$ and $\phi(\cdot; \vartheta_2)$, that represent dynamic and seasonal behavior, respectively. The function $\varphi(u, v)$ nests a great variety of models. $\varphi(u, v) = (u \times v)$ forms an

ACD-type of model whereas $\varphi(u, v) = \exp(u + v) = \exp(u) \times \exp(v)$ represents a Log-ACD type of model with

$$\psi(\bar{d}_{i-1}; \vartheta_1) = \omega + \sum_{j=1}^{J} \alpha_j \ln d_{i-j} + \sum_{\ell=1}^{L} \beta_\ell \psi_{i-\ell}. \tag{4}$$

With respect to the seasonal component, $\phi(\bar{d}_{i-1}, \bar{y}_{i-1}, \vartheta_2)$, several alternatives are available. In this class of models it is usually assumed that the seasonal term is somehow related to t_i. In order to make this dependence more explicit, we define a rescaled time variable, t_i', such that

$$t_i' = \begin{cases} t_i - \left\lfloor \frac{t_i - t_o}{t_c - t_o} \right\rfloor (t_c - t_o) & \text{if } t_i > t_c, \\ t_i & \text{otherwise,} \end{cases} \tag{5}$$

where $\lfloor x \rfloor$ is the integer part of x. If the seasonal frequency is daily, t_o and t_c stand for the stock market opening and closing (in seconds), respectively; i.e. $t_o = 9.5 \times 60 \times 60$ and $t_c = 16 \times 60 \times 60$ if it opens at 09:30 and closes at 16:00. Note that defined in this way, $t_i' \in [t_o, t_c]$ is of bounded support and is the time-of-the-day (in seconds) at which the event has occurred. By changing the values of t_o and t_c, other possible types of seasonality (hourly, weekly, monthly, etc.) can be considered.

In the context of regularly spaced variables, several functional forms for $\phi(\cdot; \theta_2)$ have been proposed in literature. If the function is assumed to fall within a known class of parametric functions then, by substituting (3) into (2), we obtain the following log-likelihood function

$$L_n(d; \vartheta_1, \vartheta_2, \xi) = \sum_{i=1}^{n} \log p\left(d_i | \bar{d}_{i-1}, \bar{y}_{i-1}, t_{i-1}'; \vartheta_1, \vartheta_2, \xi\right), \tag{6}$$

and the parameters ϑ_1, ϑ_2, and ξ can be estimated as in the one component case (2). If the error density is correctly specified then standard ML estimation methods apply.

But if very little information is available about the functional form that relates seasonality and time-of-the-day, the risk of misspecification in choosing $\phi(\cdot; \vartheta_2)$ is high. Consequently, it is worth the use of nonparametric methods to approximate the unknown seasonal function. Let $\phi(t_i')$ be the deterministic seasonal component at time t_i'. Given the proposed specification for the components, (3) becomes

$$E\left[d_i | \bar{d}_{i-1}, \bar{y}_{i-1}, t_{i-1}'\right] = \varphi\left(\psi(\bar{d}_{i-1}, \bar{y}_{i-1}; \vartheta_1), \phi(t_{i-1}')\right), \tag{7}$$

where ϑ_1 and the function $\phi(\cdot)$, evaluated at time points $t_1', \ldots, t_n'$, have to be estimated. Following the standard approach (as in (2) or (6)), one would be tempted to obtain estimators for $\vartheta_1, \xi, \phi(t_1'), \ldots, \phi(t_n')$, by choosing the values that maximize the following log-likelihood function

$$L_n(d; \vartheta_1, \xi) = \sum_{i=1}^{n} \log p\left(d_i | \bar{d}_{i-1}, \bar{y}_{i-1}; \vartheta_1, \phi\left(t_{i-1}'\right), \xi\right). \tag{8}$$

Standard ML techniques do not apply directly since the estimation of the parameter vector, ϑ_1 and ξ, does not necessarily provide consistent estimators in the presence of infinite dimensional nuisance parameters. In order to implement valid inference not only on the parameter estimators of the dynamic component but on the estimated seasonal curve as well, we propose the generalized profile likelihood approach. It has been introduced, in an i.i.d. context, by Severini and Wong (1992). The basic idea of this method is to estimate the nonparametric function $\phi(\cdot)$ by maximizing a local (and hence smoothed) likelihood function (see Staniswalis (1989)), and simultaneously estimate the parameter vector ϑ_1 and ξ by maximizing the unsmoothed likelihood function. For a given value of the time-of-the-day, $t_0' \in [t_o, t_c]$, and fixed values of ϑ_1 and ξ, we estimate $\phi(t_0')$ as the solution of the problem

$$\hat{\phi}_{\vartheta_1,\xi}(t_0') = \arg\sup_{\phi\in\Phi} \frac{1}{nh}\sum_{i=1}^{n} K\left(\frac{t_0'-t_i'}{h}\right) \log p\left(d_i|\bar{d}_{i-1}, \bar{y}_{i-1}; \vartheta_1, \phi, \xi\right), \quad (9)$$

where $K(\cdot)$ is a kernel function and h is the corresponding bandwidth. Note also that all estimators depend on ϑ_1 and ξ. Then, $\hat{\phi}_{\vartheta_1,\xi}(t_0')$ must fulfill the first order condition

$$\frac{1}{nh}\sum_{i=1}^{n} K\left(\frac{t_0'-t_i'}{h}\right) \frac{\partial}{\partial\phi}\log p\left(d_i|\bar{d}_{i-1}, \bar{y}_{i-1}; \vartheta_1, \hat{\phi}_{\vartheta_1,\xi}(t_0'), \xi\right) = 0. \quad (10)$$

Given the above estimates for the nonparametric part, a simple ML estimation for ϑ_1 and ξ is performed

$$\left(\hat{\vartheta}_{1n}\ \hat{\xi}_n\right)^T = \arg\sup_{\vartheta_1\in\Theta}\sup_{\xi\in\Xi}\sum_{i=1}^{n}\log p\left(d_i|\bar{d}_{i-1}, \bar{y}_{i-1}; \vartheta_1, \hat{\phi}_{\vartheta_1,\xi}(t_{i-1}'), \xi\right) \quad (11)$$

and $\left(\hat{\vartheta}_{1n}\ \hat{\xi}_n\right)$ must fulfill the first order condition

$$\sum_{i=1}^{n}\frac{\partial}{\partial\left(\vartheta_1\ \xi\right)^T}\log p\left(d_i|\bar{d}_{i-1}, \bar{y}_{i-1}; \hat{\vartheta}_{1n}, \hat{\phi}_{\hat{\vartheta}_{1n},\hat{\xi}_n}(t_{i-1}'), \hat{\xi}_n\right) = 0. \quad (12)$$

The procedure is implemented as follows: (1) For a given t_0', fix the values of ϑ and ξ, and find the $\hat{\phi}_{\vartheta_1,\xi}(t_0')$ that fulfills condition (9). Repeat it for all t_i'. (2) Plug the vector $\hat{\phi}_{\vartheta_1,\xi}(t_i')$ into (11). The log-likelihood is hence concentrated on ϑ and ξ, which can be easily estimated. (3) Given the estimators $\hat{\vartheta}_n$ and $\hat{\xi}_n$ come back to (1). (4) Iterate until (10) and (12) are fulfilled. This procedure is computationally intensive as the optimization in (1) has to be done n times. It would be significantly alleviated if we would have a closed form expression for $\hat{\phi}_{\vartheta_1,\xi}(t_0')$. This is the case for $\log p$ being one of the log-likelihoods that are typically assumed for financial durations.

As an example, assume that $p(\cdot)$ in the log-likelihood function (2) is the generalized gamma density function, i.e., $\varepsilon_i \sim GG(1, \gamma, \nu)$ then

$$d_i \sim GG\left(\varphi\left(\psi\left(\bar{d}_{i-1}, \bar{y}_{i-1}; \vartheta_1\right), \phi_{\vartheta_1,\xi}\left(t_{i-1}'\right)\right)^{-1}, \gamma, \nu\right).$$

Assuming that $\varphi(u, v) = \exp(u + v)$, the seasonal component is estimated through the following expression

$$\hat{\phi}_{\vartheta_1,\xi}(t_0') = \frac{1}{\gamma\nu}\log\left\{\frac{\frac{1}{nh}\sum_{i=1}^{n} K\left(\frac{t_0'-t_i'}{h}\right)\left(\frac{d_i}{\exp\{\psi(\bar{d}_{i-1},\bar{y}_{i-1};\vartheta_1)\}}\right)^{\gamma}}{\frac{1}{nh}\sum_{i=1}^{n} K\left(\frac{t_0'-t_i'}{h}\right)}\right\}. \tag{13}$$

This closed form estimator can be plugged into (11), reducing the simultaneous optimization problem to a standard ML procedure. Two useful particular cases are nested in this estimator. If $\nu = 1$, we obtain the nonparametric estimator when $p(\cdot)$ is a Weibull density function. And $\nu = \gamma = 1$ corresponds to the exponential density. In all cases, the nonparametric seasonal curve is estimated by a transformation of the Nadaraya–Watson nonparametric regression estimator of the duration—adjusted by the dynamic component—on the time-of-the-day at time t_0'.

The results available in the earlier literature were obtained for independent observations and hence do not hold for tick-by-tick data. The following Theorem shows the equivalent statistical results that allows us to make correct inference about the unknown parameters of the Log-ACD model (the proof is given in the Appendix and Eq. (8) is simplified to $\log p\,(d;\phi,\eta)$):

Theorem 1: *Let $\eta = (\vartheta_1\ \xi)^T$, and $\hat{\eta}_n$ be the vector of corresponding parametric estimates. Under conditions (L.1), (L.2), (A.1) to (A.3), (B.1), and (B.3) to (B.6) stated in the Appendix*

$$\sqrt{n}\left(\hat{\eta}_n - \eta\right) \rightarrow_d \mathbf{N}\left(0, I_\eta^{-1}(\phi,\eta)\right), \tag{14}$$

where

$$\begin{aligned} I_\eta(\phi,\eta) = {} & E\left[\frac{\partial}{\partial\eta}\log p\,(d;\phi,\eta)\frac{\partial}{\partial\eta^T}\log p\,(d;\phi,\eta)\right] \\ & - E\left[\frac{\partial}{\partial\eta}\log p\,(d;\phi,\eta)\frac{\partial}{\partial\phi}\log p\,(d;\phi,\eta)\right] \\ & \times E\left[\frac{\partial^2}{\partial\phi^2}\log p\,(d;\phi,\eta)\right]^{-1} \\ & \times E\left[\frac{\partial}{\partial\phi}\log p\,(d;\phi,\eta)\frac{\partial}{\partial\eta^T}\log p\,(d;\phi,\eta)\right] \end{aligned}$$

and

$$\sqrt{nh}\left(\hat{\phi}_{\hat{\eta}_n}(t_0') - \phi(t_0')\right) \rightarrow_d \mathbf{N}\left(0, V\left(t_0',\eta\right)\right), \tag{15}$$

where

$$V\left(t_0', \eta\right) = \frac{\int K^2(u)du}{f(t_0')I\left(t_0', \eta\right)}, \tag{16}$$

$$I\left(t_0', \eta\right) = E\left[\frac{\partial}{\partial \phi} \log p\left(d; \phi, \eta\right)^2 \bigg| t' = t_0'\right] \tag{17}$$

and $f(t_0')$ is the marginal density function of t', as n tends to infinity.

This theorem is very appealing since it allows us to make inference on the parametric and nonparametric components. However, the result depends on the correct specification of the conditional density function of the error term. In Section 2.1 we weaken certain assumptions about the error density function and show that our result remains valid.

2.1 The GLM Approach

As pointed out in Engle and Russell (1998) and Engle (2000), it is of interest to have estimation techniques available that do not rely on the knowledge of the functional form of the conditional density function. Two alternative approaches that allow for consistent estimation of the parameters of interest without specifying the conditional density are quasi maximum likelihood techniques, QML, (see Gouriéroux, Monfort and Trognon (1984)) and generalized linear models, GLM, (see McCullagh and Nelder (1989)). In both approaches it is assumed that d_i, conditional on $\bar{d}_{i-1}$ and $\bar{y}_{i-1}$, depends on a scalar parameter $\theta = h\left(\bar{d}_{i-1}, \bar{y}_{i-1}, t'_{i-1}; \vartheta_1, \vartheta_2\right)$, and its distribution belongs to a one-dimensional exponential family with conditional density

$$q\left(d_i | \bar{d}_{i-1}, \bar{y}_{i-1}; \theta\right) = \exp\left(d_i\theta - b(\theta) + c(d_i)\right),$$

where $b(\cdot)$ and $c(\cdot)$ are known functions. The main difference between the QML and the GLM approaches is simply a different parameterization. We adopt the GLM approach for the sake of convenience. By adopting the GLM parametrization, it is straightforward to see that the ML estimator of θ solves the first order conditions $\sum_{i=1}^{n}\left\{d_i - b'(\theta)\right\} = 0$. The ML estimator of θ can also be obtained from the solution of the following equation

$$\sum_{i=1}^{n} \frac{\left\{d_i - \varphi\left(\psi(\bar{d}_{i-1}, \bar{y}_{i-1}; \vartheta_1), \phi(t'_{i-1}; \vartheta_2)\right)\right\} \varphi'\left(\psi(\bar{d}_{i-1}, \bar{y}_{i-1}; \vartheta_1), \phi(t'_{i-1}; \vartheta_2)\right)}{V\left\{\varphi\left(\psi(\bar{d}_{i-1}, \bar{y}_{i-1}; \vartheta_1), \phi(t'_{i-1}; \vartheta_2)\right)\right\}} = 0. \tag{18}$$

The parameter of interest θ (the so-called canonical parameter) can be estimated without specifying the whole conditional distribution function. It is only necessary to specify the functional form of the conditional mean, $\varphi(\cdot)$, and of the conditional variance $V(\cdot)$, but not the whole distribution.

The relationship between the predictors in Eq. (7) and the canonical parameter is given by the *link function*. This function depends on the member of the exponential family that we use. For the exponential distribution the link function is

$$\theta = -\frac{1}{\varphi\left(\psi(\bar{d}_{i-1}, \bar{y}_{i-1}; \vartheta_1), \phi(t'_{i-1}; \vartheta_2)\right)}.$$

Since under this distribution $E\left[d_i|\bar{d}_{i-1}, \bar{y}_{i-1}, t'_{i-1}\right] = -\theta^{-1}$ and $\mathrm{V}\left[d_i|\bar{d}_{i-1}, \bar{y}_{i-1}, t'_{i-1}\right] = \theta^2$, Eq. (18) specifies the first order conditions for the maximization of the log-likelihood function for exponentially distributed random variables.

In order to estimate the parameters of interest we maximize the quasi-log-likelihood function

$$Q_n(d, \varphi) = \sum_{i=1}^{n} \log q\left(d_i, \varphi\left(\psi(\bar{d}_{i-1}, \bar{y}_{i-1}; \vartheta_1), \phi(t'_{i-1}; \vartheta_2)\right)\right)$$

with respect to ϑ_1 and ϑ_2. As already indicated in the likelihood context, if the seasonal component is assumed to fall within the class of known parametric function, $\phi(t_{i-1}; \vartheta_2)$, the properties of the QML estimators of ϑ_1 and ϑ_2 are well known (see Engle and Russell (1998) and Engle (2000)). But if the seasonal component is approximated nonparametrically, then standard quasi-likelihood arguments do not hold. For the standard i.i.d. data case, Severini and Staniswalis (1994) and Fan et al. (1995) propose consistent estimators of both parametric and nonparametric parts. The statistical results presented in these papers do not apply directly in our case since they assume independence of observations but at the end of the subsection equivalent statistical results are shown for the dependent case.

As in the likelihood case, let us define $\hat{\phi}_{\vartheta_1}(t'_0)$, for fixed values of ϑ_1, as the solution to the following (smoothed) optimization problem

$$\hat{\phi}_{\vartheta_1}(t'_0) = \arg\sup_{\phi\in\Phi} \frac{1}{nh} \sum_{i=1}^{n} K\left(\frac{t'_0 - t'_i}{h}\right) \log q\left(d_i, \varphi\left(\psi(\bar{d}_{i-1}, \bar{y}_{i-1}; \vartheta_1), \phi\right)\right) \tag{19}$$

for $t'_0 \in [t_o, t_c]$. Then $\hat{\phi}_{\vartheta_1}(t'_0)$ must fulfill the following first order condition

$$\frac{1}{nh} \sum_{i=1}^{n} K\left(\frac{t'_0 - t'_i}{h}\right) \frac{\partial}{\partial\phi} \log q\left(d_i, \varphi\left(\psi(\bar{d}_{i-1}, \bar{y}_{i-1}; \vartheta_1), \hat{\phi}_{\vartheta_1}(t'_0)\right)\right) = 0.$$

The estimator of ϑ_1 is obtained as the solution to the following (unsmoothed) optimization problem

$$\hat{\vartheta}_{1n} = \arg\sup_{\vartheta_1\in\Theta} \sum_{i=1}^{n} \log q\left(d_i, \varphi\left(\psi(\bar{d}_{i-1}, \bar{y}_{i-1}; \vartheta_1), \hat{\phi}_{\vartheta_1}(t'_{i-1})\right)\right),$$

and $\hat{\vartheta}_{1n}$ must fulfill the following first order condition

$$\sum_{i=1}^{n} \frac{\partial}{\partial\vartheta_1} \log q\left(d_i, \varphi\left(\psi(\bar{d}_{i-1}, \bar{y}_{i-1}; \hat{\vartheta}_{1n}), \hat{\phi}_{\hat{\vartheta}_{1n}}(t'_{i-1})\right)\right) = 0.$$

As an example, set $\varphi(u, v) = \exp(u + v)$. Then, the quasi-likelihood function corresponds to the log-likelihood function from an exponential distribution and a Log-ACD representation, i.e.,

$$-\sum_{i=1}^{n} \left[\left\{ \psi(\bar{d}_{i-1}, \bar{y}_{i-1}; \vartheta_1) + \phi_{\vartheta_1}(t'_{i-1}) \right\} + \frac{d_i}{\exp\left\{ \psi(\bar{d}_{i-1}, \bar{y}_{i-1}; \vartheta_1) + \phi_{\vartheta_1}(t'_{i-1}) \right\}} \right],$$

and, after some derivations, the estimator (19) takes the explicit form

$$\hat{\phi}_{\vartheta_1}(t'_0) = \log \left\{ \frac{\frac{1}{nh} \sum_{i=1}^{n} K\left(\frac{t'_0 - t'_i}{h}\right) \frac{d_i}{\exp\{\psi(\bar{d}_{i-1}, \bar{y}_{i-1}; \vartheta_1)\}}}{\frac{1}{nh} \sum_{i=1}^{n} K\left(\frac{t'_0 - t'_i}{h}\right)} \right\}.$$

The above expressions are obtained by assuming a seasonal component for a given period (e.g., daily). But it is also possible to extend this method to cover several seasonal effects. For example, we may be interested in looking at whether the seasonal patterns for each day of the week are different and test the differences using confidence bands. For sth day of the week, we have for the exponential density and the Log-ACD representation,

$$\hat{\phi}_s(t'_0) = \log \left\{ \frac{\frac{1}{nh} \sum_{j=1}^{\lfloor n/5 \rfloor} \sum_{i=1}^{n} K\left(\frac{t'_0 - t'_i}{h}\right) I\left(\lfloor \frac{t - t_o}{t_c - t_o} + 1 \rfloor = js\right) \frac{d_i}{\exp\{\psi(\bar{d}_{i-1}, \bar{y}_{i-1}; \vartheta_1)\}}}{\frac{1}{nh} \sum_{j=1}^{\lfloor n/5 \rfloor} \sum_{i=1}^{n} K\left(\frac{t_0 - t_i}{h}\right) I\left(\lfloor \frac{t - t_o}{t_c - t_o} + 1 \rfloor = js\right)} \right\}$$

for $s = 1, \ldots, 5$. $I(\cdot)$ is the indicator function and $\lfloor x \rfloor$ is the integer part of x.

The following Theorem shows the equivalence of statistical results for making correct inference (notation is simplified, as in Theorem 1, and the proof is given in the Appendix):

Theorem 2: *Under conditions (A.1) to (A.5), and (B.1) to (B.6), provided in the Appendix then,*

$$\sqrt{n}\left(\hat{\vartheta}_{1n} - \vartheta_1\right) \rightarrow_d \mathbf{N}\left(0, \Sigma_{\vartheta_1}^{-1}\right), \tag{20}$$

where

$$\Sigma_{\vartheta_1} = -E\left(\frac{\partial^2}{\partial \vartheta_1 \partial \vartheta_1^T} \log q\left(d, \varphi\left(\psi(\bar{d}, \bar{y}; \vartheta_1), \phi(t')\right)\right)\right),$$

and

$$\sqrt{nh}\left(\hat{\phi}_{\hat{\vartheta}_1}(t'_0) - \phi(t'_0)\right) \rightarrow_d \mathbf{N}\left(0, V\left(t'_0; \eta\right)\right), \tag{21}$$

where

$$V\left(t'_0; \eta\right) = \frac{\int K^2(u) du}{f(t'_0)} \times \frac{E\left[\left\{\frac{d - \varphi(\psi(\vartheta_1), \phi)}{V(\varphi(\psi(\vartheta_1), \phi))} \frac{\partial}{\partial \phi} \varphi\left(\psi(\vartheta_1), \phi\right)\right\}^2 \middle| t' = t'_0\right]}{E\left[\frac{1}{V_0(\varphi(\psi(\vartheta_1), \phi))} \times \frac{\partial}{\partial \phi} \varphi\left(\psi(\vartheta_1), \phi\right)^2 \middle| t' = t'_0\right]^2},$$

and $f(t'_0)$ is the marginal density function of t', as n tends to infinity.

2.2 Predictibility

We analyze the predictibility and specification of the model via density forecast, introduced by Diebold et al. (1998) in the context of GARCH models and extensively used by, among others, Bauwens et al. (2004) to compare different financial duration models. Density forecast is more accurate than point, or even interval prediction, since it relies on the forecasting performance of all the moments. The behavior of the out-of-sample density function is evaluated via the probability integral transform. If the prediction is correct, the probability integral transform over the out-of-sample should be i.i.d and uniformly distributed.

Let $p\left(d_j \left| \bar{d}_{j-1}, \bar{y}_{j-1}; \widehat{\vartheta}_{1n}, \widehat{\phi}_{\widehat{\vartheta}_{1n},\widehat{\xi}_n}\left(t'_{j-1}\right), \widehat{\xi}_n\right.\right)$ for $j = n+1, \ldots, m$ be a sequence of one-step-ahead density forecasts given by the estimated model, and let $f\left(d_j | \bar{d}_{j-1}, \bar{y}_{j-1}\right)$ be the sequence of densities defining the data generating process of the duration process d_j. n and m are the number of observations in-sample and out-of-sample, respectively. Diebold et al. (1998) show that if the density is correctly specified

$$p\left(d_j \left| \bar{d}_{j-1}, \bar{y}_{j-1}; \widehat{\vartheta}_{1n}, \widehat{\phi}_{\widehat{\vartheta}_{1n},\widehat{\xi}_n}\left(t'_{j-1}\right), \widehat{\xi}_n\right.\right) = f\left(d_j | \bar{d}_{j-1}, \bar{y}_{j-1}\right).$$

To test this equality we use the probability integral transform

$$z_j = \int_{-\infty}^{d_j} p\left(u \left| \bar{d}_{j-1}, \bar{y}_{j-1}; \widehat{\vartheta}_{1n}, \widehat{\phi}_{\widehat{\vartheta}_{1n},\widehat{\xi}_n}\left(t'_{j-1}\right), \widehat{\xi}_n\right.\right) du.$$

If the one-step-ahead density forecast equals the density defining the data generating process, z must be independent and uniformly distributed. This happens if (1) the predicted density is correctly specified, (2) the dynamics are well captured, and (3) the estimated seasonal component fits the out-of-sample seasonal pattern. If any of these three elements fails, the integral probability transform is not independent and uniformly distributed. Uniformity can be tested by using histograms based on the computed z sequence. If the density is correctly specified, the histogram should be flat. Additionally, the Spearman's ρ of various centered moments of the z sequence may reveal some dependency in z.

Last, a note on how we predict seasonality: Assuming a generalized gamma density, the seasonal estimator is (13). It depends on (1) the current time-of-the-day t'_0 and duration d_i that come from the same arrival times—see Eq. (5)—and (2) the estimated parameters $\hat{\vartheta}_1, \hat{\xi}$. When forecasting, the parameters are fixed as they have been estimated using the in-sample. But durations and the time-of-the-day are those of the out-of-sample. All this translates into a forecasted seasonality that changes with the out-of-sample information and adapts to changes in the intensity of the arrival times.

3 Illustration

3.1 Data and transformations

In this section we illustrate the method estimating the model for two duration processes—price and volume—pertaining to two different stocks traded at NYSE. A price duration is the minimum time interval required to witness a cumulative price change greater than a certain threshold. We define the threshold as \$0.125. Following Engle and Russell (1998), the price is defined as the mid-quote. A volume duration is defined as the time interval needed to observe an accumulated change in volume greater than a given number of shares. The threshold is defined to be 25,000 shares. The data sample covers September, October, and November 1996, and the stocks are Boeing for price durations and Disney for volume durations. The choice of these two stocks and its time horizon is motivated by the fact that Bauwens et al. (2004) show, in a density forecast analysis, that they are difficult to model, both in terms of expectation and density.

Price and volume durations provide an instantaneous overview of two important market features. Price durations are strongly linked to the instantaneous volatility process, see Engle and Russell (1998) and Gerhard and Hautsch (2002). By computing the conditional expectation of the duration, it is possible to compute the instantaneous volatility process. Volume durations are appealing as they convey information about two of the three dimensions of the liquidity: time and volume. Therefore, price and volume durations provide measures of the instantaneous volatility and the liquidity of the market. Prior to estimation we follow Engle and Russell (1998) and two transformations are performed: (1) Trades and bid/ask quotes recorded before 09:30 am and after 4 pm are ignored and (2) the time gaps between market closing and opening and weekends are also ignored.

3.2 Descriptive analysis

Table 1 shows some basic statistics. There are 1,778 and 2,160 price and volume durations, respectively. The first two-thirds of the sample are used for estimation (in-sample), while the last third of the sample is used for diagnosis and prediction (out-of-sample). Price and volume durations show different properties. Price durations are overdispersed (the ratio of standard deviation to mean is 1.49) while volume durations are underdispersed (0.81). The proportions of observations below the mode (% $<$ mod) are 11% and 24% for price and volume durations, respectively. These high proportions translate into humps in the density, even very close to the origin, as it is the case of price durations (see Fig. 1). Furthermore, there are large durations, specially for price durations, inducing a long right tail. A hump close to the origin and long right tails are important insights that help to choose the density function.

Regarding seasonality, we estimate a preliminary nonparametric regression, allowing for differences between the days of the week. For the sth day we define

$$\hat{\phi}_s(t_0') = \frac{\frac{1}{nh}\sum_{j=1}^{\lfloor n/5\rfloor}\sum_{i=1}^{n} K\left(\frac{t_0'-t_i'}{h}\right) I\left(\lfloor \frac{t-t_o}{t_c-t_o}+1\rfloor = js\right) d_i}{\frac{1}{nh}\sum_{j=1}^{\lfloor n/5\rfloor}\sum_{i=1}^{n} K\left(\frac{t_0-t_i}{h}\right) I\left(\lfloor \frac{t-t_o}{t_c-t_o}+1\rfloor = js\right)}, \quad s = 1, \cdots, 5,$$

Table 1 Information on duration data

	n	n_{in}	n_{out}	Mean	sd	mode	% $<$ mod	min	max
Price (Boeing)	2,160	1,426	734	647	966	70.6	0.11	3	9,739
Volume (Disney)	1,778	1,173	605	801	648	304	0.24	6	4,621

Durations are measured in seconds. n denotes the total number of observations, n_{in} the in-sample number of observations, n_{out} the out-of-sample number of observations, sd the standard deviation, % $<$mod the proportion of observations smaller than the mode, min and max are the smallest and the largest durations.

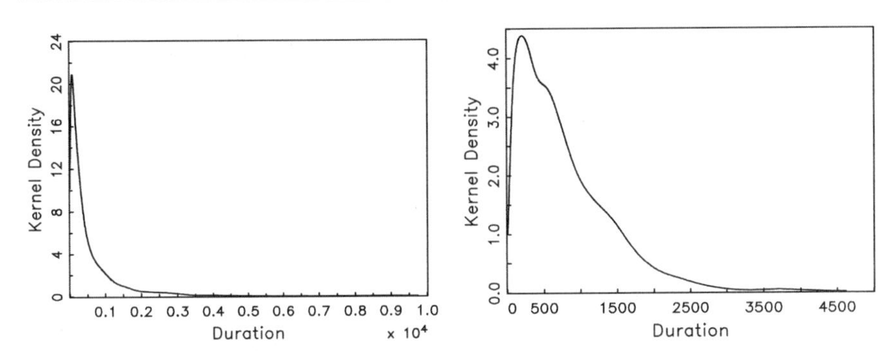

Fig. 1 Kernel marginal densities for Boeing price durations (*left box*), and Disney volume durations (*right box*). The kernel is gamma, see Chen (2000), with optimal bandwidth $(0.9sN^{-0.2})^2$ where N is the number of observations and s is the sample standard deviation.

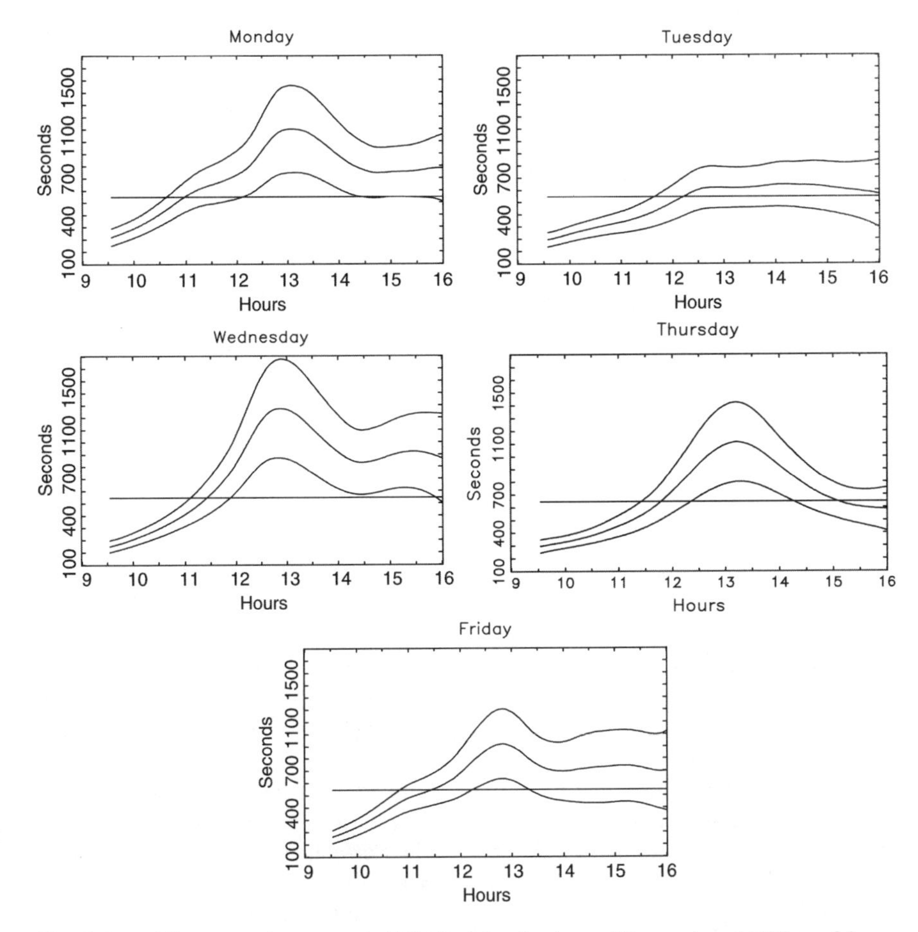

Fig. 2 Intradaily seasonal patterns (*middle line*) for the days of the week and 95% confidence bands (*side lines*) for Boeing price durations. The *straight line* represents the mean of each curve.

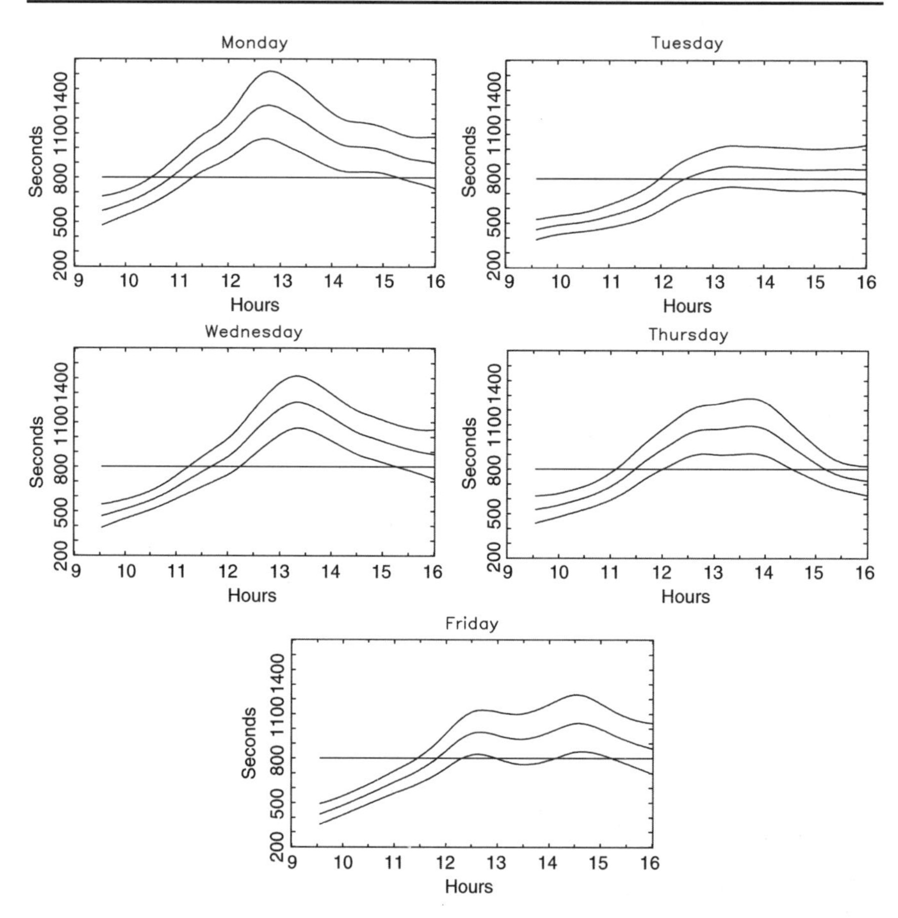

Fig. 3 Intradaily seasonal patterns (*middle line*) for the days of the week and 95% confidence bands (*side lines*) for Disney volume durations. The *straight line* represents the mean of each curve

where $I(\cdot)$ is the indicator function and $\lfloor x \rfloor$ is the integer part of x. We use a quartic kernel with bandwidth $2.78sn^{-1/5}$, where s is the sample standard deviation.

Figures 2 and 3 show the intradaily seasonality for each day of the week. They also include pointwise confidence bands (see Bosq (1998), Theorem 3.1, p. 70) for the different curves and, for comparison purposes, a straight line representing the mean of each curve. Although a more formal test is needed, the hypothesis of differences between the seasonal behavior over the days of the week is not supported. Confidence bands increase through the day: Early in the morning, the bands are tighter than near the closing. This indicates that the variance of the durations evolves through the day. Indeed, Fig. 4 presents rolling means and standard deviations, in intervals of half an hour, through the day. Each point represents the mean and standard deviation over half an hour. They show an inverted U shape, i.e., as the day goes on, the variance increases (with the exception of the half hour 13:00–13:30), widening the confidence bands with a slight tightness near the closing. Incidentally, notice that for price durations the standard deviation is above the mean while it is the opposite for volume durations. This is due to the over and under dispersion, respectively.

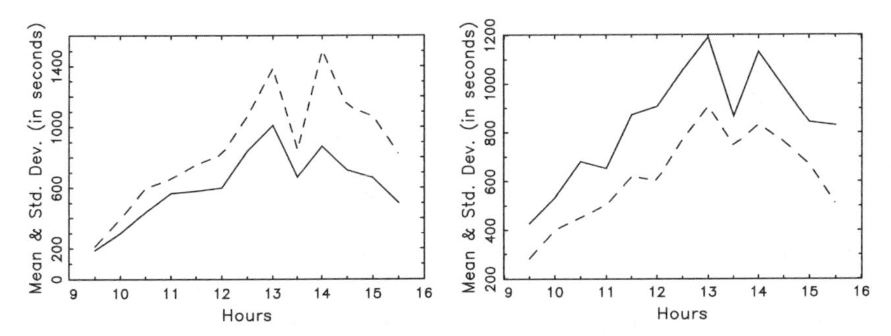

Fig. 4 Intradaily rolling means (*solid lines*) and standard deviations (*dashed lines*) for half hour intervals. Boeing price durations in the *left* box and Disney volume durations in the *right* box

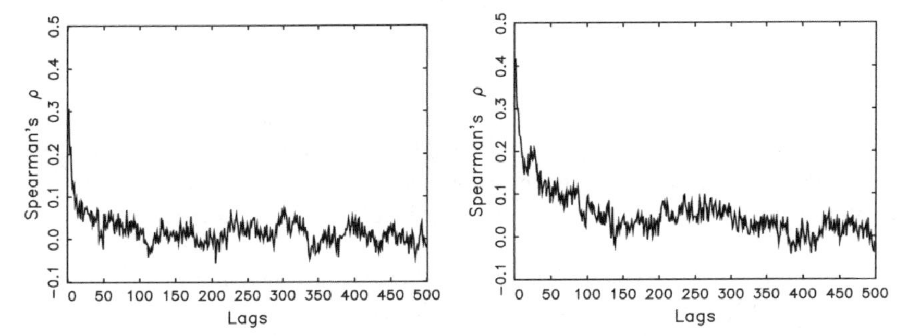

Fig. 5 Spearman's ρ coefficients for serial dependence for price durations (*left box*), and volume durations (*right box*)

Finally, Fig. 5 contains the Spearman's ρ coefficients of serial dependence. They indicate the presence of dependencies, justifying the dynamic component in the model.

3.3 Estimation results

Prior to estimation, we need to specify $\varphi(u, v)$, the lag orders, and the form of $g(\cdot)$ in (4), the conditional density of the error term and the nonparametric estimator $\phi(t^{'}_{i-1})$. We opt for a Log-ACD(1,1), which has been successfully used in the literature. As for the density, we chose a generalized gamma as it is able to reproduce the features highlighted earlier. We also provide the quasi maximum likelihood estimates using the exponential density.

We estimate $\phi(t^{'}_{i-1})$ in four different ways. Two of them are joint estimators, in the sense that estimation is performed jointly with the parameters. One is Eq. (13) that we denote by UniNW—standing for one step Nadaraya–Watson. The other is the polynomial spline used by Engle and Russell (1995 and 1997) that we denote by UniSp—standing for one step spline. We do not use their parametrization but

$$E\left[d_i|\bar{d}_{i-1}, t'_{i-1}\right] = \psi_i = \exp\left\{\omega + \alpha \ln \frac{d_{i-1}}{\phi_{\theta,\delta}(t'_{i-1})} + \beta\psi_{i-1}\right\},$$

$$\phi_{\theta,\delta}(t'_{i-1}) = \sum_{j=1}^{4} \theta_j t'^{\,j-1}_{i-1} + \sum_{g=1}^{G} \delta_g \left(t'_{i-1} - \pi_g\right)^3_+, \tag{22}$$

where $(\cdot)_+ = \max(\cdot, 0)$, G is the number of knots, and π_g is the gth knot. θ_j, $j = 1, \cdots, 4$, and δ_g, $g = 1, \cdots, G$, are the parameters of the spline to be estimated. Further details on the derivation of this form can be found in Eubank (1988, p 354). Knots are set at every hour with an additional knot in the last half hour, as in Engle and Russell (1998).

The two remaining estimators are two steps estimators, in the sense that they are not estimated jointly with the finite dimensional parameters but independently: First the seasonal curve is estimated, then durations are adjusted by seasonality and finally the parameters are estimated using the deseasonalized durations. This procedure is often followed in the literature (see, for instance, Engle and Russell (1998) and Bauwens and Giot (2000)). The first of these two estimators is the plain Nadaraya–Watson estimator

$$\hat{\phi}(t_0') = \frac{\frac{1}{nh}\sum_{i=1}^n K\left(\frac{t_0'-t_i'}{h}\right) d_i}{\frac{1}{nh}\sum_{i=1}^n K\left(\frac{t_0'-t_i'}{h}\right)}, \tag{23}$$

which we denote by BiNW—standing for two steps and Nadaraya–Watson. The second two-step estimator is the one used by Engle and Russell (1998); it is the seasonal component that is estimated by averaging the durations over 30 min intervals, and smoothing the resulting piece-wise constant function via cubic splines. We denote this estimator by BiSp.

To construct pointwise confidence bands for the seasonal curve in UniNW, we need to estimate the empirical counterparts of Eqs. (16) and (17):

$$\widehat{\phi}_{\widehat{\vartheta}_1,\widehat{\xi}}(t_0') \pm z_{1-\frac{\alpha}{2}} \sqrt{\frac{\|K\|_2^2}{\widehat{f}(t_0')\widehat{I}(\widehat{\phi}_{\widehat{\vartheta}_1,\widehat{\xi}}, t_0')}}, \tag{24}$$

where

$$\widehat{I}(\widehat{\phi}_{\widehat{\vartheta}_1,\widehat{\xi}}, t_0') = \frac{\frac{1}{nh}\sum_{i=1}^n K\left(\frac{t_0'-t_i'}{h}\right)\left[\widehat{\gamma}\left(\left(\frac{d_i}{\exp\left\{\psi(\widehat{\vartheta}_1)+\widehat{\phi}_{\widehat{\vartheta}_1,\widehat{\xi}}(t_0')\right\}}\right)^{\widehat{\gamma}} - \widehat{\nu}\right)\right]^2}{\frac{1}{nh}\sum_{i=1}^n K\left(\frac{t_0'-t_i'}{h}\right)},$$

$$\widehat{f}(t_0') = \frac{1}{nh}\sum_{i=1}^n K\left(\frac{t_0'-t_i'}{h}\right),$$

$z_{1-\frac{\alpha}{2}}$ is the $\frac{\alpha}{2}$-quantile of the standard normal distribution, and $\|K\|_2^2$ is a known constant that depends on the kernel. For the quartic kernel, $\|K\|_2^2 = \frac{5}{7}$. We can also compute consistent estimators of the variance–covariance matrix of the parameters. For the other three cases, no results in this direction are available and, therefore, the standard errors we present for these three cases have unknown properties. Nonetheless, we present them for the sake of comparison.

Table 2 presents the estimation results of the four specifications for price and volume durations. Comparing UniNW with the other specifications we conclude: The estimated parameters of the dynamic component are very similar under the exponential and the generalized gamma distribution but the standard deviations are smaller under the generalized gamma distribution. This supports the theory of ML and GLM in the sense that the

Table 2 Estimation results

	Boeing (price)				Disney (volume)			
	BiNW	BiSp	UniNW	UniSp	BiNW	BiSp	UniNW	UniSp
$\widehat{\omega}$	0.0954	0.2375	–	0.5867	0.0309	0.0894	–	1.5461
	[0.0171]	[0.0457]		[0.0738]	[0.0103]	[0.0233]		[0.0490]
$\widehat{\alpha}$	0.1436	0.2418	0.1598	0.1657	0.1054	0.1799	0.0875	0.1226
	[0.0185]	[0.0258]	[0.0203]	[0.0187]	[0.0279]	[0.0341]	[0.0260]	[0.0931]
$\widehat{\beta}$	0.7591	0.5290	0.6309	0.5144	0.8255	0.6471	0.8875	0.6886
	[0.0735]	[0.0942]	[0.0581]	[0.0901]	[0.0543]	[0.0829]	[0.0451]	[0.0710]
$\widehat{\alpha}+\widehat{\beta}$	0.9027	0.7708	0.7907	0.6801	0.9309	0.8270	0.9750	0.8112
	[0.0820]	[0.1183]	[0.0801]	[0.0997]	[0.0824]	[0.1154]	[0.0711]	[0.1638]
$\widehat{\omega}$	0.0501	0.2115	–	1.4121	0.0306	0.0896	–	1.5361
	[0.0135]	[0.0211]		[0.0654]	[0.0069]	[0.0163]		[0.0244]
$\widehat{\alpha}$	0.0611	0.1596	0.1761	0.1401	0.1003	0.1799	0.0875	0.1216
	[0.0185]	[0.0175]	[0.0128]	[0.0141]	[0.0183]	[0.0237]	[0.0185]	[0.0910]
$\widehat{\beta}$	0.8039	0.5672	0.6834	0.5899	0.8236	0.6459	0.8798	0.7786
	[0.0417]	[0.0404]	[0.0297]	[0.0494]	[0.0388]	[0.0583]	[0.0379]	[0.0545]
$\widehat{\gamma}$	0.2693	0.2900	0.2527	0.0675	1.4356	1.0180	1.3959	1.0653
	[0.0605]	[0.0579]	[0.0518]	[0.0401]	[0.1328]	[0.0972]	[0.1328]	[0.2376]
$\widehat{\nu}$	10.807	8.7329	10.332	17.852	1.2468	1.9952	1.2774	1.7702
	[3.7263]	[3.3737]	[3.1208]	[2.4329]	[0.1912]	[0.3373]	[0.2020]	[0.2453]
$\widehat{\alpha}+\widehat{\beta}$	0.8650	0.7268	0.8595	0.7300	0.9239	0.8258	0.9673	0.9002
	[0.0611]	[0.0590]	[0.0415]	[0.0538]	[0.0561]	[0.0818]	[0.0549]	[0.1459]

Entries are GLM estimates—using the exponential distribution—(top part of the table) and ML estimates—using a generalized gamma distribution—(bottom part) for the Log-ACD. Numbers in brackets are heteroskedastic-consistent standard errors.

exponential density cannot fit the empirical density, which implies inefficient estimates with respect to the ML ones. Second, the estimated parameters under BiSp and UniSp are very similar, confirming the results shown in Engle and Russell (1998, p 1137), who do not find very different results for BiSp and UniSp. By contrast, there are substantial differences between the NW (UniNW and BiNW) and the Sp (UniSp and BiSp) groups, meaning that the parameters and the nonparametric curve are not orthogonal when estimating with kernels. Third, volume durations estimates have, in general, smaller $\widehat{\alpha}$ and bigger $\widehat{\beta}$ than price durations estimates. This is due to the persistence (see Spearman's ρ in Fig. 5) and the underdispersion of volume durations. Note that the constant of the dynamic component is not present in UniNW as it is replaced by the seasonal curve.

As for the nonparametric curves (see Fig. 6), results in terms of smoothness are rather different for UniSp and UniNW. UniSp curves are sharper with small humps and, for volume durations, we also observe a rough increase (decrease) at the beginning (end) of the day. This finding can be explained by the fact that UniNW provides a data-driven method to compute the bandwidth, whereas the UniSp does not (location and number of

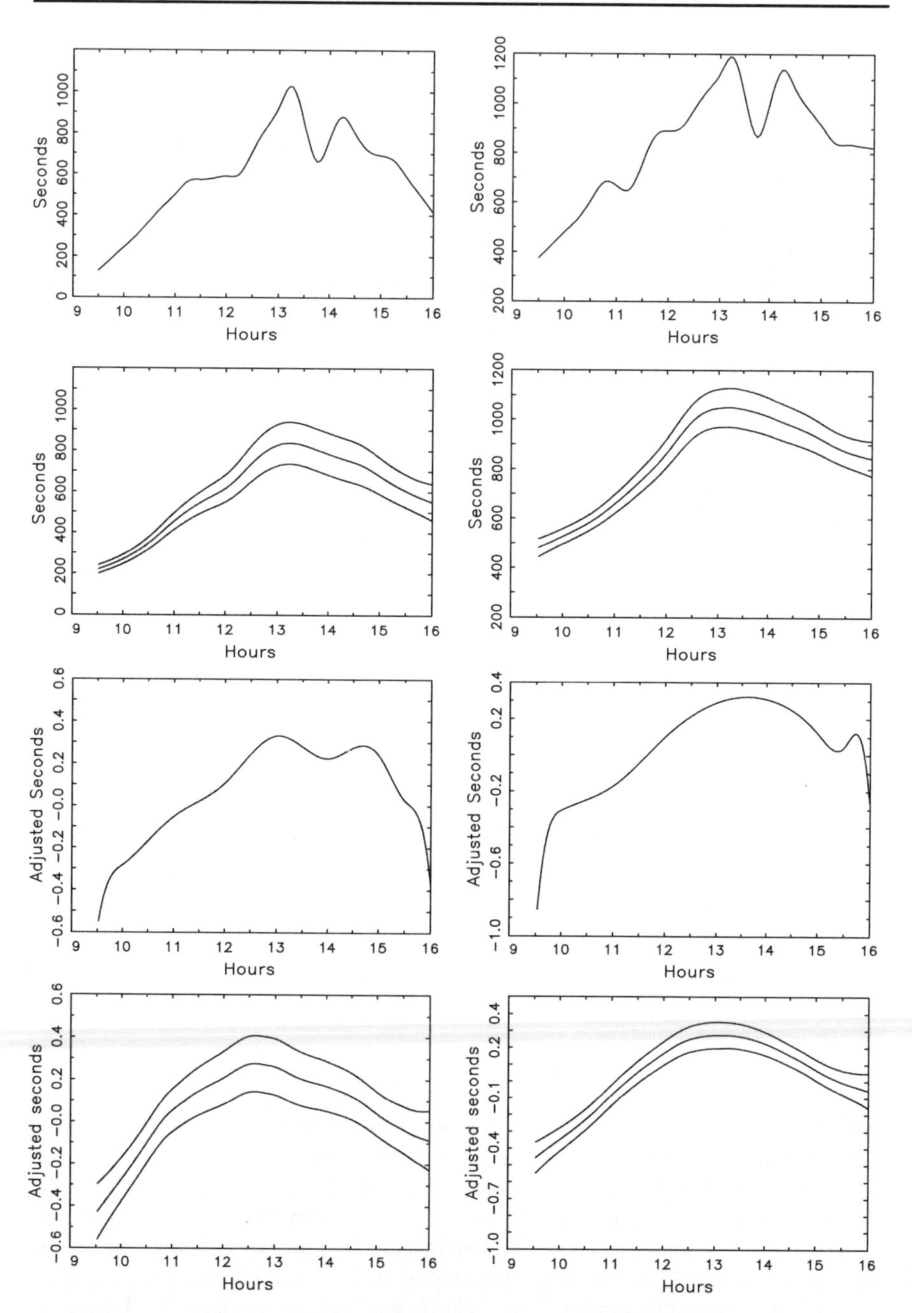

Fig. 6 Estimated seasonal curves. Price durations in *left* column and volume durations in *right* column. Rows from top to bottom: BiSp, BiNW, UniSP, and UniNW

knots are chosen ad hoc). For UniNW and UniSp we show the seasonal curve $\hat{\phi}_{\vartheta_1,\xi}(t_0')$ at a different scale than the other seasonal estimators. It is not possible to present them at the same scale since it is a logarithmic function, see Eq. (13), and the argument of the function is not the observed duration but the duration adjusted by the dynamics.

3.4 Predictibility

So far estimation results dovetail with the theory. We now check how well they predict the probability distribution. Since density forecast is strongly dependent on the distributional assumption, we only show results for ML (the exponential density does it worse than the generalized gamma). Another reason for sticking to ML is because we want to study the predictive capabilities of the model for different estimators of seasonality. Since the density and the parametric component are the same, differences in density forecast may exclusively come from the different specification of the seasonal component.

If $\varepsilon_j \sim GG\left(1, \hat{\gamma}_n, \hat{\nu}_n\right)$ then

$$d_j \sim GG\left(\varphi\left(\psi\left(\bar{d}_{j-1}, \bar{y}_{j-1}; \hat{\vartheta}_{1n}\right), \hat{\phi}_{\hat{\vartheta}_{1n},\hat{\xi}_n}\left(t_{j-1}'\right)\right)^{-1}, \hat{\gamma}_n, \hat{\nu}_n\right),$$

for UniNW and UniSp. For the two-step methods, since $\tilde{d}_j = d_j \hat{\phi}\left(t_{j-1}'\right)^{-1}$, the seasonal curve is outside the function $\varphi(\cdot)$, that is

$$\tilde{d}_j = \frac{d_j}{\hat{\phi}\left(t_{j-1}'\right)} \sim GG\left(\varphi\left(\psi\left(\bar{d}_{j-1}, \bar{y}_{j-1}; \hat{\vartheta}_{1n}\right)\right)^{-1}, \hat{\gamma}_n, \hat{\nu}_n\right)$$

or

$$d_j \sim GG\left(\varphi\left(\psi\left(\bar{d}_{j-1}, \bar{y}_{j-1}; \hat{\vartheta}_{1n}\right)\right)^{-1} \hat{\phi}\left(t_{j-1}'\right)^{-1}, \hat{\gamma}_n, \hat{\nu}_n\right).$$

Figures 7 and 8 show the density forecast results. The histograms for UniNW are significantly better than for any other estimator. Although formal tests can be performed to assess the accuracy of this statement, visual inspection of the figures reveals that estimation in two steps (BiSp and BiNW) results in misspecification in the conditional distribution of the durations—in line with Bauwens et al. (2004). Moreover, estimation in one step with splines, UniSp, leads to similar results as the two step estimators. This fits with the conclusions on estimation results that found similar estimates under BiSp and UniSp. Second, for volume durations none of the specifications correctly captures the dynamic component. This is expected as the dynamics are modeled in the same way for all specifications and Bauwens et al. (2004) already show that a simple Log-ACD(1,1) is unable to fit the dynamics.

The better performance of UniNW is a result of its flexibility, as already mentioned. UniNW adapts to changes in the seasonal pattern. We can interpret the seasonal estimator UniNW as a time-varying intercept that adapts the conditional expectation of the durations to changes in seasonality. This is because UniNW is a pure nonparametric estimator and follows the principle of *let the data speak*. This is not the case of the polynomial spline (22). In fact, Eq. (22) not only depends implicitly on $\hat{\vartheta}_{1n}$ and $\hat{\xi}_n$ but also depends on $\hat{\theta}_n$ and $\hat{\delta}_n$. And these parameters are estimated using the in-sample to fit the in-sample

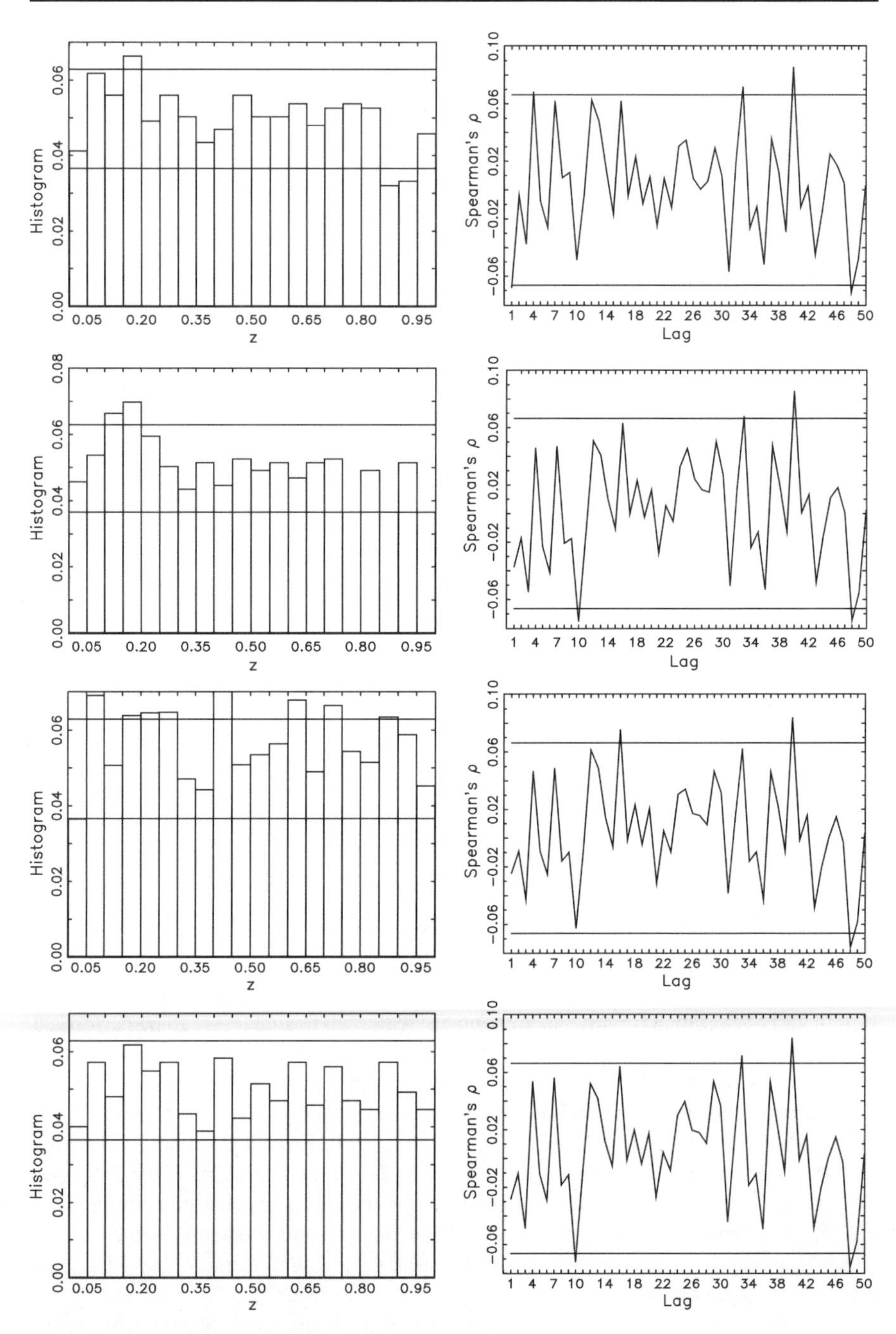

Fig. 7 Out-of-sample density forecast evaluation for price durations. Rows from top to bottom: BiSP, BiNW, UniSp, and UniNW

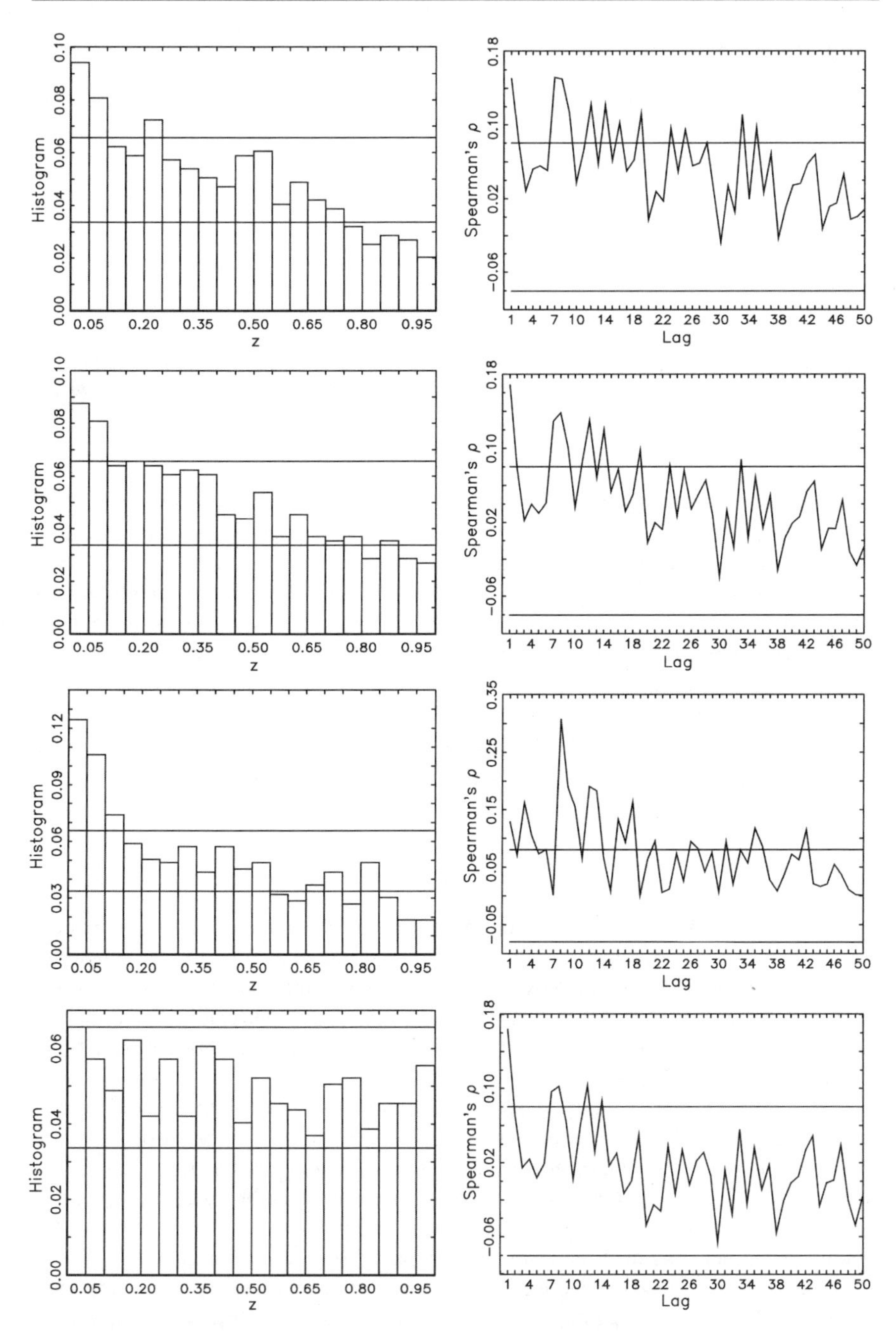

Fig. 8 Out-of-sample density forecast evaluation for volume durations. Rows from top to bottom: BiSP, BiNW, UniSp, and UniNW

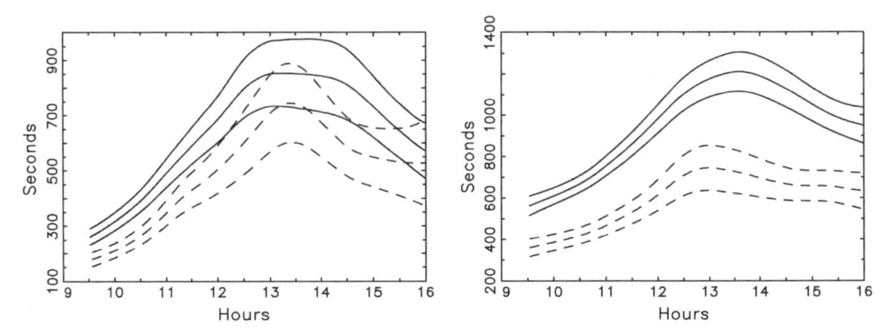

Fig. 9 Intradaily seasonal curves, and confidence bands, for in-sample (*solid lines*) and out-of-sample (*dashed lines*). Price durations in the *left* box and volume durations in the *right* box

seasonality. Hence the spline will fail to fit the out-of-sample seasonality if its pattern has changed. This is not the case of UniNW as it does not depend on parameters, other than $\hat{\vartheta}_{1n}$ and $\hat{\xi}_n$, to capture the seasonality.

The empirical evidence supports this intuition. Figure 9 shows the in-sample and out-of-sample intradaily seasonalities. We observe that the out-of-sample seasonal patterns shift down compared with the in-sample patterns. This is especially visible for volume durations. If we use BiSp, BiNW, or UniSp to evaluate the out-of-sample density forecast, they all fail because none may capture changes in seasonality. By contrast, UniNW adapts quickly to changes in the out-of-sample and this explains why the density forecast produced by UniNW is better than for any other estimator.

4 Conclusions

We propose a component model for the analysis of financial durations. The components are dynamics and seasonality. The latter is left unspecified and the former is assumed to fall within the class of (Log-)ACD models. Joint estimation of the parameters of interest and the smooth curve is performed through a local (quasi-)likelihood method. The resulting nonparametric estimator of the seasonal component shows a closed form expression.

Although the methodology is applied to intradaily seasonality, it could also be used to measure relations between any other two variables. This is particularly useful when the relation is nonlinear and we are not sure about the functional link. So, for instance, the time-of-the-day may be replaced by some volatility measure, spread, or any other microstructure variables. Alternatively, the nonparametric estimator could be multivariate, including the time-of-the-day and any other variable of interest.

The model is applied to the price and volume duration processes of two stocks traded on the NYSE. We show that the proposed method produces better predictions, in terms of densities, since it adapts quickly to changes in the seasonal pattern.

Acknowledgements The authors acknowledge financial support from the Université catholique de Louvain (project 11000131), the Institute of Statistics at the Université Catholique de Louvain, the Dirección General de Investigación del Ministerio de Ciencia y Tecnología under research grants BEC2001-1121 and SEJ2005-05549/ECON as well as the Spanish Ministry of Education (project PB98-0140), respectively. Work supported in part by the European Community's Human

Potential Programme under contract HPRN-CT-2002-00232, [MICFINMA]. Thanks are due also to Luc Bauwens, Joachim Grammig, M. Paz Moral, Wendelin Schnedler, Laura Spierdijk, Jorge Yzaguirre for useful remarks, and to the seminar participants at CEMFI, University Carlos III de Madrid, the PAI Conference in Financial Econometrics in Leuven, and at the CAF conference on Market Microstructure and High-Frequency data in Finance in Aarhaus. Last, we are also grateful to the co-editor Winfried Pohlmeier, a referee for insightful remarks.

Appendix

A. Definitions and assumptions

In order to prove the results claimed in Theorems 1 and 2, we need to establish some definitions and assumptions.

(L.1) For fixed but arbitrary η_1, ϕ^+, where $\eta \in \Theta$, and $\phi^+ \in \Phi$, let

$$\rho\left(\phi^+, \eta\right) = \int \log p\left(\varphi(\psi, \phi^+); \xi\right) dF\left(\varphi(\psi, \phi^+); \xi\right).$$

If $\eta \neq \eta_1$, then

$$\rho\left(\phi, \eta\right) < \rho\left(\phi^+, \eta\right)$$

Let $I_\eta\left(\phi, \eta\right)$, denote the marginal Fisher information for η in the parametric model. Then we assume that the matrix $I_\eta\left(\phi, \eta\right)$ is positive definite for all $\eta \in \Theta$ and $\phi \in \Phi$.

(L.2) Assume that for all $r, s = 0, \ldots, 4$, $r + s \leq 4$, the derivative

$$\frac{\partial^{r+s}}{\partial \eta^r \partial \eta^s} \log p\left(\varphi(\psi, \phi^+); \xi\right)$$

exists for almost all d and that

$$E\left\{\sup_\eta \sup_\phi \left|\frac{\partial^{r+s}}{\partial \eta^r \partial \eta^s} \log p\left(\varphi(\psi, \phi^+); \xi\right)\right|^2\right\} < \infty.$$

(A.1) The marks y take values in a compact set $\mathbf{Y} \subset R^p$.

(A.2) The observations $\{(d_i, y_i)\}_{i=1,\ldots,n}$ are a sequence of stationary and ergodic random vectors.

(A.3) ϑ_1 takes the values in the interior of Θ, a compact subset in R^p and ϕ takes the values in the interior of Φ, a compact subset of R:

$$\Phi = \left\{g \in C^2[t_o, t_c] : g(t') \in \text{int}\left(\Phi\right) \quad \text{for } \forall t' \in [t_o, t_c]\right\}.$$

(A.4) Let Ξ be a compact subset of R such that $\varphi\left(\psi(\bar{d}, \bar{y}; \vartheta_1), \phi(t')\right) \in \Xi$ for all $t' \in [t_o, t_c]$, $y \in \mathrm{Y}$, $\vartheta_1 \in \Theta$, and $\phi \in \Phi$.

(A.5) The matrix

$$\Sigma_{\vartheta_1} = E\left(\frac{\partial^2}{\partial \vartheta_1 \partial \vartheta_1^T} \log q\left(d, \varphi\left(\psi(\bar{d}, \bar{y}; \vartheta_1), \phi(t')\right)\right)\right)$$

is positive definite.

(B.1) The kernel function $K(\cdot)$ is of order $k > 3/2$ with support $[-1, 1]$ and it has bounded $k+2$ derivatives.
(B.2) For $r = 1, \ldots, 10+k$ the functions $\partial^r \varphi(m)/\partial m^r$ and $\partial^r V(\mu)/\partial \mu^r$ exist and they are bounded in their respective supports.
(B.3) d is a strong mixing process where the mixing coefficients must satisfy for some $p > 2$ and r being a positive integer

$$\sum_{i=1}^{\infty} i^{r-1} \alpha(i)^{1-2/p} < \infty.$$

Furthermore, for some even integer q satisfying $\frac{(k+2)(3+2k)}{(2k-3)} \leq q \leq 2r$

$$E\left\{|d|^q\right\} < \nu,$$

where ν is a constant not depending on t'.
(B.4) Let f denote the marginal density of t', and let $f(\cdot|t')$ denote the conditional density function of d given t'. f and $f(\cdot|t')$ has $k+2$ bounded derivatives uniformly in $[t_o, t_c]$.
(B.5) Let

$$M\left(\phi; \vartheta_1, t'\right) = E\left\{\frac{\partial}{\partial \eta} \log q\left(d, \varphi\left(\psi(\bar{d}, \bar{y}; \vartheta_1), \phi\right)\right)\bigg| t'\right\}.$$

For each fixed ϑ_1 and t', let $\phi_{\vartheta_1}(t')$ the unique solution to $M\left(\phi; \vartheta_1, t'\right) = 0$. Then for any $\epsilon > 0$ there exists a $\delta > 0$ such that

$$\sup_{\vartheta_1 \in \Theta} \sup_{t' \in [t_I, t_F]} \left|\phi_{\vartheta_1}(t') - \phi(t')\right| < \epsilon$$

whenever

$$\sup_{\vartheta_1 \in \Theta} \sup_{t' \in [t_I, t_F]} \left|M\left(\phi(t'); \vartheta_1, t\right)\right| < \delta.$$

(B.6) The sequence of bandwidths must satisfy $h = O(n^{-\alpha})$ where

$$\frac{1}{4k} < \alpha < \frac{1}{4} \frac{q - (2+p)}{q + (2+p)}.$$

The following result is also needed to prove Theorems 1 and 2,

Lemma A.1: Consider the following expression

$$\frac{1}{nh} \sum_{i=1}^{n} \left[K\left(\frac{\tau - t_i}{h}\right) \varphi\left(\psi(\bar{d}, \bar{y}; \vartheta_1), \phi\right) - E\left\{K\left(\frac{\tau - t}{h}\right) \varphi\left(\psi(\bar{d}, \bar{y}; \vartheta_1), \phi\right)\right\}\right]$$

and define

$$W_i = \frac{1}{h} K\left(\frac{\tau - t_i}{h}\right) \varphi\left(\psi(\bar{d}, \bar{y}; \vartheta_1), \phi\right) - E\left\{K\left(\frac{\tau - t}{h}\right) \varphi\left(\psi(\bar{d}, \bar{y}; \vartheta_1), \phi\right)\right\}$$

Then, for $\epsilon > 0$

$$P\left\{\left|\frac{1}{n}\sum W_j\right| > \epsilon\right\} \leq \frac{E\left[(\sum W_i)^q\right]}{n^q \epsilon^q}$$

$$\leq \frac{1}{n^q \epsilon^q} C \left\{ n^{q/2} \sum_{i=P}^{\infty} i^{q/2-1} \alpha(i)^{1-2/p} + \sum_{j=1}^{q/2} n^j P^{q-j} \nu^j \right\}$$

for any integers n and P with $0 < P < n$.

B. Proof of Lemma A.1

Under assumptions (A.2) and (B.3) the process $W_1, \ldots, W_n$ is strong mixing and therefore Theorem 1 from Cox and Kim (1995) applies and the sequence of inequalities holds.

C. Proof of Theorem 1

The proof is in two parts. First we show Eq. (14). In order to do this, we claim

$$\sup_{\vartheta_1} \sup_{t} |\hat{\phi}_{\vartheta_1}(t) - \phi(t)| = o_p\left(n^{-1/4}\right) \tag{25}$$

The proof of this expression follows the same steps as the proof of Lemma 5 from Severini and Wong (1992), p. 1784. The bias term must be treated in the same way as it is there. With respect to the variance term, an additional result must be included to account for the dependence. In fact, under assumptions (A.2) and (B.3), Lemma A.1 applies and, proceeding as for Severini and Wong (1992) in the proof of Lemma 8, the proof of Eq. (25) is completed.

Finally, the proof of Eq. (14) consists of verifying conditions I (Identification), S (Smoothness), and NP (Nuissance Parameter) from Severini and Wong (1992). Condition NP(a) is the result already shown in Eq. (25). Condition NP(b) (least favorable curve) is immediate from Lemma 6 of Severini and Wong (1992). By assuming (L.2) the smoothness condition holds. Finally, assumption (L.1) implies I. Then, under assumptions (A.1) to (A.3), a Uniform Weak Law of Large Numbers and a Central Limit Theorem for stationary and ergodic processes (see for example Wooldridge (1994)) holds, propositions 1 and 2 from Severini and Wong (1992) apply and the proof is completed.

The proof of Eq. (15) consists of taking a Taylor expansion around (ϑ_1, ξ). This can be done because of assumption (L.2). Then apply result (25), and the rest of the proof follows the same lines as the proof of Theorem 1 in Staniswalis (1989). Note that instead of using a standard Liapunov CLT here under assumption (A.2), we need to use one for for stationary and ergodic processes.

D. Proof of Theorem 2

The proof of Eq. (20) follows exactly the same lines as the proof of Eq. (14) in Theorem 1, but instead of using (L.1) for identification and (L.2) for smoothness, we use respectively

(A.5) and (A.1) to (A.4). Condition NP(b) (least favorable curve) is immediate from Lemma 6 of Severini and Wong (1992). This is due to the fact that we assume that the conditional density function belongs to the exponential family.

The proof of Eq. (21) is equal to the proof of Eq. (15) except for that (L.2) is replaced by (B.2).

References

Andersen T, Bollerslev T (1997) Heterogenous information arrivals and return volatility dynamics: unrecovering the long-run in high frequency returns. J Finance 52:975–1005
Andersen T, Bollerslev T (1998) Deustche mark-dollar volatility: intraday activity patterns, macroeconomic announcements, and longer run dependencies. J Finance 53:219–265
Baillie R, Bollerslev T (1990) Intra-day and inter-market volatility in foreign exchange rates. Rev Econ Stud 58:565–585
Bauwens L, Giot P (2000) The logarithmic ACD model: an application to the bid-ask quote process of three NYSE stocks. Annales d'Economie et de Statistique 60:117–149
Bauwens L, Giot P, Grammig J, Veredas D (2004) A comparison of financial duration models via density forecast. Int J Forecast 20:589–604
Bauwens L, Veredas D (2004) The stochastic conditional duration model: a latent factor model for the analysis of financial durations. J Econ 119(2):381–412
Beltratti A, Morana C (1999) Computing value at risk with high frequency data. J Empir Finance 6:431–455
Birgé L, Massart P (1993) Rate of convergence for minimum contrast estimators. Probab Theory Relat Field 97:113–150
Bollerslev T, Domowitz I (1993) Trading patterns and prices in the interbank foreign exchange market. J Finance 48:1421–1443
Bosq D (1998) Nonparametric statistics for stochastic processes. Estimation and prediction, 2nd edition. Lecture Notes in Statistics, 110. Springer-Verlag, New York
Chen SX (2000) A beta kernel estimation for density functions. Comput Stat Data Anal 31:131–145
Chen SX, Shen SX (1998) Sieve extremum estimates for weakly dependent data. Econometrica 66:289–314
Cox DR, Kim TY (1995) Moment bounds for mixing random variables useful in nonparametric function estimation. Stochastic Processes and their Applications 56:151–158
Diebold FX, Gunther TA, Tay AS (1998) Evaluating density forecasts, with applications to financial risk management. Int Econ Rev 39:863–883
Drost FC, Werker BJM (2004) Efficient estimation in semiparametric time series: the ACD model. J Bus Econ Stat 22:40–50
Engle RF, Russell JR (1995) Autoregressive conditional duration: a new model for irregularly spaced data. unpublished manuscript , University of California, San Diego
Engle RF, Russell JR (1997) Forecasting the frequency of changes in quoted foreign exchange prices with the ACD model. J Empir Finance 4:187–212
Engle RF, Russell JR (1998) Autoregressive conditional duration: a new approach for irregularly spaced transaction data. Econometrica 66:1127–1162
Engle RF (2000) The econometrics of ultra high frequency data. Econometrica 68:1–22
Eubank RL (1988) Spline smoothing and nonparametric regression. Marcell Decker, New York
Fan J, Heckman NE, Wand MP (1995) Local polynomial kernel regression for generalized linear models and quasi-likelihood functions. J Am Stat Assoc 90:141–150
Fernandes M, Grammig J (2001) A family of autoregressive conditional duration models. CORE DP 2001/36, Université catholique de Louvain. Forthcoming J Econ
Gerhard F, Haustch N (2002) Volatility estimation on the basis of price intensities. J Empir Finance 9:57–89
Ghysels E, Gouriéroux C, Jasiak J (1997) Stochastic volatility duration models. Working paper 9746. J Econ 119(2):413–435
Gouriéroux C, Monfort A, Trognon A (1984) Pseudo maximum likelihood methods: theory. Econometrica 52:681–700
Grammig J, Maurer KO (2000) Non-monotonic hazard functions and the autoregressive conditional duration model. Econ J 3:16–38

Grenander U (1981) Abstract inference. Wiley, New York
Kiefer J, Wolfowitz J (1956) Consistency of the maximum likelihood estimator in the presence of infinetely many incidental parameters. Ann Math Stat 28:887–906
McCullagh P, Nelder JA (1983) Generalized linear models. Chapman and Hall, London
Meitz M, Terasvirta T (2004) Evaluating models of autoregressive conditional duration. Working Paper Series in Economics and Finance 557, Stockholm School of Economics. Forthcoming in J Bus Econ Stat
Severini TA, Staniswalis JG (1994) Quasi-likelihood estimation in semiparametric models. J Am Stat Assoc 89:501–511
Severini TA, Wong WH (1992) Profile likelihood and conditionally parametric models. Ann Stat 20:1768–1802
Shen X, Wong WH (1994) Convergence rates for sieve estimates. Ann Stat 22:580–615
Spierdijk L, Nijman TE, van Soest AHO (2004) Price dynamics and trading volume: a semiparametric approach. TW Memorandum, University of Twente
Staniswalis JG (1989) On the kernel estimate of a regression function in likelihood based models. J Am Stat Assoc 84:276–283
Wooldridge JM (1994) Estimation and inference for dependent processes. In: Engle RF, McFadden DL (eds) Handbook of econometrics, vol. 4. Elsevier Science, New York
Zhang MY, Russell JR, Tsay RT (2001) A nonlinear autoregressive conditional duration model with applications to financial transaction data. J Econ 104:179–207

Anthony S. Tay · Christopher Ting

Intraday stock prices, volume, and duration: a nonparametric conditional density analysis

Abstract We investigate the distribution of high-frequency price changes, conditional on trading volume and duration between trades, on four stocks traded on the New York Stock Exchange. The conditional probabilities are estimated nonparametrically using local polynomial regression methods. We find substantial skewness in the distribution of price changes, with the direction of skewness dependent on the sign of trade. We also find that the probability of larger price changes increases with volume, but only for trades that occur with longer durations. The distribution of price changes vary with duration primarily when volume is high.

1 Introduction

Time—in the form of the duration between trades—matters in the formation of stock prices. This has been demonstrated from both theoretical and empirical perspectives. Durations may be negatively related to prices because short-selling constraints prevent trading on private bad news whereas there are no similar constraints to prevent trading on private good news (Diamond and Verrecchia 1987). Durations may also be negatively related to volatility of price changes (Easley and O'Hara 1992). The connection between price change and duration has been verified empirically (Engle 2000; Grammig and Wellner 2002). Further investigations into the relationship between durations and prices have yielded interesting insights. For instance, the size and speed of price movements increase with decreasing duration (Dufour and Engle 2000), reflecting a link between duration and market liquidity. There is also an interesting dynamic relationship

A. S. Tay (✉)
School of Economics and Social Sciences, Singapore Management University,
90 Stamford Road, Singapore, 178903, Singapore
E-mail: anthonytay@smu.edu.sg

C. Ting
Lee Kong Chian School of Business, Singapore Management University, Singapore, Singapore

between price changes and duration (Russell and Engle 2004) and within duration itself (Engle and Russell 1998). Grammig and Wellner (2002) explore the interdependence of transaction intensity and volatility, and find that lagged volatility has significant negative effect on volatility in the secondary equity market following a initial public offering. The economic interpretation of the (contemporaneous) role of duration in the distribution, particularly the volatility, of price changes is the subject of a study by Renault and Werker (2004) where, using a structural model, the effect of duration on price is decomposed into a temporal effect and an informational effect. This enables them to disentangle the effects of instantaneous causality from Granger causality in the relationship between duration to prices.

As in Renault and Werker (2004) we are interested in the contemporaneous relationship between duration and price changes. We report estimates of the entire distribution of price change conditional on duration, and investigate the role of volume and trade sign on these distributions. We explore these relationships from a fully nonparametric perspective. In particular, we estimate nonparametrically the probabilities of various price changes conditional on duration, volume, lagged values of these three variables, and trade sign.

As the conditional distribution contains all probabilistic information regarding the statistical behavior of a variable given values of a set of explanatory variables, analyzing estimates of conditional distributions may reveal interesting and useful structure in the data, and provide insights that would complement studies that focus on the conditional mean or variance. The purpose for adopting a nonparametric approach to estimating the conditional distribution is to allow the data to speak for itself. There are numerous applications in the literature that highlight the value of incorporating nonparametric density estimation into an analysis. For example, Gallant et al. (1992) study the bivariate distribution of daily returns and volume conditional on lagged values of these variables. Among other findings, their analysis indicates a positive correlation between risk and return after conditioning on lagged volume. Other studies that highlight the usefulness of non-parametric methods include Engle and Gonzalez-Rivera (1991), and Gallant et al. (1991).

We estimate the conditional distributions for four stocks traded on the NYSE using intraday data from TAQ spanning a period of 1 year, from Jan 2, 2002 to Dec 30, 2002. We find substantial skewness in the distribution of price changes, with the direction of skewness dependent on trade sign. On the whole, the relationship between price changes and volume is much weaker than the relationship between price changes and duration, and shows up most clearly at long durations. When durations are long, the probability of large price changes increases with volume.

In the following section we describe the data used in this study, and explain precisely all the adjustments made to the data to get it into a form suitable for analysis. In Section 3, we describe the nonparametric conditional distribution estimation technique used in the paper, and discuss the practical issues we had to address in order to the implement the technique. The results of our study are presented in Section 4, followed by concluding comments.

2 Data and data adjustments

We estimate the conditional distributions of price changes for four NYSE stocks: IBM (International Business Machines), GE (General Electric), BA (Boeing), and

MO (the Altria Group, formerly Philip Morris.) The data are obtained from the TAQ database, and cover the period Jan 2, 2002 to Dec 30, 2002. We extract, for each stock, the time of trade of the tth transaction τ_t, from which we obtain the duration of the tth transaction $d_t=\tau_t-\tau_{t-1}(\tau_0=34{,}200$ s after midnight, i.e., 9:30 am), the transacted price p_t , from which we compute the price change $\Delta p_t=p_t-p_{t-1}$, and volume v_t in lots of 100 shares. Clearly, these are all discrete variables. Each trade is then signed as in Lee and Ready (1991) to indicate if the transaction is buyer (+) or seller (−) initiated, but without modifying the reported times of quotes as developments in the NYSE trading procedures no longer warrant this. There are newly proposed methods (e.g., Vergote 2005) that aim to determine the appropriate adjustments to the times of quotes. We do not apply these techniques, but as the analysis in Vergote (2005) suggests that a delay of 1 or 2 s may be appropriate in the sample period that we work with, we check if adding 2 s to the quote time-stamp affects our results.

We use data from 0930 to 1600 h, deleting all trades that occur outside of these hours. Data from 4 days with unusual market openings and closings are dropped from our sample. These are July 5, September 11, November 29, and December 24, 2002. The first of these dates is an early closing for Independence Day, the second is a late opening, in respect of memorial events commemorating the 1-year anniversary of the attacks on World Trade Center. The latter 2 days are early closings for Thanksgiving and Christmas.

There are several noteworthy characteristics of the data set in our sample period. One is that by the start of this sample period the NYSE had already completed the move to decimal pricing. One benefit of this is that the bid-ask spreads are small so that that the bid-ask bounce is less of an issue for our estimates. Perhaps the more important characteristic of this sample period is that, in general, trading is so active that for each stock in our study there are large numbers of trades with zero duration. These may be trades that occur almost simultaneously, but are recorded as having occurred at the same time because the TAQ database records time of trade to an accuracy of 1 s. Some of these trades may also reflect large trades that are broken up into smaller simultaneous trades. The exact number of such trades for the stocks we analyze are presented in Table 1. The lowest proportion of zero-duration trades is BA at 3.7%. About 5% of the IBM and MO observations have zero duration, and almost 10% of the observations for GE have zero duration. We aggregate in standard fashion all trades that occur with the same time-stamp and consider the aggregate as a single trade. The price of the first trade in the aggregate is taken as the price of the aggregate trade. Signed volume is

Table 1 Number of zero duration trades

	IBM	GE	BA	MO
Total number of observations	923,577	1,292,532	594,186	797,373
Number of zero durations	54,799	121,924	21,925	41,651
	(5.9%)	(9.4%)	(3.7%)	(5.2%)
Number of observations after aggregation	868,778	1,170,608	572,261	755,722

simply aggregated. Even after aggregation we have a large number of observations for each stock, ranging from 572,261 (BA) to 1,170,608 (GE). Finally, note that, unlike many studies that work with intraday stock prices, we do not remove the diurnal patterns that are present in durations. This is a more critical issue if the focus of the study involves the dynamics in duration data, but our focus is on the contemporaneous relationship between durations and price changes, and so we choose to examine the data with as few adjustments as possible. In addition, we do not estimate the distributions at a fine enough grid on durations for the removal of diurnal patterns to affect our results in any important way.

In Fig. 1 we show histograms of price changes, duration, and signed volume for IBM. Also displayed is the histogram for signed duration (duration multiplied by the sign of the trade). The distribution of price change appears to be very symmetric. The distribution of signed durations is symmetric, so the distribution of durations is similar for both buyer- and seller-initiated trades. The histogram for trading volume is not informative except to indicate the presence of a few outliers. However, apart from these the distribution is also fairly symmetric. A more detailed picture can be obtained from Table 2. Here we show every 10th percentile of price change, (unsigned) duration, duration for buyer- and seller-initiated trades, (unsigned) volume, and volume for buyer- and seller-initiated trades. For all stocks the distribution of duration and volume across buyer- and seller-initiated trades are very similar to the overall distribution. In fact there is almost no correlation between the sign of trade and duration. Volume is substantially larger for the more

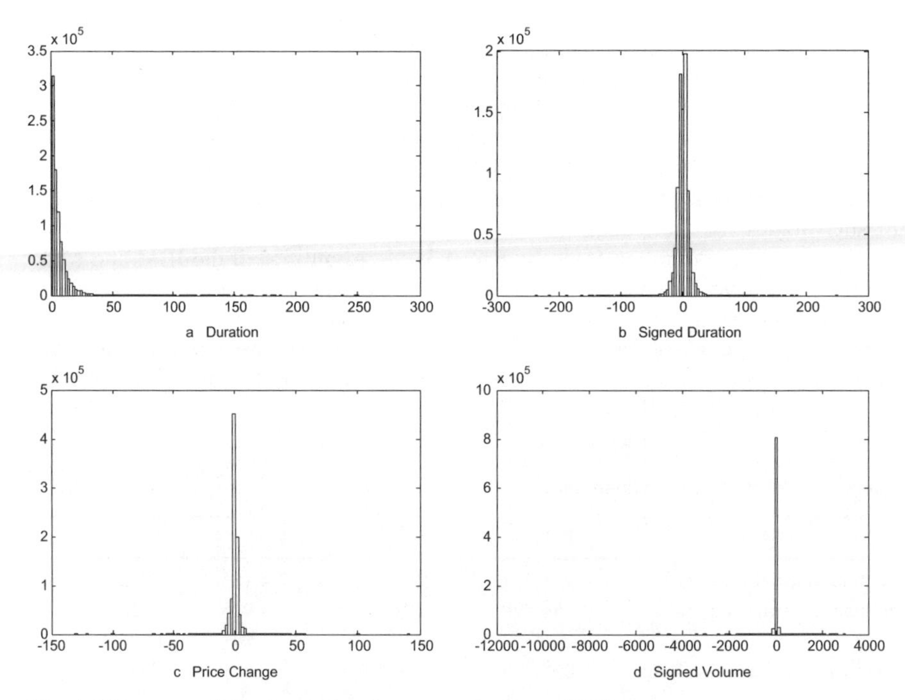

Fig. 1 Histograms of duration, price changes, and volume (*IBM*)

Table 2 Distribution of prices, duration, and volume

		Percentiles								
		10	20	30	40	50	60	70	80	90
IBM	Price change	−3	−1	0	0	0	0	1	1	3
	Duration	1	2	3	4	5	6	7	10	14
	Duration (sell)	2	2	3	4	5	6	7	10	14
	Duration (buy)	1	2	3	4	5	6	7	10	14
	Volume	1	2	4	5	7	10	14	23	48
	Volume (sell)	1	2	4	5	7	10	13	22	45
	Volume (buy)	1	2	4	5	7	10	15	24	50
GE	Price change	−1	−1	0	0	0	0	0	1	1
	Duration	1	1	2	2	3	4	5	7	11
	Duration (sell)	1	1	2	2	3	4	5	7	11
	Duration (buy)	1	1	2	2	3	4	5	7	11
	Volume	1	3	5	8	10	15	24	42	94
	Volume (sell)	1	3	5	8	10	15	24	40	90
	Volume (buy)	1	3	5	8	10	15	24	43	99
BA	Price change	−2	−1	0	0	0	0	0	1	2
	Duration	1	2	3	4	5	7	10	15	25
	Duration (sell)	1	2	3	4	5	7	10	15	25
	Duration (buy)	1	2	3	4	5	7	10	15	25
	Volume	1	2	2	3	5	7	10	13	24
	Volume (sell)	1	2	2	3	5	6	10	12	23
	Volume (buy)	1	2	2	4	5	7	10	13	25
MO	Price change	−2	−1	0	0	0	0	0	1	2
	Duration	1	2	3	3	5	6	8	11	18
	Duration (sell)	1	2	3	3	5	6	8	11	18
	Duration (buy)	1	2	3	3	5	6	8	11	17
	Volume	1	2	3	5	6	9	12	20	42
	Volume (sell)	1	2	3	5	6	9	12	20	41
	Volume (buy)	1	2	3	5	6	9	12	20	43

frequently traded of the four stocks. The distribution of price changes is mostly symmetric. Table 2 also provides a useful reference for interpreting our conditional distribution plots, as our conditional probabilities will be estimated and plotted at various *percentile* values of the conditioning information set.

3 Nonparametric estimation of conditional distribution functions

Let $\{\mathbf{X}_t, Y_t\}_{t=1}^{T}$ be observations from a strictly stationary process where Y_t is a scalar and $\mathbf{X}_t = (x_{1t}, x_{2t}, \ldots x_{Mt})$. In our first application, $Y_t = \Delta p_t$ and $\mathbf{X}_t = (d_t, sv_t)$. One non-parametric approach to estimating the conditional distribution function

$$F(y|\mathbf{x}) = \Pr(Y_t \leq y|\mathbf{X}_t = \mathbf{x})$$

is to make use of the fact that if $Z_t = I(Y_t \leq y)$, where $I(.)$ is the indicator function, then $E[Z_t \mid \mathbf{X}_t = \mathbf{x}] = F(y \mid \mathbf{x})$. The particular technique we use is the Adjusted Nadaraya–Watson estimator (Hall et al. 1999; Hall and Presnell 1999) which we state in its multivariate ($M \geq 1$) form.

3.1 The Adjusted Nadaraya–Watson estimator

The Adjusted Nadaraya–Watson estimator of $F(y|\mathbf{x})$ is given by

$$\tilde{F}(y|\mathbf{x}) = \frac{\sum_{t=1}^{T} Z_t w_t K_{\mathbf{H}}(\mathbf{X}_t - \mathbf{x})}{\sum_{t=1}^{T} w_t K_{\mathbf{H}}(\mathbf{X}_t - \mathbf{x})} \tag{1}$$

where $\{w_t\}_{t=1}^{T} = \arg\max \prod_{t=1}^{T} w_t$, with $\{w_t\}_{t=1}^{T}$ satisfying the conditions (1) $w_t \geq 0$ for all t, (2) $\sum_{t=1}^{T} w_t = 1$, and (3) $\sum_{t=1}^{T} w_t (X_{mt} - x_m) K_{\mathrm{H}}(\mathbf{X}_t - \mathbf{x}) = 0$ for all $m = 1, \ldots, M$, and $K_{\mathrm{H}}(.)$ is a multivariate kernel with bandwidth matrix $\mathbf{H}$.

Although the Adjusted Nadaraya–Watson estimator is based on the biased bootstrap idea of Hall and Presnell (1999), it is useful, as noted in Hall et al. (1999), to view the estimator as the local linear estimator of $F(y|\mathbf{x})$ with weights $K_{\mathbf{H}}(\mathbf{X}_t - \mathbf{x})$ replaced by $w_t K_{\mathbf{H}}(\mathbf{X}_t - \mathbf{x})$, i.e., $\tilde{F}(y|\mathbf{x}) = \hat{a}$ where $\hat{a}$ is obtained from the solution of

$$\max_{a,\,\mathbf{b}} \sum_{t=1}^{T} (Z_t - a - (\mathbf{X}_t - \mathbf{x})\mathbf{b})^2 w_t K_{\mathbf{H}}(\mathbf{X}_t - \mathbf{x})$$

(see Fan and Gijbels 1996, for an authoritative introduction to local linear and local polynomial regression methods). It is easy to see from the first-order condition

$$\frac{\partial}{\partial a} \sum_{t=1}^{T} (Z_t - a - (\mathbf{X}_t - \mathbf{x})\mathbf{b})^2 w_t K_{\mathbf{H}}(\mathbf{X}_t - \mathbf{x}) = 0$$

that $\hat{a}$ reduces to $\tilde{F}(y|\mathbf{x})$ under condition (3).

Using the unmodified version of the local linear approach ($w_t = 1$) may result in estimates of conditional distributions that are not monotonic in y, or that do not lie always between 0 and 1. The Adjusted Nadaraya–Watson estimates, on the other hand, always lie between 0 and 1, is monotonic in y, and yet share the superior bias properties as estimates from local linear methods (Hall et al. 1999), and also automatic adaptation to estimation at the boundaries (see e.g., Fan and Gijbels 1996). There is no requirement for the conditional distribution to be continuous in y. Another justification for using the adjusted Nadaraya–Watson estimator is provided by Cai (2002) who establish asymptotic normality and weak consistency of the estimator for time series data under conditions more general than in Hall et al. (1999).

3.2 Practical issues

Implementation of the estimator $\tilde{F}(y|\mathbf{x})$ requires a number of practical issues to be addressed. In particular, $\{p_t\}$ has to be computed, and $K_{\mathbf{H}}(.) = |\mathbf{H}|^{-1} K(\mathbf{H}^{-1}\mathbf{x})$ has to be chosen. The choice of K in smoothing problems is usually not crucial (see e.g., Wand and Jones 1995) but the choice of $\mathbf{H}$ is important. For K we use the standard M-variate normal distribution

$$K(\mathbf{x}) = (2\pi)^{-1/2} \exp\left(-\|\mathbf{x}\|^2/2\right).$$

We take $\mathbf{H}=h\mathbf{I}_M$ where $\mathbf{I}_M$ is the M-dimensional identity matrix. Other, possibly non-symmetric kernels may be useful here (for instance, the gamma kernels in Chen 2000), although we stay with the Gaussian kernel as the theoretical properties for this specific method has been established for symmetric kernels only. As we are effectively using only one bandwidth for multiple regressors, our regressors are always scaled to a common variance before the estimator is implemented.

To obtain the optimal value of h, we adapt the bootstrap bandwidth selection method suggested by Hall et al. (1999). This approach exploits the fact that, as we are estimating distribution functions, there is limited scope for highly complicated behavior. First, a simple parametric model is fitted to the data and used to obtain an estimate $\hat{F}(y|\mathbf{x})$. We use

$$Y_t = a_0 + a_1 X_{1t} + \ldots + a_M X_{Mt} + a_{M+1} X_{1t}^2 + \ldots + a_{2M} X_{Mt}^2 + \varepsilon_t \tag{2}$$

and assume that ε_t is heteroskedastic, depending on the square of lagged price changes. We then simulate from this model to obtain a bootstrap sample $\{Y_1^*, Y_2^*,...,Y_T^*\}$ using the actual observations $\{\mathbf{x}_1,\mathbf{x}_2,...\mathbf{x}_T\}$. For each bootstrap sample (and for a given value of h), we compute a bootstrap estimate $\tilde{F}_h^*(y|\mathbf{x})$. We choose h to minimize

$$\sum \left|\tilde{F}_h^*(y|\mathbf{x}) - \hat{F}(y|\mathbf{x})\right|$$

where the summation is over all bootstrap replications and over values of y for any given $\mathbf{x}$. We checked the sensitivity of our estimates to the choice of the bandwidth, and we note that we obtained very similar results over a wide range of bandwidth values.

Computation of the weights w_t is carried out using the Lagrange Multiplier method. The Lagrangian is

$$L = \sum_{t=1}^{T} \log(w_t) - \lambda_0 \left(\sum_{t=1}^{T} w_t - 1\right) - \sum_{m=1}^{M} \lambda_m \sum_{t=1}^{T} w_t (X_{m,t} - x_m) K_{\mathbf{H}}(\mathbf{X}_t - \mathbf{x}),$$

which gives the first-order conditions

$$\frac{1}{w_t} - \lambda_0 - \sum_{m=1}^{M} \lambda_m (X_{m,t} - x_m) K_{\mathbf{H}}(\mathbf{X}_t - \mathbf{x}) = 0, \ \forall m = 1, \ldots, M \tag{3}$$

together with the restrictions $\sum_{t=1}^{T} w_t = 1$ and $\sum_{t=1}^{T} w_t (X_{mt} - x_m)\ K_{\mathbf{H}}(\mathbf{X}_t - \mathbf{x}) = 0$ $\forall m = 1,...,M$. Solving these equations, we get $\lambda_0 = T$ with $\{\lambda_m\}_{m=1}^{M}$ satisfying the equations

$$\sum_{t=1}^{T} \frac{(X_{mt} - x_m)K_{\mathbf{H}}(\mathbf{X}_t - \mathbf{x})}{T + \sum_{m=1}^{M} \lambda_m (X_{mt} - x_m) K_{\mathbf{H}}(\mathbf{X}_t - \mathbf{x})} = 0, \quad \forall m = 1, \ldots, M \tag{4}$$

We obtain $\{\lambda_m\}_{m=1}^{M}$ by solving Eq. (4) numerically (we use the MATLAB function *fsolve* to do this.) The weights $\{w_t\}_{t=1}^{T}$ are then computed from Eq. (3).

4 Empirical results

For each of the four stocks in our sample, we estimate two sets of the conditional distribution $F(\Delta p_t | d_t, v_t)$, one set for seller-initiated trades, and one for buyer-initiated trades. For each set, the distribution is estimated at the 20th, 40th, 60th, and 80th percentiles of d_t and v_t. That is, for each stock, we estimate 16 conditional distributions. It is not worth the computational burden to estimate the conditional distributions at a finer grid: it would be difficult to present that much information clearly and simply, and the chosen values of the conditioning variables are sufficient for highlighting important empirical regularities. For each pair (d_t, v_t) we estimate the conditional cumulative distribution at values of Δp_t from −10 to 9. We report the conditional probabilities $\Pr(\Delta p_t \leq -10 | d_t, v_t)$, $\Pr(\Delta p_t = i | d_t, v_t)$ for $i = -9, -8, ..., 8, 9$, and $\Pr(\Delta p_t \geq 10 | d_t, v_t)$. These conditional probabilities are obtained by taking the difference of the estimated cumulative distribution between adjacent points of the grid over Δp_t. It is possible to go further and obtain estimates of the bivariate conditional distribution $\Pr(\Delta p_t, d_t | v_t)$ by first estimating $\Pr(d_t | v_t)$, and getting the distribution $\Pr(\Delta p_t, d_t | v_t)$ by multiplying the estimates of $\Pr(d_t | v_t)$ with the estimates of $\Pr(\Delta p_t | d_t, v_t)$ obtained previously. Our estimates of $\Pr(d_t | v_t)$ show that more trades occur at short durations than at long durations, regardless of volume, and as a result, much of the interesting structure in the conditional probabilities $\Pr(\Delta p_t | d_t, v_t)$ do not show up well in plots of the bivariate distributions. We therefore discuss only our estimates of $\Pr(\Delta p_t | d_t, v_t)$.

The results for IBM are presented in Fig. 2(a) and (b). In Fig. 2(a), we show the distributions of Δp_t conditional on d_t and v_t for seller-initiated trades, organized into four panels. In each panel we have the conditional distribution at four percentile levels of duration (20th, 40th, 60th, and 80th percentiles—see Table 2 for actual values.) The four panels correspond to four different levels of volume, with volume at the 20th, 40th, 60th, and 80th percentile levels starting from the top left panel and proceeding clockwise.) In Fig. 2(b) we have the same plots for buyer-initiated trades. The most obvious pattern is that the distributions in Fig. 2(a) are negatively skewed, whereas the distributions for buyer-initiated trades in Fig. 2(b) are positively skewed. This skewness is also clear in the estimated probabilities provided later in Table 3. As security prices tend to move in the direction of trade sign, the skewness observed is not surprising. Buyer-initiated trades tend to move the security price higher, especially when these trades exhaust the prevailing depth in the limit-order book and NYSE specialists revise the quotes higher to reflect the higher demand for the stock.

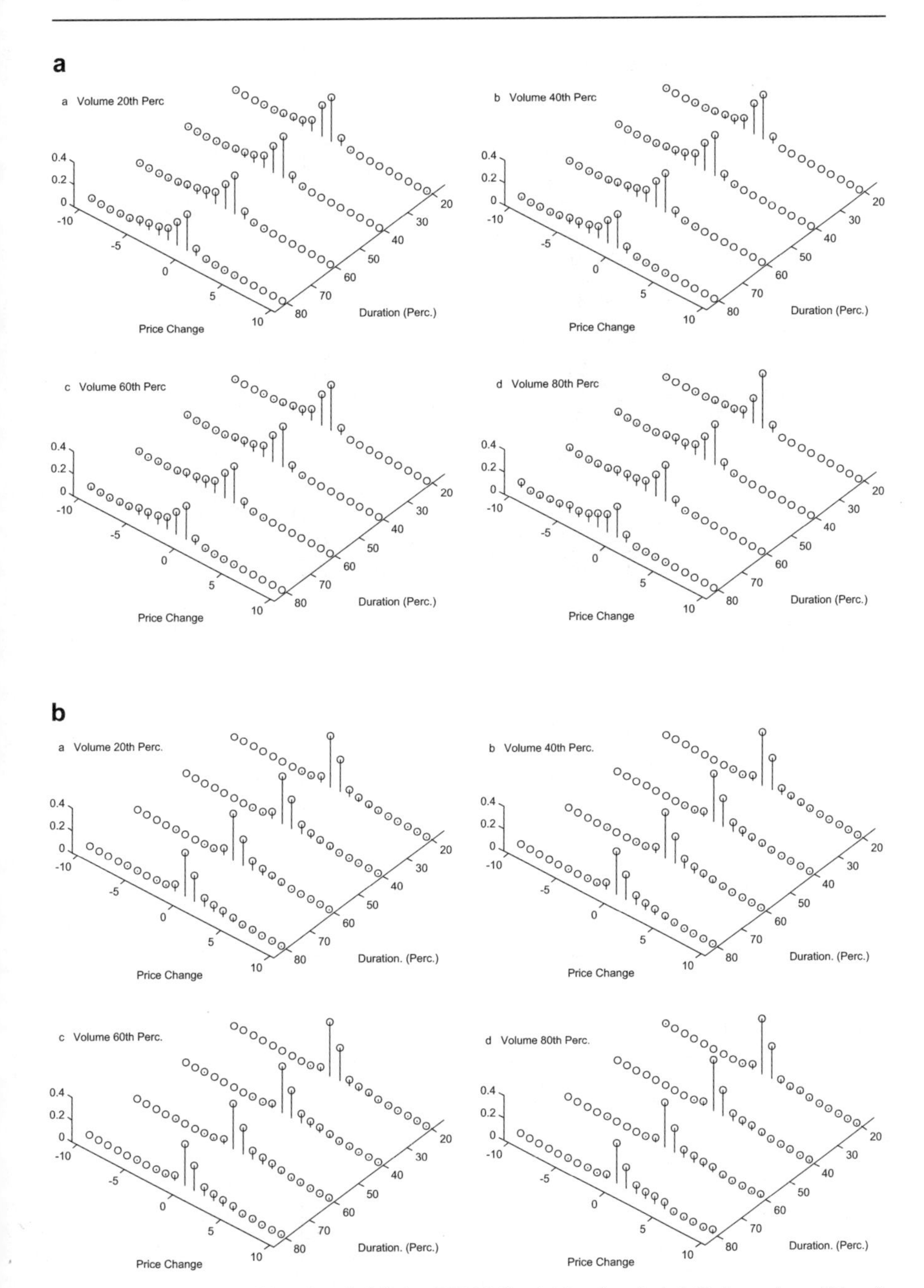

Fig. 2 **a** Estimated conditional probabilities (*IBM* Seller-initiated trades). **b** Estimated conditional probabilities (*IBM* Buyer-initiated trades)

Table 3 Estimated probabilities of price change conditional on volume and duration

	Volume (percentile)	Duration (percentile)	$\Delta p_t \leq -2$				$-1 \leq \Delta p_t \leq 1$				$\Delta p_t \geq -2$			
			IBM	GE	BA	MO	IBM	GE	BA	MO	IBM	GE	BA	MO
(a) Seller-initiated trades	20th	20th	0.243	0.119	0.139	0.148	0.742	0.875	0.848	0.839	0.016	0.006	0.014	0.013
		40th	0.298	0.142	0.203	0.184	0.677	0.850	0.775	0.799	0.025	0.008	0.022	0.017
		60th	0.327	0.157	0.222	0.199	0.645	0.834	0.751	0.784	0.027	0.008	0.027	0.017
		80th	0.374	0.163	0.226	0.210	0.601	0.831	0.738	0.774	0.025	0.006	0.036	0.016
	40th	20th	0.246	0.125	0.141	0.158	0.742	0.869	0.846	0.832	0.012	0.006	0.013	0.010
		40th	0.307	0.146	0.205	0.192	0.673	0.846	0.775	0.796	0.020	0.008	0.021	0.012
		60th	0.341	0.163	0.227	0.215	0.636	0.829	0.748	0.772	0.024	0.008	0.025	0.013
		80th	0.390	0.168	0.253	0.224	0.583	0.825	0.717	0.763	0.027	0.007	0.030	0.014
	60th	20th	0.241	0.128	0.142	0.163	0.749	0.866	0.847	0.829	0.011	0.007	0.012	0.009
		40th	0.320	0.153	0.212	0.200	0.661	0.839	0.769	0.790	0.018	0.008	0.019	0.011
		60th	0.366	0.174	0.241	0.231	0.612	0.818	0.737	0.757	0.022	0.008	0.023	0.012
		80th	0.411	0.181	0.269	0.245	0.564	0.812	0.704	0.746	0.026	0.007	0.027	0.010
	80th	20th	0.202	0.120	0.133	0.161	0.789	0.874	0.858	0.831	0.009	0.005	0.009	0.008
		40th	0.341	0.153	0.223	0.202	0.640	0.840	0.759	0.788	0.018	0.007	0.018	0.010
		60th	0.397	0.202	0.267	0.264	0.582	0.791	0.712	0.723	0.021	0.006	0.021	0.013
		80th	0.472	0.215	0.313	0.269	0.509	0.777	0.663	0.720	0.018	0.008	0.024	0.011

Table 3 (continued)

	Volume (percentile)	Duration (percentile)	$\Delta p_t \leq -2$				$-1 \leq \Delta p_t \leq 1$				$\Delta p_t \geq -2$			
			IBM	GE	BA	MO	IBM	GE	BA	MO	IBM	GE	BA	MO
(b) Buyer-initiated trades	20th	20th	0.021	0.006	0.016	0.015	0.804	0.895	0.868	0.843	0.176	0.099	0.116	0.142
		40th	0.030	0.008	0.023	0.019	0.743	0.868	0.802	0.809	0.227	0.124	0.175	0.171
		60th	0.033	0.007	0.022	0.015	0.717	0.854	0.799	0.834	0.250	0.140	0.179	0.151
		80th	0.037	0.007	0.031	0.013	0.695	0.855	0.783	0.822	0.269	0.138	0.187	0.165
	40th	20th	0.015	0.006	0.014	0.012	0.813	0.891	0.857	0.841	0.171	0.103	0.129	0.148
		40th	0.026	0.008	0.021	0.015	0.729	0.861	0.786	0.806	0.245	0.131	0.193	0.179
		60th	0.030	0.008	0.022	0.012	0.700	0.846	0.784	0.819	0.269	0.146	0.194	0.169
		80th	0.035	0.007	0.027	0.012	0.676	0.852	0.776	0.821	0.288	0.141	0.198	0.168
	60th	20th	0.013	0.006	0.011	0.010	0.816	0.886	0.890	0.837	0.171	0.108	0.100	0.154
		40th	0.024	0.008	0.019	0.012	0.701	0.848	0.776	0.793	0.275	0.145	0.206	0.195
		60th	0.027	0.008	0.021	0.011	0.694	0.831	0.783	0.807	0.279	0.161	0.197	0.182
		80th	0.033	0.007	0.025	0.010	0.650	0.838	0.766	0.798	0.317	0.155	0.210	0.192
	80th	20th	0.015	0.005	0.012	0.007	0.844	0.888	0.866	0.844	0.142	0.107	0.123	0.149
		40th	0.019	0.009	0.017	0.010	0.719	0.866	0.729	0.767	0.261	0.125	0.254	0.223
		60th	0.024	0.007	0.020	0.011	0.657	0.795	0.765	0.789	0.319	0.199	0.215	0.200
		80th	0.030	0.008	0.021	0.007	0.602	0.804	0.745	0.743	0.368	0.188	0.234	0.250

More interesting is the observation, for both buyer- and seller-initiated trades, that duration has a much stronger influence on the distribution of price change at high volumes than at low volumes. Looking at panel (a) of Fig. 2(a) and (b) we see a slight change in the distribution of price changes between the four duration levels at low volume. Panel (d) of both figures on the other hand show that at high volume, the probability of larger price changes increases substantially with duration. The influence of volume on price change also depends on duration. At short durations (i.e., when trading is very active), volume does not have very much influence on the distribution of price changes; moving from panel (a) to panel (d), in both diagrams, the distributions at the 20th percentile level of duration are very similar. At long durations, volume matters. Again moving from panel (a) to panel (d), the distribution of price change for duration at the 80th percentile shows that the probability of larger prices changes (negative for seller-initiated trades, positive for buyer-initiated ones) increases as volume increases. Duration between trades influences the distribution of price changes, but the degree of influence appears to depend on the volume.

This result suggests that at the intraday frequency, volume may have some influence on the distribution of price changes, even after controlling for duration. This contrasts with the result in Jones et al. (1994) that, at the daily frequency, the relationship between volume and the volatility of stock returns actually reflects the relationship between volatility and the number of transactions (thus, average duration). The result is consistent with research on high-frequency stock prices (e.g., Easley et al. 1997) and suggests the importance of the interaction between volume and duration.

We repeat this exercise with the other three stocks in our sample. The figures are qualitatively the same as Fig. 2, so we do not display them. Instead, the estimated probabilities $\Pr(\Delta p_t \leq -2|d_t,v_t)$, $\Pr(-1 \leq \Delta p_t \leq 1|d_t,v_t)$, and $\Pr(\Delta p_t \geq -2|d_t,v_t)$ are presented in Table 3. The top half of the table shows estimates for seller-initiated trades, while the lower half shows estimates for buyer-initiated trades.

We see that for IBM at the 20th percentile of duration, $\Pr(\Delta p_t, \leq -2|d_t,v_t)$ remains around 0.2 for seller-initiated trades, and $\Pr(\Delta p_t \geq -2|d_t,v_t)$ lies mostly around 0.17 for buyer-initiated trades, as volume increases. At the 80th percentile level of duration, both probabilities increase substantially as volume increases. Focusing on the 20th percentile level of volume, $\Pr(\Delta p_t \leq -2|d_t,v_t)$ increases from 0.243 to 0.374 as duration increases from the 20th percentile level to the 80th. At the 80th percentile level of volume, $\Pr(\Delta p_t \leq -2|d_t,v_t)$ increases from 0.202 to 0.472 as duration increases from the 20th to 80th percentile levels.

Looking over the estimates for other three stocks, we see that similar comments can be made, with differences only in degree. These patterns appear to be strongest in the case of IBM, and weakest for GE and BA. In addition to the estimates for the full sample period, we also compute the estimates for two subsamples (Jan to June, and July to December) to check if our results are robust across different sample periods. The estimates for these two sample periods are very similar to the estimates for the full sample. We also checked if adding a 2 s delay to the reported quote time when signing trades affects our results. Again, the estimates in this case are very similar to what is reported here.

We have also computed 95% confidence intervals around the probability estimates in Table 3 using formulas that exploit the local linear nature of the estimates (see e.g. Fan and Yao 2003, Section 6.3.4). We do not display the intervals, because it suffices to note that they are all very narrow-the boundary of the intervals differ from the estimated probabilities only in the third or fourth decimal places-and therefore indicate that the difference between the estimated probabilities conditional on long versus short durations are statistically significant. For instance, the 95% confidence interval for Prob($\Delta p_t \leq -2$) for IBM (volume at 80th perc., duration 80th perc.) is (0.471, 0.474) whereas the corresponding interval for IBM (volume at 80th perc., duration 20th perc.) is (0.201, 0.202).

It seems therefore that there are substantial interaction effects between duration and volume, and the sign of trade. Any parametric analysis of the relationship between duration, volume, and prices should take these interaction effects into account. In addition, the high degree of skewness in the distributions indicates that care should be taken in interpreting results concerning the volatility of price changes. We interpret our results for seller-initiated trades as evidence in support of the Diamond and Verrecchia (1987) analysis, where we expect larger probability of price falls with higher levels of duration.

As a further robustness check, we re-estimate the conditional probabilities, including lagged values of duration, price changes, and volume, in addition to contemporaneous duration and volume. We continue to report the estimated probabilities at the 20th, 40th, 60th, and 80th percentile levels of duration and volume, but only at the median values of the lagged variables. Estimates of Pr($\Delta p_t \leq -2$) for seller-initiated trades, and Pr($\Delta p_t \geq -2$) for buyer-initiated trades are listed in Table 4. We note that by conditioning on the median value of lagged price change (which is zero), the estimated probabilities of price changes of -1 to 1 ticks increase substantially; this is because trades at the same price tend to be following by more trades at the same price. Nonetheless, it is still the case that among the probabilities Pr($\Delta p_t \leq -2$) for seller-initiated trades, and Pr($\Delta p_t \geq -2$) for buyer-initiated trades, the largest estimates occur when duration and volume are both at the 60th or 80th percentiles.

The focus of this paper is on the contemporaneous relationships between duration, volume, and price changes. Nonetheless, it is of interest to extend the analysis to the dynamic interrelationships between these variables (beyond the robustness check reported in the previous paragraph) as much of the interest in these variables arise from the effect of information arrival, and changes over time of durations are indicative of this. For instance, it is of interest to explore and compare the densities of price changes when trading is becoming more frequent, with the densities when trading is becoming less frequent.

In Fig. 3, we show the estimated probabilities for GE in these two situations. The densities for the other three stocks are similar, and are omitted. The top row shows the estimated price densities for seller-initiated trades, whereas the bottom row shows the estimates for buyer initiated trades. The left column shows the estimates for the situation when trades are getting more frequent (current duration at the 20th percentile, lagged duration at the 80th percentile). Again the skewness in the distributions is clear, although in the case of GE, the modal price change is now -1 and 1 for seller and buyer initiated trades, respectively. In the other three stocks, the probability of a price change also increases relative to the probability of no price change.

In the right column, we show the estimated probabilities when current duration is long relative to lagged duration (current duration at the 80th percentile, lagged duration at the 20th percentile). Here the modal probability is clearly at zero price

Table 4 Estimated probabilities of price change conditional on volume and duration and lagged information

	Volume (Perc.)	Duration (Perc.)	$\Delta p_t \leq -2$					Volume (Perc.)	Duration (Perc.)	$\Delta p_t \geq -2$			
			IBM	GE	BA	MO				IBM	GE	BA	MO
(a) Seller-initiated trades	20th	20th	0.070	0.026	0.074	0.096	(b) Buyer-initiated trades	20th	20th	0.073	0.025	0.044	0.048
		40th	0.026	0.046	0.141	0.067			40th	0.129	0.036	0.167	0.067
		60th	0.245	0.053	0.116	0.059			60th	0.084	0.029	0.070	0.040
		80th	0.156	0.107	0.111	0.033			80th	0.119	0.023	0.092	0.066
	40th	20th	0.175	0.034	0.066	0.135		40th	20th	0.097	0.031	0.055	0.051
		40th	0.118	0.052	0.117	0.087			40th	0.120	0.044	0.097	0.074
		60th	0.151	0.118	0.113	0.098			60th	0.138	0.041	0.114	0.079
		80th	0.211	0.046	0.109	0.059			80th	0.172	0.040	0.102	0.065
	60th	20th	0.113	0.054	0.065	0.064		60th	20th	0.088	0.036	0.054	0.057
		40th	0.202	0.067	0.136	0.095			40th	0.148	0.048	0.115	0.086
		60th	0.427	0.077	0.183	0.104			60th	0.199	0.055	0.095	0.081
		80th	0.324	0.087	0.189	0.082			80th	0.195	0.057	0.089	0.061
	80th	20th	0.082	0.038	0.084	0.086		80th	20th	0.076	0.022	0.156	0.085
		40th	0.235	0.062	0.176	0.146			40th	0.243	0.049	0.126	0.112
		60th	0.254	0.362	0.192	0.108			60th	0.447	0.136	0.260	0.132
		80th	0.215	0.088	0.130	0.108			80th	0.272	0.285	0.117	0.086

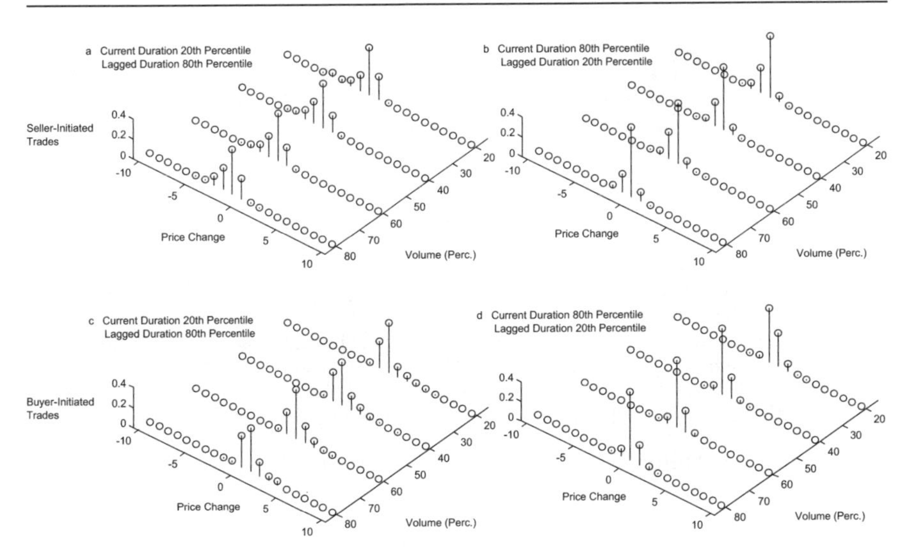

Fig. 3 Estimated conditional probabilities (*GE* with lagged duration)

change. The distributions here are also much tighter than the corresponding distributions in Fig. 2. This suggests that our results reported earlier are more appropriate at median values of the lagged variables, or when current durations are short relative to lagged durations. These results are merely indicative; a more extensive study of the dynamic relationships in the framework used here is left for future research.

5 Concluding comments

We investigate the distribution, conditional on trading volume, duration between trades, and the sign of trades of high-frequency price changes on four stocks traded on the New York Stock Exchange. The conditional probabilities are estimated nonparametrically using local polynomial regression methods. We find substantial skewness in the distribution of price changes, with the direction of skewness dependent on trade sign. We also find that the probability of larger price changes increases with volume, but only for trades that occur with longer durations. Durations affect prices, with a stronger effect when volume is high.

The evidence suggests substantial interaction effects between duration and volume with respect to their effect on prices; parametric analyses of the relationship between duration, volume, and prices should take these into account, such as in Tay et al. (2004). The high degree of skewness in the distributions indicate that it is also important to distinguish between seller- and buyer-initiated trades, and that care should be taken in interpreting results concerning the volatility of price changes. The results are consistent with theoretical research. Our findings for seller-initiated trades provide direct evidence in support of the Diamond and Verrecchia (1987) analysis, where larger probability of price falls is associated with a higher level of duration.

Acknowledgements Tay gratefully acknowledges research support from the Singapore Management University's Wharton-SMU Research Centre. We thank participants at the 2005 North American Meeting of the Econometric Society for helpful questions and comments. All errors remain our own. Zhang Feng provided able research assistance.

References

Cai Z (2002) Regression quantiles for time series. Econom Theory 18:169–192
Chen SX (2000) Probability density function estimation using gamma kernels. Ann Inst Stat Math 52:471–480
Diamond DW, Verrecchia RE (1987) Constraints on short-selling and asset price adjustment to private information. J Financ Econ 18:277–311
Dufour A, Engle RF (2000) Time and the price impact of a trade. J Finance 55(6):2467–2498
Easley D, O'Hara M (1992) Time and the process of security price adjustment. J Finance 47 (2):577–605
Easley D, Kiefer NM, O'Hara M (1997) The information content of the trading process. J Empir Finance 4:159–186
Engle RF (2000) The econometrics of ultra-high frequency data. Econometrica 68(1):1–22
Engle RF, Gonzalez-Rivera G (1991) Semiparametric ARCH models. J Bus Econ Stat 9:345–360
Engle RF, Russell JR (1998) Autoregressive conditional duration: a new model for irregularly spaced transaction data. Econometrica 66(5):1127–1162
Fan J, Gijbels I (1996) Local polynomial modelling and its application, monographs on statistics and applied probability 66, Chapman & Hall/CRC
Fan J, Yao Q (2003), Nonlinear time series: nonparametric and parametric methods, Springer
Gallant AR, Hsieh DA, Tauchen G (1991) On fitting a recalcitrant series: the Pound/Dollar exchange rate, 1974–83. In: Barnett WA, Powell J, Tauchen G (eds) Nonparametric and semiparametric methods in econometrics and statistics: proceedings of the fifth international symposium in economic theory and econometrics, Cambridge University Press
Gallant AR, Rossi PE, Tauchen G (1992) Stock prices and volume. Rev Financ Stud 5(2):199–242
Grammig J, Wellner M (2002) Modeling the interdependence of volatility and inter-transaction duration processes. J Econom 106:369–400
Hall P, Presnell B, Intentionally Biased Bootstrap Methods. J R Stat Soc B 61:143–158
Hall P, Wolff RCL, Yao Q (1999) Methods for estimating a conditional distribution function. J Am Stat Assoc 94(445)154–163
Jones CM, Kaul G, Lipson ML (1994) Transactions, volume, and volatility. Rev Financ Stud 7 (4):631–651
Lee CMC, Ready MJ (1991) Inferring trade direction from intraday data. J Finance 46(2)733–746
Renault E, Werker BJM (2004) Stochastic Volatility Models with Transaction Time Risk, manuscript, Universite de Montreal and Tilburg University
Russell JR, Engle RF (2004) A Discrete-state continuous-time model of financial transactions prices: the ACM-ACD Model, forthcoming in J Bus Econo Stat
Tay AS, Ting C, Tse Y, Warachka M (2004) Transaction-data analysis of marked durations and their implication for market microstructure, Singapore Management University SESS Working Paper
Vergote O (2005) How to match trades and quotes for NYSE stocks? Center for economic studies discussion paper, K.U.Leuven, DPS 05.10
Wand MP, Jones MC (1995) Kernel smoothing, Monographs on statistics and applied probability 60, Chapman & Hall/CRC

David Veredas

Macroeconomic surprises and short-term behaviour in bond futures

Abstract This paper analyses the effect of macroeconomic news on the price of the ten year USA Treasure bond future. We consider 15 fundamentals and we analyse the effect of their forecasting errors conditional upon their sign and the momentum of the business cycle. To obtain a smooth effect of the news arrival we estimate a Polynomial Distributed Lag model. Using 10 minutes sampled data during 9 years, we conclude that 1) releases affect the bond future for only few hours, 2) their effect depends on the sign of the forecast error, 3) their effect also depends on the business cycle and 4) the timeliness of the releases is significant.

Keywords US bonds · PDL model · Business cycle · Macroeconomic announcements

JEL Classification C22 · G14

1 Introduction

Researchers and security price analysts are in fundamental disagreement about the causes of price movements. Both groups, economic theorists and econometrics practitioners, have produced a plethora of articles explaining the reasons for volatility, trends, clustering, long memory and other price movements. But one phenomenon in which they express common agreement is that when prices are sampled at a sufficiently high frequency (i.e. at least intradaily), short term movements can partly be attributed to news arrivals. In particular, to macroeconomic announcements.

D. Veredas (✉)
ECARES-Université Libre de Bruxelles and CORE, 50, Av Roosevelt CP114, B1050 Brussels, Belgium
E-mail: dveredas@ulb.ac.be

This paper explores the effect of macroeconomic announcements on the returns of bond futures. The literature in the field tracks back to Berkman (1978) and a series of articles published until mid 1980s (e.g. Grossman 1981; Roley 1983) that focus on money supply announcements. Goodhart et al. (1993) analyse the effect of the announcements of two macroeconomic numbers on the mean and variance, using a GARCH-M and three months of tick-by-tick US-pound exchange rates. McQueen and Roley (1993) study the effect of eight fundamentals on the mean during eleven years, on a daily basis, of S&P500 price movements. Instead of using the news itself, they use the forecasting error, i.e. the difference between the released and the expected number. Moreover, they control for the momentum of the business cycle (hereafter we will refer indistinctively to the economic cycle as the business cycle or economic cycle). Flemming and Remolona (1999) study the effect of 25 macroeconomic numbers on the US bond market (see Table 1 of this article for a summary of the research done in macroeconomic announcements prior to 1997). They employ standard regression using five minutes returns in absolute value, the number of transactions per hour as market measures and the macroeconomic forecasting errors. DeGennaro and Shrieves (1997) look at the effect of contemporaneous and expost announcements of 27 USA and Japan fundamentals on the volatility of US-Yen exchange rates using GARCH models. Li and Engle (1998) work on the daily volatility of the 30 years note future (T-bond), via GARCH and GARCH-M models, and how it is affected by the announcements of just two fundamentals. Andersen and Bollerslev (1998) also study the effect of announcements on volatility but using a slightly different approach to Engle and Li's. They analyse the Deutsche Mark-US exchange rate in a two-step procedure and using weighted least squares. First, they estimate a standard GARCH model and, second, they standardise the absolute residuals centered with the estimated volatility.

Table 1 Information on macro numbers

Name	Acronymic	Rel. hour	Details
Business inventories	BI	08:30 NY	%chg m/m
Consumer confidence	CC	10:00 NY	Index SA
Consumer price index	CPI	08:30 NY	%chg m/m SA
Durable goods orders	DG	08:30 NY	%chg m/m SA
Gross domestic product	GDP	08:30 NY	%chg q/q SAAR
Housing starts	HS	08:30 NY	Millions of units SA
Industrial production	IP	09:15 NY	%chg m/m SA
Industrial Production	IP	09:15 NY	%chg m/m SA
Institute for Supply Management	ISM	10:00 NY	Index
Non-Farms Payrolls	NFP	08:30 NY	*K* persons chg SA
Personal Income	PI	08:30 NY	%chg m/m SA
Producer Price Index	PPI	08:30 NY	%chg m/m SA
Retail Sales	RS	08:30 NY	%chg m/m SA
Trade Balance	TB	08:30 NY	$billions
Unemployment claims	UNEMW	08:30 NY	*K* persons as reported
Unemployment rate	UNEM	08:30 NY	Rate as reported

NY NY time, *Rel.* released, *%chg* percent change, *SA* seasonally adjusted, *SAAR* seasonally adjusted annual rate, *m/m* month to month, *q/q* quarter to quarter

They assume that this variable captures the intradaily seasonality and the news effect that standard GARCH models are not able to capture. Finally, two papers focus on the mean of the returns of the Treasury bond future and several exchange rates. Andersen et al. (2003) analyse, in a seven year period, the forecasting errors of 40 fundamentals using a very similar approach to Andersen and Bollerslev (1998) but for the mean. Hautsch and Hess (2002) use the revision unemployment report numbers that include important variables such as the unemployment rate and non-farms payrolls.

In this paper 1) the financial asset on which we base our analysis is the US Treasury 10-year bond future, 2) we consider 15 fundamentals representing different sectors of the economy (inflation, real economy, supply and demand confidence indexes and export–import measures), 3) we are interested in the effect of news announcements on the mean rather than on volatility, 4) we consider the forecasting errors rather than the released numbers, 5) we differentiate between positive and negative forecasting errors, 6) we assume that the effect of the fundamentals on the bond market differs depending on the business cycle (top, bottom, expansion and contraction), and 7) we use an econometric model that estimates contemporaneous and ex-post effects of the macroeconomic news smoothly, that is, the parameters that measure the news effect vary smoothly through time.

One of the main features of the paper is the interaction between 5 and 6. That is, we not only permit good and bad news to affect differently the bond future, but we also investigate whether these effects are robust within the economic cycle. If there are variations, we may conclude that market's behaviour depends on the mood of the economy or how the market perceives the state of the economy. This is related to the work in behavioural economics and, in particular, behavioural finance. Since the introduction of prospect theory (Kanehman and Tversky 1979), it is well known that individuals suffer from an asymmetry between the way they make decisions involving gains (good news) and decisions involving losses (bad news). This leads to a failure of invariance, i.e. inconsistent choices (or reactions to good and bad news) when the same problem (a good or a bad news item) appears in different frames (top, bottom, expansion and contraction of the business cycle). An alternative means of explaining why the market responds differently to good and bad news in different phases of the business cycle is by means of uncertainty (Veronesi 1999): The market is uncertain about the state of the economy. For example, after a period of economic boom, good news will have no effect as traders are already aware of the strength of the economy. If, by contrast, bad news start to arrive, the market reacts because it is not no longer confident about the phase of the business cycle. This introduces more uncertainty and causes an asymmetry in the response to good and bad news. On this basis, Conrad et al. (2002) test this asymmetric reaction to the news sign and the state of the stock market (the equivalent to the business cycle in our paper). In a firm-specific study, and using earnings announcements, they conclude that price responses to negative earnings surprises increase as the difference between the market value at the moment of the announcement and the average market value during the last 12 months, increases. In other words, the effect of a negative earnings surprise is higher if the market feels itself to be in expansion with respect to the previous 12 months, than in contraction. Or, put differently, markets respond more strongly to bad news in good times. By contrast, price responses to positive earnings surprises do not necessarily increase in bad times with respect to good times.

In this paper we find similar results to the above studies and more, as our set of news is much richer. The basic conclusions we reach are: First, the signs of the responses are all intuitive and the fundamentals that cause most the bond future changes are (see Table 1 for the acronyms) CC, ISM, UNEM, NFP, PPI, CPI, RS and IP. On the other hand, HS, GDP, PI, BI and TB have little or no statistical significance on the bond future. Second, the responses to positive and negative forecast errors are statistically different. This is the case, for instance, for UNEM, NFP, CPI and RS. Third, and more importantly, we observe an asymmetry in responses to positive and negative forecasting errors at different phases of the business cycle. In fact, our findings agree with Conrad et al. (2002). Bad news has the strongest effect when the business cycle is at the top and in contraction. This suggests that when the economy is at the top of the cycle, traders know that sooner or later the downward part of the cycle will start, and hence bad news may be a signal of the beginning of the contraction. For equivalent (but inverse) reasons, the smallest effect of bad news is when the economy is in expansion. Overconfidence in the state of the economy may be an explanation. Our findings regarding positive news also agree with Conrad et al. (2002): They are ambiguous. When the economy is expanding good news has barely any effect. This is again a sign that the market is overconfident. However, when the economy is contracting, positive and negative news have similar effect. This suggests that, regardless of the sign of the error, news increases uncertainty. Fourth, timeliness matter, i.e. the sooner the fundamental is released, the more it influences the bond future. This is strongly related to the first conclusion. For example, CC is the first number released, at the end of the month it is reporting on, and it has an important influence. On the contrary, BI is released two months after the month it is covering and has barely any influence. These findings are also found by Flemming and Remolona (1999) and Hess (2004).

The rest of the paper is organized as follows; Section 2 shows the financial model, Section 3 explains the econometric methodology, Section 4 discusses the data, Section 5 shows and discusses the results and Section 6 draws conclusions.

2 On the components of short run price movements

The hypothesis underlying this paper is that temporary jumps observed in the pricing of financial assets reflect: 1) the market expectation of fundamental factors driving the asset valuation (e.g. inflation or unemployment) and 2) the time in the business cycle where these expectations are formed (and potential signs of reversal of the cycle) i.e. the idea of looking past the immediate macroeconomic release.

To measure the effect on market expectation of fundamentals driving asset valuation, we need an asset and a set of fundamentals. Among all the financial markets, we chose the bond market. Bonds are the most widely used of all the financial instruments. In particular, we chose the 10 year Treasury Note 6% day session Future (TY) because it is a liquid and important contract and reflects the general response of the yield curve to news. The ten year future is a reflection of the state of the US bond market with maturities between seven and ten years. It is an efficient way of constructing a long time series that is not subject to the problems encountered by taking a particular bond (for instance, when a bond matures its price represents shorter and shorter yields). Regarding the fundamentals, we chose 15 different numbers, representing wide areas of the economy such as real economy, inflation, confidence indexes and export-import measures (we refer to Table 1 and Section 4 for further details).

The second component that affects short run jumps in asset prices is the momentum of the business cycle. Following recent literature on behavioural finance and Veronesi (1999), we conjecture that the effect of the fundamentals on TY depends on the general economic situation of the economy. For example, it is intuitive to believe that the effect of a positive surprise in the unemployment rate on the TY is not the same when the business cycle is in contraction as when it is in expansion. As measure of the business cycle we use the Institute for Supply Management Survey (ISM) index. The difference between ISM and other measures of the business cycle, such as GDP, is that it is the most forward-looking measure available of the market since it is based on expectations. The ISM index is the result of a survey among 300 people selected from 20 manufacturing industries. The survey includes questions related to new orders, production, employment, supplier deliveries and inventories (see Niemira and Zukowsky (1998, chapter 19) for further details). In order to see the effect of the fundamentals on the TY, we divide the ISM according to two criteria: 1) if it is expanding or contracting and 2) if it is above, below or in between (differentiating as well expansions and contractions) upper and a lower thresholds (further details are given in Section 4).

Ideally, we would like to estimate a model like

$$\begin{aligned}(1-L)TY_t &= [\alpha_1. + \alpha_2(\text{trend strength of the business cycle}_t) \\ &\quad + \alpha_3(\text{trend strength of } TY_t)]\,(N_t - E[N_t]) + \varepsilon_t, \qquad (1)\end{aligned}$$

where N_t is the fundamental, $E[N_t]$ its expectation, and hence $N_t - E[N_t]$ is the forecasting error (or surprise or news).

The effect of the news in TY is caused by the news itself as well as by the trend strength of the business cycle and the trend strength of the future bond. At each point in time, the market has a working hypothesis as to where the economy is and where it is going. Simplistically, the macroeconomic releases are testing the market hypothesis. No drastic price response should be seen if the number falls in a reasonable range around the expectations. If the macroeconomic releases fall beyond the *reasonable* bounds, measured by the forecasting error, the re-pricing effect should be proportional to the forecast error (coefficient α_1).

In reality, more factors are at work and they should be evident mostly around the changes in macroeconomic trends. In general, agents are not very good at calling market tops or bottoms and when the economy is perceived to be turning, one usually sees a rush to close or reverse existing positions. Agents are unsure of whether a turn is occurring or whether the number is a statistical fluke in the macroeconomic trend. So we expect an asymmetry in the market's response to positive or negative forecast errors at perceived phases of the economic cycle. This is represented by coefficient α_2.

This asymmetry can be exacerbated by the net positioning of the speculator and hedgers community. The speculators community are usually trend followers and

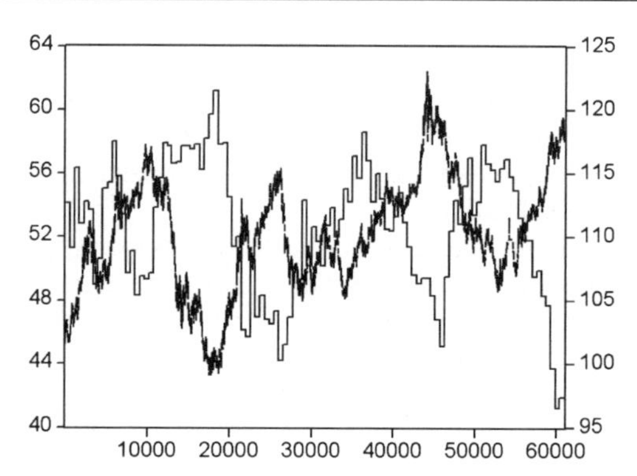

Fig. 1 Continuous line is the TY future (*scale in the right axis*). Piecewise line is ISM (*scale in the left axis*)

gradually accumulate positions in the direction of the trend whilst hedgers are counter-trend followers. Trends by themselves tend to last in time and the correction to our equation coming from this momentum is represented by α_3.

We can simplify model (1) above in the following way. First we notice that the ISM and the TY price are inversely correlated: −0.71 (see also Fig. 1). TY moves up when ISM goes down, peaks close to when ISM bottoms then sells off when ISM starts recovering again, until it reaches a top. So the price trend and the macroeconomic trend are strongly related and, in a way, represent the same explanatory variable. In other words, we can assume that $\alpha_1 = -\alpha_2$. We therefore decide to divide the business cycle into different phases, explained below, and estimate the following equation:

$$(1 - L)TY_t = \sum_{i=0}^{p} \beta_{i,C,S(N)}(N_{t-i} - E[N_{t-i}]) + \varepsilon_t, \tag{2}$$

where i is the time measured since the moment of the release, C is the variable representing where we are in the business cycle and taking values 'Top, Bottom, Expansion, Contraction', and $S(N)$ is the sign of the forecast error $N - E(N)$.

The fact that parameters depend on the moment of the business cycle and the sign of the forecasting errors, will permit us to discover hidden relations that otherwise are not possible to discern. For example, in an analysis of non-farms payrolls for β independent of C and $S(N)$ we find a very strong relation, as expected. However, if β depends on the business cycle, non-farms payrolls affect TY only when the cycle is in the top and the bottom range and matters little when the cycle is in expansion.

3 Methodology

Given the previous arguments, our interest is to model the effect of news arrivals in the bond future market not only contemporaneously but also through time and depending on the momentum of the business cycle and the sign of the forecast error. We want to answer questions like: For how long does employment news affect the bond future market? Or how does the effect of a CPI release in the bond future market change throughout the day?

Under the econometric viewpoint, several alternatives are available. We could consider that the news arrivals and the bond future price are pure stationary time series and hence, in a univariate context, build a transfer function model, where the transfer function is the ratio of two finite polynomials yielding to an infinite polynomial and hence an infinite response. In a more simplistic context, we could consider a standard regression model with the news arrival and all its lags as exogenous variables and returns of TY as endogenous variables, yielding a model similar to (2)

$$(1-L)TY_t = \sum_{i=0}^{p} \beta_i (N_{t-i} - E[N_{t-i}]) + \varepsilon_t. \tag{3}$$

Here we assume that just one news item, i.e. one macroeconomic number, arrives to the market. Notice that p should not be fixed and should vary through the macroeconomic number, i.e. the shock effect does not last for the same length of time for all macroeconomic numbers. The differences in p through fundamentals will be tested empirically. This model, although very simple, is not adequate for several reasons. First, we would like to have a smooth effect of the news shock through time and, hence, some smoothness constraints between parameters instead of allowing them to vary freely. Second, the number of parameters can become large if our sample frequency is high and the effect of the fundamental on the TY stands for long.

In order to avoid these problems we use a Polynomial Distributed Lag (PDL) model, also known as Almon's model (1965). PDL models were introduced for a different reason than the two aforementioned: often when contemporaneous and past values of exogenous variables are introduced, the model may suffer of multicollinearity. This does not happen in our case since the exogenous variables take zero value everywhere except when the news is released. The Almon's model is based on the assumption that the coefficients are represented by a polynomial of small degree K

$$\beta_i = \alpha_0 + i\alpha_1 + i^2\alpha_2 + \ldots + i^K\alpha_K, i = 0, \ldots, p > K, \tag{4}$$

which can be expressed in matrix terms as $\beta{=}H\alpha$ where

$$H= \begin{pmatrix} 1 & 0 & 0 & \cdots & 0 \\ 1 & 1 & 1 & \cdots & 1 \\ 1 & 2 & 4 & \cdots & 2^K \\ 1 & 3 & 9 & \cdots & 3^K \\ \vdots & \vdots & \vdots & & \vdots \\ 1 & p & p^2 & & p^K \end{pmatrix}. \tag{5}$$

This specification permits us to calculate the p coefficients estimating only K coefficients. K is an integer number, usually between three and four. The degree of the polynomial will determine its flexibility. For degree zero, all the β's will be equal and hence they form an horizontal straight line. For degree one, the β's decrease uniformly. For K=*2* the β's form a concave or convex bell. For K=*3* the β's can form a convex shape during some periods and concave during others, yielding a sort of wave with decreasing amplitude. Since a priori we believe that the news release will affect the bond future price smoothly and with periods of positive effect and periods of negative effect, we set three as the degree of the polynomial.

Substituting (4) into (3) we obtain

$$\begin{aligned}(1-L)TY_t &= \alpha_0\left(\sum_{i=0}^{p} N_{t-i} - E[N_{t-i}]\right) + \alpha_1\left(\sum_{i=0}^{p} i(N_{t-i} - E[N_{t-i}])\right)\\ &\quad + \ldots + \alpha_3\left(\sum_{i=0}^{p} i^3(N_{t-i} - E[N_{t-i}])\right) + \varepsilon_t \\ &= \alpha_0 z_{t,0} + \alpha_1 z_{t,1} + \ldots + \alpha_3 z_{t,3} + \varepsilon_t = AZ_t + \varepsilon_t,\end{aligned} \tag{6}$$

The matrix of parameters A can be estimated by OLS if the error term fulfills the classical assumptions. The variance/covariance matrix of $\hat{A}$, $\Sigma_{\hat{A}}$, is estimated with the White estimator.

It is worth making a note on stationarity and variance effects. The bond future has a unit root. Hence $(1-L)TY_t = \Delta TY_t$ will be stationary. The exogenous variables are stationary since we are not analyzing the effect of the macroeconomic number on the future prices but on the news. That is, the forecasting error is always stationary. Finally, conditioning the matrix of coefficients to the business cycle and the sign of the forecast error is immediate and similar to (2).

4 Data

TY and the macroeconomic numbers are sampled in 10 minutes segments from 08:20 to 12:30 NY time and from April 1992 to April 2001. The sample size is 61048 observations. TY is traded at the Chicago Board of Trade (CBOT) from 08:20 am to 03:00 pm, NY time. We use 15 fundamentals, summarized in Table 1. The forecast of the fundamentals is generated (from surveys conducted before the announcement) and released by MMS International.

The choice of the fundamentals is such that they represent the real economy (UNEM, UNEMW, NFP, RS, HS and BI), inflation (PPI, CPI), supply and demand confidence (ISM and CC) and export-import (TB). We do not consider monetary fundamentals such as M1 and the Fed interest rates. The former does not affect the TY and the latter is not a proper fundamental but rather a consequence of the fundamentals, i.e. the Fed acts according to the fundamental releases.

For the futures contracts, some transformation is needed since they mature every three months and hence often have different pricing, producing a discontinuity. We rollover the future contracts when the open interest rate becomes greater in the next contract (this usually means that the following contract is the more liquid contract) using linear interpolation.

Further, notice that data are sampled from 08:20 a.m. to 12:30 p.m. NY time, while the market closes at 03:00 p.m. We discard all the observations from 12:30 p.m. to 03:00 p.m. There are three reasons to consider the 08:20–12:30 interval. First, the aim of the paper is to analyze the news impact of macroeconomic numbers on the bond future prices. If the news impact stands only for a few hours, as we believe, and as most of the news is released at 08:30 a.m. NY time, the news

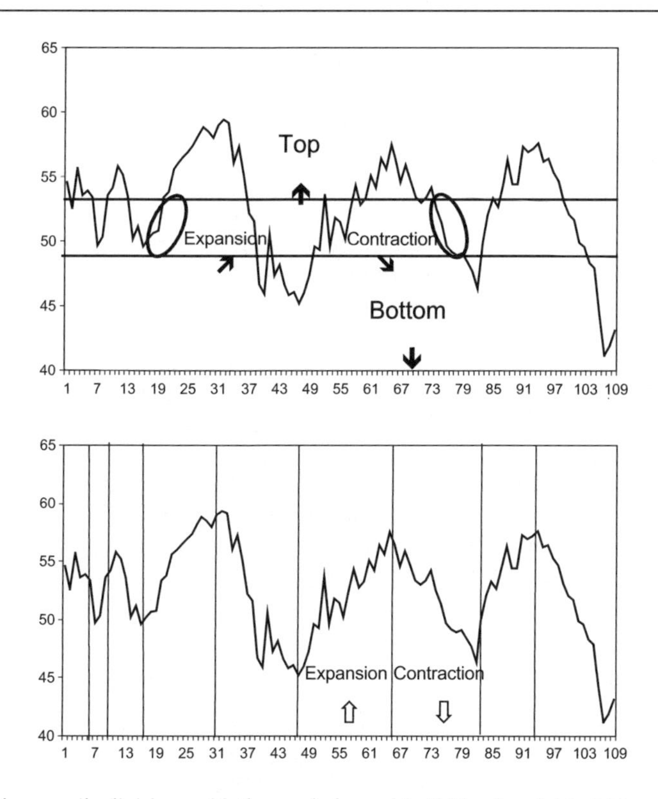

Fig. 2 Business cycle divisions with the symbols used in Tables 2 and A1–A12

impact on TY should be hardly distinguishable after 12:30 p.m. Second, lunchtime typically starts at 12:00, and hence the trading activity slows down sharply, implying that necessarily the news effect vanishes. Finally, after lunch and before 03:00 p.m. (close) the price action may be somewhat distorted by the activity of the *bond pit locals* whose activity is not related to the fundamentals releases but they take intraday positions and close them towards the end of the session.

Where do an expansion and a contraction begin? A well known measure of the momentum of the economy is the NBER dates of recessions and expansions. However, as noticed by McQueen and Roley (1993), the NBER turning points only classify the direction of the cycle rather than the level. Instead, we consider the ISM. It is more widely used among fixed income traders than the NBER indicator since it represents more accurately the cycles of the bonds and it has been proven to provide a forward-looking indication of the business cycle direction. We consider two different divisions of the business cycle: 1) expansions and contractions and 2) top and bottom.

With respect to the first division, a change of trend is produced when it stands for at least three periods, as commonly accepted among bank's analysts (see top plot of Fig. 2). The second division requires longer explanation. McQueen and Roley (1993) analyse the news effect on the S&P500 in different phases of the business cycle. They measure the business cycle with IP. The levels of high, medium and low economic activity are determined by estimating a trend and then fixing some intervals around the trend. This is a purely statistical method subject to the choice of bandwidth. By contrast, we believe that the right determination of the business cycle phases would be the one commonly accepted by the fixed income traders since they are the ones who generate the TY process.

The ISM is largely used among traders as a measure of the business cycle. It is constructed from a survey that asks about the state of the economy. It has only three possible answers: better, worse or equal than the previous period. With these answers a percentage is built. A percentage equal to 50 means that half of the respondents think that the business conditions are good, while the other half believes the contrary. A value below 50 is a sign of a weak economy. On the other hand, historical data shows that an ISM above 54.5 is an indicator of expansion. Therefore, we consider the following classification: Top if ISM is above 55, bottom if it is below 50, expansion if it is between 50 and 55 and rising, and contraction if it is between 50 and 55 and falling. A graphical explanation is shown in the bottom plot of Fig. 2.

Finally, some autocorrelation is found on the TY and one lag suffices to take it into account. This effect is usually found in stock prices. It does not necessarily mean predictive capabilities but could be due to microstructure effects (see Andersen and Bollerslev, 1998, footnote 10).

5 Results

Results are shown in Table 2 and A1–A12 in the Appendix and Figs. 3, 4, 5. Tables A1–A5 show full quantitative results for the fundamentals that have a higher effect in TY, that is CC, ISM, UNEM and UNEMW. Results are full in the sense that we consider a benchmark case (no business cycle division) and the aforementioned divisions. Results in Tables A6–A11 are for a second set of fundamentals (PPI, CPI, RS, IP, HS and DG). They are qualitative in the sense that we do not present the exact value of the effect but rather whether it is significant or not. For this second set of fundamentals we do not present results for all the divisions of the business cycle but only for the most exhaustive. We present very few results for GDP, BI, TB and PI in Table A12 for reasons that will become clear. Last, Table 2 summarizes and gives further insights into the asymmetry of responses depending on the sign of the forecasting error and the business cycle.

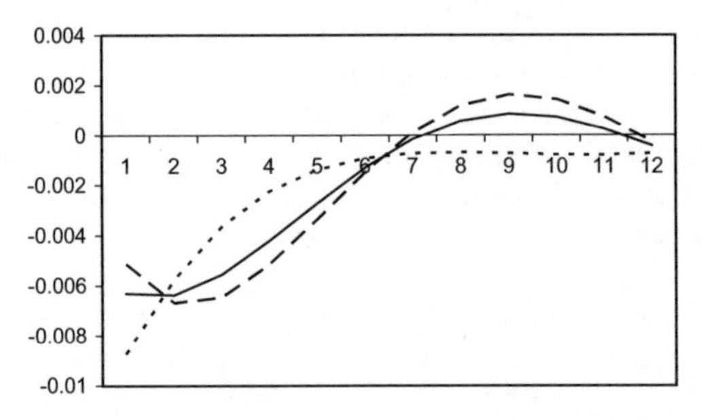

Fig. 3 Impulse respond function of a shock in CC on TY in division 1. *Solid line* considering all forecasting errors. *Dotted line* when only considering negative forecasting errors. *Dashed line* when considering positive forecasting errors. *Vertical axis* is the coefficient, *horizontal* are the minutes (divided by ten) after the release

Table 2 Summary

•↑	•↓	•↘	•↗	✓↑	✓↓	✓↘	✓↗	×↑	×↓	×↘	×↗
46	37	42	27	30	21	33	18	30	22	27	17

Total number of significant coefficients for all the fundamentals. See legend Table A1

Full results for all the fundamentals are available at core.ucl.ac.be, discussion paper 2002–37. In the sequel, we first present a detailed comment for one fundamental, Consumer Confidence. We then provide an analysis based on more general results.

5.1 A case study: consumer confidence

Consumer confidence is one of the fundamentals affecting the most bond futures. It is an interesting fundamental to analyse because it is not an economic measure but is rather an index of economic sentiment, not based on quantitative measures but rather on feelings.

Results are in Table A1. The signs of the coefficients are intuitively correct. Higher confidence than expected yields bond market sell-offs (column •). On average, and especially during the first hour, the effect of a positive forecast error is larger than the effect of a negative forecast error. See Fig. 3 and columns ✓ and

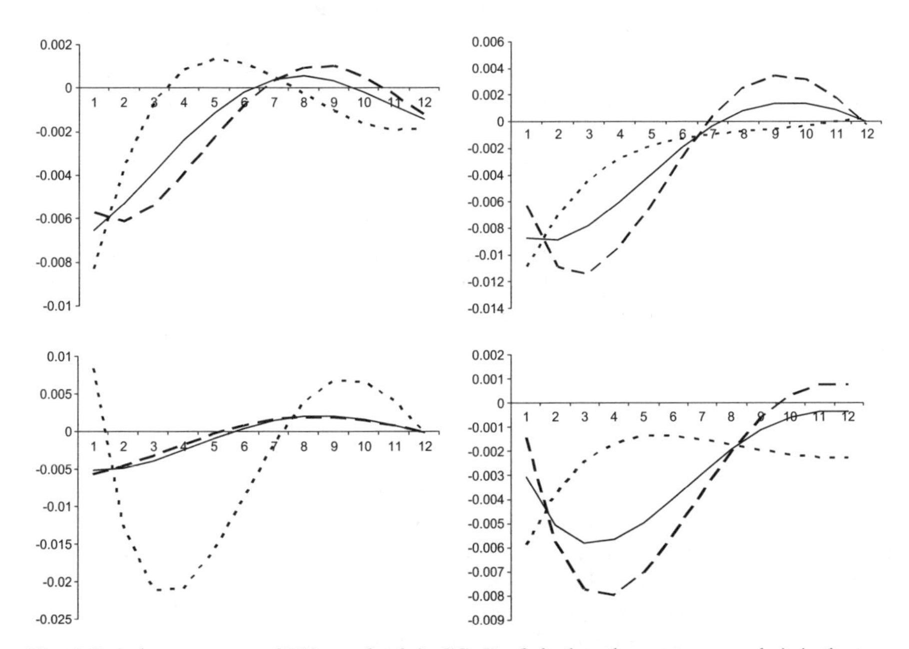

Fig. 4 Impulse responses of TY to a shock in CC. *Top left* when the economy cycle is in the top. *Top right* when the cycle is in the bottom. *Bottom left* when it is in expansion and *bottom right* when it is in contraction. *Solid, dotted and dashed lines* are the same as in Fig. 3. *Vertical axis* is the coefficient, *horizontal* are the minutes (divided by ten) after the release

×. That is for a given absolute value of forecast error, the bond sell-off is more violent than a bond rally. And it holds for all the economic environments (i.e. ✓⇧ and ✓⇩ vs. ×⇧ and ×⇩) except in an economic expansion (✓➚ vs. ×➚) where it seems that negative forecast errors (i.e. a weaker confidence than expected) generate larger bond moves than a positive error (see Fig. 4). This is an illustration that when CC comes 'against the trend', it tends to get more publicity in the bond market. Also, when the economy is contracting, positive forecast errors tend to impact on the market more than negative ones.

5.2 General results

5.2.1 Tables A2 to A5

All the signs, except in some pathological cases, are intuitively correct. ISM and Non-farm payrolls higher than expected yield bond market sell-offs. And UNEM and UNEMW higher than expected yields bond market rallies. Additionally, asymmetries and results are robust across divisions of the economic cycle.

NFP, UNEM and UNEMW represent the labour market. But NFP is the variable that has the stronger effect. NPF has tendency to exacerbate the peaks and bottoms of the business cycle (see ✓⇧, ×⇩, ✓⬆, ✓⬇). By contrast, UNEM has a lower and different effect. It has a tendency to accelerate the bond trend (see ✓➘, ×➚). Yet, as they are both released at the same time we would expect similar effects on TY. The reasons for this different reaction are: 1) UNEM is not seasonally adjusted by the US Census Bureau whilst the Bureau of Labour Statistics adjusts NFP. We could adjust UNEM by seasonality but it is not what the market observes. 2) The NFP survey has a larger sample size (390,000 establishments) than UNEM (60,000 households) and 3) in 1994 the questionnaire and the collection method for UNEM changed, introducing some disturbances on the sample. Regarding UNEMW, it is the only weekly released fundamental. It is not based on a survey but on a complete register announced every week. It has therefore a short term ongoing view of the labour market and can provide the first signals of some future change in the economy, i.e. it is a sort of leading indicator of the state of the economy. But, although it can be very useful for discover ongoing hidden problems, it is very erratic.

5.2.2 Tables A6–A11

These fundamentals have a smaller effect on the bond future than the previous ones. Nevertheless, sings (not reported here) are intuitively correct and results are robust to all economic environments. This is especially true for PPI, CPI, RS and IP. By contrast, the effects of HS and DG are very limited. They last no longer than half an hour with some rebound in some cases after one hour and a half. PPI and CPI deserve a special comment. These two fundamentals are inflation measures but PPI has a weak impact in TY, contrary to CPI. PPI is not significant because ISM, released slightly earlier, is informative enough about the manufacturing sector. Traders still keep in mind the ISM results as a benchmark of the manufacturing and industrial sector but wait for the CPI before gaining a precise idea of the inflation.

5.2.3 Table A12

Even in the less exhaustive case, PI, GDP, BI and TB do not have any effect on TY except, very vaguely, PI. The statistical significance of the coefficients is below acceptance and we will not push the analysis further.

In summary, the market also seems to have a general bias when it approaches certain numbers. We call a *Bearish Bias* a situation where the expected sell-off is larger in magnitude than the expected rally for the same absolute value of the forecast error and the opposite sign. Also the *Bullish Bias* is where the expected rally is larger in magnitude than the expected sell-off for the same absolute value of the forecast error and the opposite sign. Across the sample we draw the conclusions that there is a *Bearish Bias* for news coming from CC, CPI, and IP, and a *Bullish Bias* for news coming from UNEM, ISM, NFP, RS and UNEMW.

Table 2 summarizes the effect of all the fundamentals in TY for each stage of the business cycle. It shows the number of significant coefficients for each phase of the business cycle. The left side of the Table (first four columns) indicates that the market reacts more when the economy is in the top of the cycle and when it is contracting. In other words, the market feels that the business cycle is at the top and therefore sooner or later the cycle will start its downward trend, but they do not know when. After the first signals of weakness, the economy starts to contract and the market, which can feel the turn in the cycle, eagers for further signals that confirm the negative trend. Any news therefore has an effect. By contrast, when the economy is expanding, the market is confident about its strength and so news, of any kind, affects the bond future much less.

The right and middle parts of the Table give a more refined history. Last four columns show that the market responds more strongly to bad news in good times than in bad times. This confirms the theories of behavioural finance and Veronesi (1999), as well as Conrad et al. (2002) findings. The market responds strongly when the economy is contracting. This is again a confirmation that when the cycle is downward, TY reacts to bad news more than when the economy is growing. Results in the middle columns are not so enlightening. In particular, good news does not have the largest impact in the bottom of the cycle, but when the economy is contracting. It may be an indication that, when the business cycle is downward, the market is uneasy and afraid that the negative trend may continue. In this context, any forecasting error, positive or negative, introduces more uncertainty to the market and in turn induces a reaction in prices. This result nicely dovetails with the results in the left columns. Conversely, when the economy is expanding, the effect of positive news is smaller than in any other phase of the cycle. This again dovetails with the equivalent result in the left columns.

In summary, bad news has a stronger effect in good times than in bad times and good news has little effect in bad times. And when the economic phase is downward, any news has a strong effect but when the phase is upward, news has barely any effect.

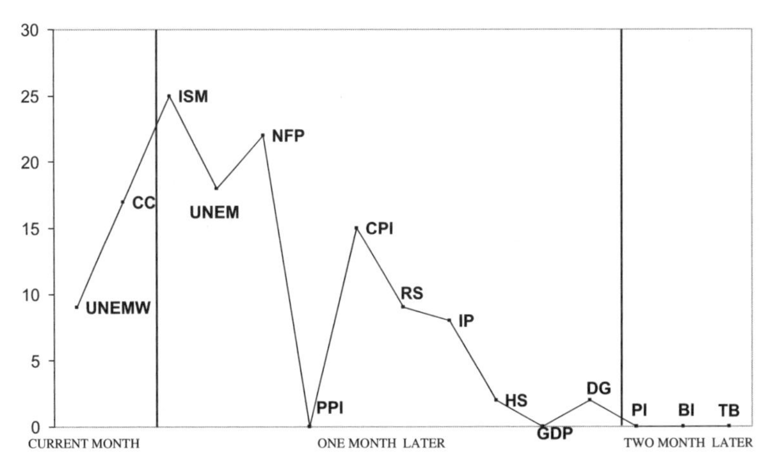

Fig. 5 Timeliness. Each point represents the number of significative coefficients in division four, the most exhaustive. The *vertical lines* represent the month when they are released. Fundamentals are temporally ordered. For example, CC is released at the end of the month it is covering and BI and TB two months later

Last, there is still an important question to answer. Some fundamentals have a definitive effect on TY while some others have no effect whatsoever. And some of these innocuous fundamentals are rather basic. Could we expect that GDP, the classical measure of the economy, is totally innocuous to TY, one of the most heavily traded long-term interest rate contracts in the world? Or that the TB, the most important indicator of import-exports, is irrelevant to the behaviour of TY? Or that CC, a fundamental based in opinions, has more effect that PI?

The answers are in Fig. 5. It shows the total number of significant parameters (in the most exhaustive division of the business cycle) for each fundamental and is ordered temporally according to when they are released. The importance of the fundamental decreases as the time interval between its release and the period it is covering increases. This effect, called timeliness, has been acknowledged previously by other authors (Flemming and Remolona 1999, and Hess 2004).

Timeliness, in this context, simply means that traders are not interested in information about the state of the economy at some point in time if it is released some months after. By contrast, the sooner the fundamental is released, the more it is perceived as important by traders and hence the greater effect it has on TY. It is the case with CC, released in the month that it is covering. Although CC is based on opinions, rather than in tangible goods, workers or prices, it conveys fresh information about the state of the economy that is rapidly absorbed by traders. The counter example is GDP. It is released around a month and a half after the period it refers. Given that it is the tenth fundamental to be released, it does not really add information and hence agents do not react when it is released. Moreover, GDP is quarterly released and continuously revised, being another reason for the lack of interest in it.

6 Conclusions

Movements in intra-day bond prices are produced by small shocks, with effects that stand in the market for few hours. These shocks can be, among others, political events, natural catastrophes, rumours, macroeconomic releases, etc. Among all, the macroeconomic releases are the only ones that arrive to the market systematically every week, month or quarter. And as their arrival time is known with certitude, agents can form systematic expectations about them. If a macroeconomic release is far from the expected number, agents will react.

Results in this paper show that traders react when the forecasting error varies from zero. The reaction to a positive or negative forecasting error is different and, most importantly, the reaction varies significantly depending on the momentum of the economic cycle. We also find that the time of the release matters and the closer it is to the covering period, the more effect it has on the bond future.

Acknowledgements I thank the participants in the CORE seminar, Sébastien Laurent, Winfried Pohlmeier and an anonymous referee for useful discussions and remarks. Special acknowledgment to Eugene Durenard, at the time proprietary fixed income trader in Credit Suisse First Boston, from who I learnt so much during my short internship at CSFB. The author acknowledges financial support from Credit Suisse Group and the Université catholique de Louvain (project 11000131) in the period this article was written. This text presents research results of the Belgian Program on Interuniversity Poles of Attraction initiated by the Belgian State, Prime Minister's Office, Science Policy Programming. The scientific responsibility is assumed by the author.

Appendix

Table A1 Consumer confidence

	•	✓	×	•⇧	•⇩	×	✓⇩	×⇧	•⬆	•⬇	•↘	•↗	✓⬆	✓⬇	✓↘	✓↗	×⬆	×⬇	•↘	×↘	×↗
10 min	**−0.06**	**−0.05**	**−0.09**	**−0.07**	**−0.06**	−0.06	**−0.04**	**−0.09**	**−0.09**	**−0.07**	−0.09	−0.03	−0.05	−0.06	−0.06	−0.01	−0.06	−0.08	−0.11	−0.06	0.08
20 min	**−0.06**	**−0.07**	**−0.06**	**−0.06**	**−0.06**	−0.06	**−0.07**	−0.07	**−0.05**	**−0.05**	−0.09	−0.05	−0.05	−0.06	−0.11	−0.06	−0.05	−0.04	−0.07	−0.04	−0.13
30 min	**−0.06**	**−0.06**	**−0.04**	**−0.05**	**−0.06**	−0.06	**−0.08**	−0.05	−0.03	−0.04	−0.08	−0.06	−0.04	−0.05	−0.11	−0.08	−0.03	−0.01	−0.04	−0.02	−0.21
40 min	**−0.04**	**−0.05**	−0.02	**−0.04**	**−0.05**	−0.04	**−0.07**	−0.03	−0.02	−0.02	−0.06	−0.06	−0.02	−0.04	−0.10	−0.08	−0.02	0.01	−0.03	0.02	0.21
50 min	**−0.03**	**−0.03**	−0.01	*−0.02*	**−0.03**	−0.02	**−0.05**	−0.02	−0.01	−0.01	−0.04	−0.05	−0.01	−0.02	−0.06	−0.07	0.00	0.01	−0.02	−0.01	−0.16
60 min	*−0.01*	*−0.01*	−0.01	−0.01	*−0.02*	−0.01	−0.03	−0.02	−0.01	0.00	−0.02	−0.04	0.00	−0.01	−0.03	−0.05	0.01	0.01	−0.01	−0.01	−0.09
70 min	0.00	0.00	−0.01	0.00	−0.01	0.01	−0.01	−0.01	−0.01	0.00	0.00	−0.03	0.01	0.00	0.00	−0.04	0.02	0.01	−0.01	−0.01	−0.02
80 min	0.01	*0.01*	−0.01	0.01	0.00	0.02	0.01	−0.01	−0.01	0.01	0.01	−0.02	0.02	0.01	0.03	−0.02	0.02	0.00	−0.01	−0.02	0.04
90 min	0.01	**0.02**	−0.01	0.01	0.01	0.02	0.01	−0.01	−0.01	0.00	0.01	−0.01	0.02	0.01	0.03	−0.01	0.02	−0.01	−0.01	−0.02	0.07
100 min	0.01	**0.01**	−0.01	0.01	0.01	0.01	0.01	−0.01	−0.01	0.00	0.01	−0.01	0.02	0.01	0.03	0.00	0.01	−0.02	0.00	−0.02	0.07
110 min	0.00	0.01	−0.01	0.00	0.00	0.01	0.01	−0.01	−0.01	−0.01	0.01	0.00	0.01	0.00	0.02	0.01	0.01	−0.02	0.00	−0.02	0.04
120 min	0.00	0.00	−0.01	−0.01	0.00	0.00	0.00	−0.01	−0.01	−0.01	0.00	0.00	0.00	−0.01	0.00	0.01	0.00	−0.02	0.00	−0.02	0.00

Estimation results using PDL models with polynomial of degree three. Standard errors estimated with the heteroskedastic consistent White estimator. Bold and italic numbers are significant at 5 and 10%, respectively. Otherwise, they are statistically equal to zero. •, ✓ and × stand for estimation results for all, positive and negative forecasting errors, respectively. ⇧ and ⇩ stand for estimation results for forecasting errors in expansions and contractions of the business cycle, respectively. ⬆, ⬇,↘ and ↗ stand for estimation results for forecasting errors in the top, bottom, expansions and contractions of the business cycle

Table A2 ISM

	•	✓	×	•⇧	•⇩	✓⇧	✓⇩	×⇧	×⇩	•🡅	•🡇	•🡖	•🡕	✓🡅	✓🡇	✓🡖	✓🡕	×🡅	×🡇	×🡖	×🡕
10 min	**−0.09**	**−0.12**	**−0.08**	*−0.07*	**−0.11**	**−0.12**	0.03	**−0.10**	**−0.11**	−0.07	**−0.12**	**−0.11**	−0.07	**−0.12**	−0.03	**−0.18**	**−0.15**	0.02	**−0.14**	−0.09	0.01
20 min	**−0.20**	**−0.19**	**−0.21**	**−0.23**	**−0.18**	**−0.20**	*−0.27*	**−0.13**	**−0.19**	**−0.18**	**−0.19**	**−0.19**	**−0.26**	**−0.26**	−0.06	**−0.15**	**−0.18**	−0.05	**−0.22**	**−0.19**	*−0.34*
30 min	**−0.23**	**−0.20**	**−0.25**	**−0.27**	**−0.20**	**−0.22**	**−0.39**	**−0.14**	**−0.21**	**−0.22**	**−0.20**	**−0.20**	**−0.31**	**−0.29**	−0.07	−0.11	**−0.17**	−0.09	**−0.23**	**−0.22**	**−0.46**
40 min	**−0.21**	**−0.17**	**−0.23**	**−0.25**	**−0.18**	**−0.19**	**−0.37**	**−0.13**	**−0.19**	**−0.20**	**−0.18**	**−0.18**	**−0.28**	**−0.26**	−0.09	−0.06	*−0.13*	−0.10	**−0.20**	**−0.20**	**−0.43**
50 min	**−0.16**	**−0.13**	**−0.17**	**−0.19**	**−0.14**	**−0.14**	**−0.29**	**−0.11**	**−0.14**	**−0.15**	**−0.14**	**−0.13**	**−0.20**	**−0.19**	**−0.09**	−0.01	−0.09	−0.09	**−0.16**	**−0.15**	*−0.32*
60 min	**−0.10**	**−0.08**	**−0.11**	**−0.11**	**−0.09**	**−0.08**	*−0.17*	**−0.09**	**−0.09**	**−0.10**	**−0.10**	**−0.08**	*−0.11*	**−0.11**	**−0.09**	0.02	−0.05	−0.07	**−0.10**	**−0.09**	−0.17
70 min	**−0.04**	−0.03	**−0.04**	−0.03	**−0.04**	−0.02	−0.05	−0.07	*−0.04*	−0.04	**−0.05**	−0.03	−0.02	−0.04	**−0.08**	0.04	−0.01	−0.05	−0.05	−0.04	−0.02
80 min	0.01	0.00	0.01	0.02	0.00	0.01	0.04	−0.05	0.00	0.00	−0.02	0.01	0.05	0.02	−0.07	0.04	0.02	−0.02	0.00	0.00	0.09
90 min	**0.03**	0.02	*0.04*	*0.05*	0.02	0.03	0.09	−0.04	0.03	0.03	0.01	0.03	**0.09**	0.04	−0.05	0.02	0.03	0.00	0.02	0.03	**0.15**
100 min	**0.04**	0.02	**0.05**	**0.05**	0.03	0.03	0.09	−0.03	0.04	0.03	0.02	0.03	**0.09**	0.04	−0.03	−0.01	0.03	0.01	0.03	0.04	**0.15**
110 min	**0.02**	0.01	**0.03**	0.03	0.02	0.02	0.06	−0.02	0.03	0.01	0.01	0.02	**0.06**	0.01	−0.01	−0.04	0.02	0.02	0.02	0.03	**0.11**
120 min	0.00	−0.01	0.00	−0.01	0.00	−0.01	0.00	−0.01	0.01	−0.01	0.00	−0.01	0.02	−0.03	0.01	−0.06	0.01	0.01	−0.01	0.00	0.02

See legend Table A1

Table A3 Non-farm payrolls

	•	✓	×	•⇧	•⇩	✓⇧	✓⇩	×⇧	×⇩	•↑	•↓	•↘	•↗	✓↑	✓↓	✓↘	✓↗	×↑	×↓	×↘	×↗
10 min	**−0.77**	**−0.08**	**−0.08**	**−0.08**	**−0.08**	**−0.08**	**−0.08**	**−0.08**	**−0.08**	**−0.08**	**−0.08**	**−0.08**	−0.08	**−0.08**	**−0.08**	*−0.08*	−0.08	**−0.08**	−0.08	**−0.08**	−0.08
20 min	**−0.79**	**−0.08**	**−0.08**	**−0.08**	**−0.08**	**−0.08**	**−0.08**	**−0.08**	**−0.08**	**−0.08**	**−0.08**	**−0.08**	−0.08	**−0.08**	**−0.08**	−0.08	−0.08	**−0.08**	**−0.08**	**−0.08**	−0.08
30 min	**−0.70**	**−0.07**	**−0.07**	**−0.07**	**−0.07**	**−0.07**	**−0.07**	−0.07	**−0.07**	**−0.07**	**−0.07**	**−0.07**	−0.07	**−0.07**	**−0.07**	−0.07	−0.07	−0.07	*−0.07*	**−0.07**	−0.07
40 min	**−0.53**	**−0.05**	**−0.05**	**−0.05**	**−0.05**	**−0.05**	−0.05	−0.05	**−0.05**	**−0.05**	**−0.05**	**−0.05**	−0.05	**−0.05**	**−0.05**	−0.05	−0.05	−0.05	*−0.05*	**−0.05**	−0.05
50 min	**−0.34**	**−0.03**	**−0.03**	**−0.03**	**−0.03**	**−0.03**	−0.03	−0.03	**−0.03**	**−0.03**	**−0.03**	−0.03	−0.03	**−0.03**	**−0.03**	−0.03	−0.03	−0.03	−0.03	*−0.03*	−0.03
60 min	**−0.15**	**−0.02**	−0.02	**−0.02**	*−0.02*	**−0.02**	−0.02	−0.02	−0.02	*−0.02*	**−0.02**	−0.02	−0.02	**−0.02**	**−0.02**	−0.02	−0.02	−0.02	−0.02	−0.02	−0.02
70 min	0.00	0.00	0.00	0.00	0.00	0.00	0.00	0.00	0.00	0.00	0.00	**0.00**	0.00	0.00	0.00	**0.00**	0.00	0.00	0.00	0.00	0.00
80 min	**0.10**	0.01	**0.01**	*0.01*	*0.01*	0.01	0.01	0.01	**0.01**	0.01	0.01	**0.01**	0.01	0.01	0.01	**0.01**	0.01	0.01	0.01	**0.01**	0.01
90 min	**0.14**	**0.01**	**0.01**	**0.01**	**0.01**	**0.01**	0.01	0.01	**0.01**	0.01	**0.01**	**0.01**	0.01	0.01	**0.01**	0.01	0.01	0.01	0.01	**0.01**	0.01
100 min	**0.13**	**0.01**	**0.01**	**0.01**	**0.01**	**0.01**	0.01	0.01	**0.01**	0.01	**0.01**	**0.01**	0.01	0.01	**0.01**	0.01	0.01	0.01	0.01	**0.01**	0.01
110 min	**0.07**	**0.01**	0.01	0.01	0.01	**0.01**	0.01	0.01	0.01	0.01	**0.01**	*0.01*	0.01	0.01	**0.01**	0.01	0.01	0.01	0.01	0.01	0.01
120 min	0.00	0.00	0.00	0.00	0.00	0.00	0.00	0.00	0.00	0.00	0.00	0.00	0.00	0.00	**0.00**	0.00	0.00	0.00	0.00	0.00	0.00

See legend Table A1

Table A4 Unemployment rate

	•	✓	×	•⇧	•⇩	✓⇧	✓⇩	×⇧	×⇩	•↑	•↓	•↘	•↗	✓↑	✓↓	✓↘	✓↗	×↑	×↓	×↘	×↗
10 min	**0.28**	0.30	**0.27**	0.12	**0.43**	0.08	*0.35*	0.12	**0.50**	**0.25**	0.09	**0.59**	0.19	0.37	−0.75	*0.53*	0.07	*0.21*	0.23	0.68	*0.22*
20 min	**0.38**	**0.50**	**0.33**	**0.26**	**0.49**	0.34	**0.53**	*0.25*	**0.45**	**0.34**	0.20	**0.69**	0.28	**0.44**	−0.59	**0.78**	0.51	**0.31**	0.33	*0.56*	**0.23**
30 min	**0.38**	**0.52**	**0.32**	**0.30**	**0.45**	0.42	**0.55**	**0.29**	**0.37**	**0.35**	0.25	**0.64**	0.29	**0.42**	−0.43	**0.78**	*0.65*	**0.32**	0.37	*0.41*	**0.21**
40 min	**0.32**	**0.44**	**0.27**	**0.28**	**0.36**	0.39	**0.46**	**0.26**	*0.28*	**0.30**	0.26	**0.48**	0.24	**0.35**	−0.28	**0.63**	*0.61*	**0.29**	0.35	0.25	**0.16**
50 min	**0.23**	**0.31**	**0.20**	**0.21**	**0.25**	0.29	**0.31**	**0.20**	*0.20*	**0.23**	*0.24*	**0.29**	0.17	**0.26**	−0.15	**0.41**	0.45	**0.22**	*0.30*	0.09	**0.10**
60 min	**0.13**	**0.16**	**0.12**	**0.13**	**0.13**	0.16	**0.16**	**0.13**	0.11	**0.15**	**0.19**	0.09	0.09	*0.16*	−0.04	**0.17**	0.25	**0.14**	**0.23**	−0.04	**0.05**
70 min	*0.04*	0.02	0.05	0.05	0.03	0.04	0.02	0.05	0.04	0.07	*0.13*	−0.07	0.02	0.08	0.04	−0.03	0.06	0.06	*0.14*	−0.13	0.01
80 min	−0.03	−0.07	−0.01	−0.01	−0.04	−0.06	−0.07	−0.01	−0.02	0.01	0.06	**−0.18**	−0.03	0.02	0.08	**−0.17**	−0.09	0.01	0.06	**−0.19**	−0.02
90 min	**−0.07**	**−0.12**	−0.05	−0.05	*−0.09*	−0.12	*−0.12*	−0.05	−0.06	−0.03	−0.01	**−0.22**	−0.06	−0.01	0.11	**−0.23**	−0.17	−0.03	−0.03	**−0.20**	−0.03
100 min	**−0.08**	**−0.11**	*−0.07*	−0.07	**−0.09**	−0.13	*−0.10*	−0.06	−0.08	−0.04	−0.07	**−0.20**	−0.05	−0.02	0.10	**−0.21**	−0.18	−0.05	−0.10	**−0.18**	−0.02
110 min	**−0.07**	−0.06	**−0.08**	*−0.07*	**−0.07**	−0.10	−0.05	*−0.06*	−0.09	−0.04	**−0.12**	**−0.13**	−0.03	0.00	0.08	−0.12	−0.13	−0.05	**0.15**	*−0.13*	−0.01
120 min	*−0.04*	0.01	**−0.07**	**−0.05**	−0.03	−0.05	0.03	−0.05	*−0.09*	−0.02	**−0.15**	−0.03	0.00	0.04	0.05	0.00	−0.04	−0.04	**−0.19**	−0.06	0.01

See legend Table A1

Table A5 Weekly unemployment claims

	•	✓	×	•⇧	•⇩	✓⇧	✓⇩	×⇧	×⇩	•↑	•↓	•↘	•↗	✓⇩	✓↑	✓↓	✓↘	×↑	×↓	×↘	×↗
10 min	**0.08**	0.07	**0.09**	**0.09**	**0.07**	0.05	**0.08**	**0.12**	**0.07**	**0.09**	**0.08**	0.06	*0.08*	0.06	**0.09**	0.05	0.06	**0.14**	0.07	0.07	0.09
20 min	**0.07**	**0.08**	**0.06**	**0.07**	**0.07**	*0.07*	**0.08**	**0.07**	*0.05*	**0.07**	**0.07**	**0.07**	0.05	0.08	**0.08**	0.07	0.09	**0.07**	*0.07*	*0.06*	0.03
30 min	**0.05**	**0.07**	*0.04*	**0.05**	**0.05**	**0.07**	**0.07**	0.04	0.03	**0.05**	**0.06**	*0.05*	0.03	**0.07**	*0.06*	0.06	0.10	0.02	0.05	0.05	0.01
40 min	**0.03**	**0.05**	0.02	0.03	**0.03**	0.05	**0.05**	0.02	0.02	0.03	*0.04*	0.04	0.02	0.06	0.04	0.04	0.08	0.00	0.04	0.03	−0.01
50 min	*0.02*	**0.03**	0.00	0.01	0.02	0.03	0.03	0.00	0.01	0.01	0.02	0.01	0.01	0.04	0.02	0.02	0.05	−0.01	0.02	0.01	−0.01
60 min	0.00	0.01	−0.01	0.00	0.00	0.01	0.01	−0.01	−0.01	0.00	0.00	−0.01	0.00	0.02	0.00	0.00	0.02	−0.02	0.00	−0.01	−0.01
70 min	−0.01	−0.01	*−0.02*	−0.01	−0.01	0.00	−0.01	−0.02	−0.01	−0.01	−0.01	−0.02	−0.01	0.00	−0.01	−0.01	−0.01	−0.01	−0.01	−0.03	−0.01
80 min	**−0.02**	−0.02	**−0.02**	*−0.02*	**−0.02**	−0.01	−0.02	−0.02	−0.02	−0.01	*−0.02*	−0.03	−0.02	−0.01	−0.02	−0.02	−0.03	−0.01	−0.02	−0.04	−0.02
90 min	**−0.02**	−0.02	**−0.02**	−0.02	**−0.02**	−0.02	−0.02	−0.02	−0.02	0.00	**−0.03**	−0.03	−0.04	−0.01	−0.02	−0.01	−0.04	0.00	−0.03	−0.05	−0.04
100 min	**−0.02**	−0.01	*−0.02*	−0.02	**−0.02**	−0.01	−0.02	−0.02	*−0.02*	0.00	**−0.03**	−0.02	−0.05	−0.01	−0.02	0.00	−0.04	0.01	−0.03	−0.05	−0.06
110 min	−0.01	0.00	−0.02	−0.01	−0.02	0.00	−0.01	−0.02	−0.02	0.01	*−0.02*	−0.01	−0.06	0.00	−0.02	0.02	−0.03	0.02	−0.02	−0.05	−0.08
120 min	−0.01	0.01	−0.02	−0.01	−0.01	0.01	0.01	−0.02	−0.02	0.02	−0.01	0.00	−0.07	0.02	−0.01	0.04	−0.02	0.02	−0.02	−0.05	−0.09

See legend Table A1

Table A6 Producer price index

	•	✓	×	•↑	•↓	•↘	•↗	✓↑	✓↓	✓↘	✓↗	×↑	×↓	×↘	×↗
10 min	■	□	■	■	□	□	■					■	■		■
20 min	■	■	■	■	■		□					□	□		□
30 min	■	□	■	□	□										□
40 min	□														
50 min															
60 min															
70 min	□	□				■								■	
80 min	■	■				■			■					■	
90 min	■	■							□					■	
100 min		■							□						
110 min		■							□						
120 min									□						

See legend Table A1. *Black* (*white*) squares stand for significant effect at 5% (10%)

Table A7 Consumer price index

	•	✓	×	•↑	•↓	•↘	•↗	✓↑	✓↓	✓↘	✓↗	×↑	×↓	×↘	×↗
10 min	■	■	■	■		■	□			■		■			
20 min	■	■	■	■		■	■			■	■	■			
30 min	■	■	□	□			■			■	■	□			
40 min	■	■					■			■	■				
50 min							■	■		□	■				
60 min				■				■							
70 min				■				■				□			
80 min	■	■		■				■				■			
90 min	■	■		■				■				■			
100 min	■	■		■								■			
110 min				□								■			
120 min															

See legend in Tables A1 and A6

Table A8 Retail sales

	•	✓	×	•↑	•↓	•↘	•↗	✓↑	✓↓	✓↘	✓↗	×↑	×↓	×↘	×↗
10 min	▪	▪	▪	▪	▪	▪		□		□		□	▪	▪	□
20 min	▪	▪	▪	▪	▪	▪			▪	▪		▪	▪	□	
30 min	▪	▪	▪	▪	▪	▪			▪	▪		▪	▪		
40 min	▪	▪	▪	▪	▪	□			▪	▪		▪	▪		
50 min	▪	▪	▪	▪	▪					▪		▪	▪		
60 min	□			□						▪					
70 min				□			▪								
80 min					▪		▪								
90 min	□				▪		□		□						
100 min	□				▪					□					
110 min										▪					
120 min										▪					

See legend in Tables A1 and A6

Table A9 Industrial production

	•	✓	×	•↑	•↓	•↘	•↗	✓↑	✓↓	✓↘	✓↗	×↑	×↓	×↘	×↗
10 min															
20 min	▪	▪		▪				▪		▪		▪		□	
30 min	▪	▪		▪		□		▪		▪		▪		▪	
40 min	▪	▪		▪				▪		▪		▪		▪	
50 min	▪	▪		▪				▪		□		▪		▪	
60 min				▪								▪			
70 min			□												
80 min		▪	□					□	□						
90 min		▪	□	▪				▪	□				▪		
100 min		▪	□	▪					□				▪		
110 min		□	□										▪		
120 min			□										▪		

See legend in Tables A1 and A6

Table A10 Housing starts

	•	✓	×	•↑	•↓	•↘	•↗	✓↑	✓↓	✓↘	✓↗	×↑	×↓	×↘	×↗
10 min	■	□	■							■					
20 min	■		■			■	■					□			■
30 min	■		■			□	■								■
40 min	□		■				■								■
50 min			□				■								
60 min															
70 min	□								□	■					
80 min	■	□	■						□			■			
90 min	■	□	■									■			
100 min	■	□	■									■			
110 min	■											■			
120 min															

See legend in Tables A1 and A6

Table A11 Durable goods

	•	✓	×	•↑	•↓	•↘	•↗	✓↑	✓↓	✓↘	✓↗	×↑	×↓	×↘	×↗
10 min	■	■	■	■	■	■	■		■		■			□	□
20 min	■	■	■			■	■							■	□
30 min	■		■			■	□							■	□
40 min			□											□	
50 min															
60 min															
70 min						■								□	
80 min	□		■			■				□				■	
90 min	□		■			■	■			□				■	
100 min	□					■	■			□				■	
110 min															
120 min		■									□				

See legend in Tables A1 and A6

Table A12 GDP, BI, TB and PI

	•			PI		
	GDP	BI	TB	•	✓	×
10 min			□			
20 min						
30 min				▪		
40 min				▪		
50 min				▪		
60 min				□		
70 min						
80 min						
90 min						
100 min	□					
110 min	□					
120 min						

See legend in Tables A1 and A6

References

Almon S (1965) The distributed lag between capital appropiations and expenditures. Econometrica 33:178–196

Andersen TG, Bollerslev T (1998) Deutsche mark-dollar volatility: intraday activity patterns, macroeconomic announcements, and longer run dependencies. J Finance LIII:219–265

Andersen TG, Bollerslev T, Diebold FX, Vega C (2003) Micro effects and macro announcements: real-time price discovery in foreign exchange. Am Econ Rev 93:38-62

Berkman NG (1978) On the significance of weekly changes in M1. New England economic review May–June 5–22

Conrad J, Cornell B, Landsman WR (2002) When is bad news really bad news? J Finance 57: 2507–2533

DeGennaro RP, Shrieves RE (1997) Public information releases, private information arrival and volatility in the foreign exchange market. J Empir Finance 4:295-315

Flemming MJ, Remolona EM (1999) What moves the bond market? J Portf Manage 25:28–38

Goodhart CAE, Hall SG, Henry SGB, Pesaran B (1993) News effects in a high frequency model of the sterling–dollar exchange rate. J Appl Econ 8:1–13

Grossman J (1981) The rationality of money supply expectations and the short-run response of interest rates to monetary surprises. J Money Credit Bank 13:409–424

Hautsch N, Hess D (2002) The processing of non-anticipated information in financial markets. Analyzing the impact of surprises in the employment report. European Finance Review 6: 133–161

Hess D (2004) Determinants of the relative price impact of unanticipated information in US macroeconomic releases. J Futures Mark 24:609–630

Kanehman D, Tversky A (1979) Prospect theory: an analysis of decision under uncertainty. Econometrica 47:263–291

Li L, Engle RF (1998) Macroeconomic announcements and volatility of treasury futures. UCSD Discussion Paper 98–27

McQueen G, Roley VV (1993) Stock prices, news and business conditions. Rev Financ Stud 6:683–707

Niemira MP, Zukowsky GF (1998) Trading the fundamentals. McGraw Hill, New York

Roley VV (1983) The response of short-term interest rates to weekly money announcements. J Money Credit Bank 15:344–354

Veronesi P (1999) Stock market overreaction to bad news in good time: a rational expectations equilibrium model. Rev Financ Stud 12:975–1007

Valeri Voev

Dynamic modelling of large-dimensional covariance matrices

Abstract Modelling and forecasting the covariance of financial return series has always been a challenge due to the so-called 'curse of dimensionality'. This paper proposes a methodology that is applicable in large-dimensional cases and is based on a time series of realized covariance matrices. Some solutions are also presented to the problem of non-positive definite forecasts. This methodology is then compared to some traditional models on the basis of its forecasting performance employing Diebold–Mariano tests. We show that our approach is better suited to capture the dynamic features of volatilities and covolatilities compared to the sample covariance based models.

1 Introduction

Modelling and forecasting the variances and covariances of asset returns is crucial for financial management and portfolio selection and re-balancing. Recently, this branch of the econometric literature has grown at a very fast pace. One of the simplest methods used is the sample covariance matrix. A stylized fact, however, is that there is a serial dependence in the second moments of returns. Thus, more sophisticated models had to be developed which incorporate this property, as well as other well-known features of financial return distributions such as leptokurtosis or the so-called 'leverage effect'. This led to the development of the univariate GARCH processes and their extension—the multivariate GARCH (MGARCH) models (for a comprehensive review see Bauwens et al. (2006)), which include also the modelling of covariances. One of the most severe drawbacks of the MGARCH models, however, is the difficulty of handling dimensions higher than 4 or 5 (or with very restrictive assumptions). Another more practically oriented field of research deals

Valeri Voev (✉)
University of Konstanz, CoFE, Konstanz, Germany
PBox D-124, University of Konstanz, 78457 Konstanz, Germany
E-mail: valeri.voev@uni-konstanz.de

with the problem of how to reduce the noise inherent in simpler covariance estimators such as the sample covariance matrix. Techniques have been developed to 'shrink' the sample covariance (SC) matrix, thereby reducing its extreme values in order to mitigate the effect of the so-called error maximization noted by Michaud (1989). One of the shrinkage estimators used among practitioners is the Black–Litterman model (Black and Litterman (1992)). This model uses a prior which reflects investor's beliefs about asset returns and combines it with implied equilibrium expected returns to obtain a posterior distribution, whose variance is a combination of the covariance matrix of implied returns and the confidence of the investor's views (which are reflected in the prior covariance). Further, Ledoit and Wolf (2003) and (2004) use shrinkage methods to combine a SC matrix with a more structured estimator (e.g. a matrix with equal pairwise correlations, or a factor model). The idea is to combine an asymptotically unbiased estimator having a large variance with a biased estimator, which is considerably less noisy. So the shrinkage actually amounts to optimizing in terms of the well-known trade-off between bias and variance.

Recently, with the availability of high-quality transaction databases, the technique of realized variance and covariance (RC) gained popularity. A very comprehensive treatment of volatility modelling with focus on forecasting appears in Andersen et al. (2006). Andersen et al. (2001a), among others, have shown that there is a long-range persistence (long memory) in daily realized volatilities, which allows one to obtain good forecasts by means of fractionally integrated ARMA processes. At the monthly level, we find that the autocorrelations decline quite quickly to zero, which led us to choose standard ARMA models for fitting and forecasting.

The aim of this paper is to compare the forecasting performance of a set of models, which are suitable to handle large-dimensional covariance matrices. Letting H denote the set of considered models, we have $H = \{s, ss, rm, rc, src, drc, dsrc\}$, where the first two models are based on the sample covariance matrix, the third model is a RiskMetrics™ exponentially weighted moving average (EWMA) estimator developed by J.P. Morgan (1996), the fourth and the fifth represent simple forecasts based on the realized and on the shrunk realized covariance matrix, and the last two models employ dynamic modelling of the RC and shrunk RC, respectively. We judge the performance of the models by looking at their ability to forecast individual variance and covariance series by employing a battery of Diebold–Mariano (Diebold and Mariano (1995)) tests. Of course, if we have good forecasts for the individual series, then the whole covariance matrix will also be well forecast. The practical relevance of a good forecast can be seen by considering an investor who faces an optimization problem to determine the weights of some portfolio constituents. One of the crucial inputs in this problem is a forecast of future movements and co-movements in asset returns. Our contribution is to propose a methodology which improves upon the sample covariance estimator and is easy to implement even for very large portfolios. We show that in some sense these models are more flexible than the MGARCH models, although this comes at the expense of some complications.

The remainder of the paper is organized as follows: Section 2 sets up the notation and describes the forecasting models, Section 3 presents the data set used to compare the forecasting performance of the models, Section 4 discusses the results on the forecast evaluation and Section 5 concludes the paper.

2 Forecasting models

In this section we describe each of the covariance forecasting models. First, we introduce some notation and description of the forecasting methodology. We concentrate on one-step ahead forecasts of covariance matrices of N stocks, and consider the monthly frequency. The information is updated every period and a new forecast is formed. Thus, each new forecast incorporates the newest information which has become available. Such a strategy might describe an active long-run investor, who revises and rebalances her portfolio every month. Let the multivariate price process be defined as $\mathbf{P} = \{\mathbf{P}_t(\omega), t \in (-\infty, \infty), \omega \in \Omega\}$, where Ω is an outcome space.[1] The portfolio is set up at $t = 0$ and updated at each $t = 1, 2, \ldots, \bar{T}$, where $\bar{T}$ is the end of the investment period. The frequency of the observations in our application is daily, which we refer to as intra-periods. In this setup, we can formally define the information set at each time $t \geq 0$ as a filtration $\mathcal{F}_t = \sigma(\mathbf{P}_s(\omega), s \in \mathcal{T})$ generated by $\mathbf{P}$, with $\mathcal{T} = \{s : s = -L + \frac{j}{M}, j = 0, 1, \ldots, (L+t)M\}$, M – the number of intra-periods within each period[2] and L – the number of periods, for which price data is available, before the investment period. It is important to note that not all information is considered in the forecasts based on the sample covariance matrix. For these models only the lower frequency monthly sampling is needed. Furthermore, we define the monthly returns as $\mathbf{r}_t = \ln(\mathbf{P}_t) - \ln(\mathbf{P}_{t-1})$, where $\mathbf{P}_t$ is the realization of the price process at time t, and the jth intra-period return by $\mathbf{r}_{t+\frac{j}{M}} = \ln\left(\mathbf{P}_{t+\frac{j}{M}}\right) - \ln\left(\mathbf{P}_{t+\frac{j-1}{M}}\right)$. The realized covariance at time $t+1$ is given by

$$\Sigma_{t+1}^{RC} = \sum_{j=1}^{M} \mathbf{r}_{t+\frac{j}{M}} \mathbf{r}'_{t+\frac{j}{M}}. \tag{1}$$

Assessing the performance of variance forecasts has been quite problematic, since the true covariance matrix Σ_t is not directly observable. This has long been a hurdle in evaluating GARCH models. Traditionally, the squared daily return was used as a measure of the daily variance. Although this is an unbiased estimator, it has a very large estimation error due to the large idiosyncratic noise component of daily returns. Thus a good model may be evaluated as poor, simply because the target is measured with a large error. In an important paper, Andersen and Bollerslev (1998) showed that GARCH models actually provide good forecasts when the target to which they are compared is estimated more precisely, by means of sum of squared intradaily returns. Since then, it has become a practice to take the realized variance as the relevant measure for comparing forecasting performance. In this spirit, we use the realized monthly covariance in place of the true matrix. Thus we will assess a given forecast $\hat{\Sigma}_{t+1|t}^{(h)}$, $h \in H$ by its deviation from Σ_{t+1}^{RC}.

[1] Of course, in reality the price process could not have started in the infinite past. Since we are interested in when the process became observable, and not in its beginning, we leave the latter unspecified.

[2] This number is not necessarily the same for all periods and should be denoted more precisely by $M(t)$. This is not done in the text to avoid cluttering of the notation.

2.1 A sample covariance forecast

In this section we describe a forecasting strategy based on the sample covariance matrix, which will serve as a benchmark. The sample covariance is a consistent estimator for the true population covariance under weak assumptions. We use a rolling window scheme and define the forecast as

$$\hat{\Sigma}^{(s)}_{t+1|t} = \frac{1}{T} \sum_{s=t-T+1}^{t} (\mathbf{r}_s - \bar{\mathbf{r}}_{t,T})(\mathbf{r}_s - \bar{\mathbf{r}}_{t,T})', \tag{2}$$

where for each t, $\bar{\mathbf{r}}_{t,T}$ is the sample mean of the return vector $\mathbf{r}$ over the last T observations. We will denote the sample covariance matrix at time t by Σ_t^{SC}. For T we choose a value of 60, which with monthly data corresponds to a time span of 5 years. As the near future is of the highest importance in volatility forecasting, this number might seem too large. Too small a number of periods, however, would lead to a large variance of the estimator; therefore other authors (e.g. Ledoit and Wolf (2004)) have also chosen 60 months as a balance between precision and relevance of the data. A problem of this approach, as simple as it is, is that new information is given the same weight as very old information. Another obvious oversimplification is that we do not account for the serial dependence present in the second moments of financial returns.

2.2 A shrinkage sample covariance forecast

In this section we briefly present the shrinkage estimator, proposed by Ledoit and Wolf (2003), in order to give an idea of the shrinkage principle.

The shrinkage estimator of the covariance matrix Σ_t is defined as a weighted linear combination of some shrinkage target F_t and the sample covariance matrix, where the weights are chosen in an optimal way. More formally, the estimator is given by

$$\Sigma_t^{SS} = \hat{\alpha}_t^* F_t + (1 - \hat{\alpha}_t^*)\Sigma_t^{SC}, \tag{3}$$

$\hat{\alpha}_t^* \in [0, 1]$ is an estimate of the optimal shrinkage constant α_t^*.

The shrinking intensity is chosen to be optimal with respect to a loss function defined as a quadratic distance between the true and the estimated covariance matrices based on the Frobenius norm. The Frobenius norm of an $N \times N$ symmetric matrix Z with elements $(z_{ij})_{i,j=1,\ldots,N}$ is defined by

$$\|Z\|^2 = \sum_{i=1}^{N} \sum_{j=1}^{N} z_{ij}^2. \tag{4}$$

The quadratic loss function is the Frobenius norm of the difference between Σ_t^{SS} and the true covariance matrix:

$$L(\alpha_t) = \left\| \alpha_t F_t + (1 - \alpha_t)\Sigma_t^{SC} - \Sigma_t \right\|^2. \tag{5}$$

The optimal shrinkage constant is defined as the value of α which minimizes the expected value of the loss function in expression (5):

$$\alpha_t^* = \underset{\alpha_t}{\operatorname{argmin}} \, \mathrm{E}\left[L(\alpha_t)\right]. \tag{6}$$

For an arbitrary shrinkage target F and a consistent covariance estimator S, Ledoit and Wolf (2003) show that

$$\alpha^* = \frac{\sum_{i=1}^N \sum_{j=1}^N \left(\mathrm{Var}\left[s_{ij}\right] - \mathrm{Cov}\left[f_{ij}, s_{ij}\right]\right)}{\sum_{i=1}^N \sum_{j=1}^N \left(\mathrm{Var}\left[f_{ij} - s_{ij}\right] + (\phi_{ij} - \sigma_{ij})^2\right)}, \tag{7}$$

where f_{ij} is a typical element of the sample shrinkage target, s_{ij} – of the covariance estimator, σ_{ij} – of the true covariance matrix, and ϕ_{ij} – of the population shrinkage target Φ. Further they prove that this optimal value is asymptotically constant over T and can be written as[3]

$$\kappa_t = \frac{\pi_t - \rho_t}{\nu_t}. \tag{8}$$

In the formula above, π_t is the sum of the asymptotic variances of the entries of the sample covariance matrix scaled by $\sqrt{T}$: $\pi_t = \sum_{i=1}^N \sum_{j=1}^N \mathrm{AVar}\left[\sqrt{T} s_{ij,t}\right]$, ρ_t is the sum of asymptotic covariances of the elements of the shrinkage target with the elements of the sample covariance matrix scaled by $\sqrt{T}$: $\rho_t = \sum_{i=1}^N \sum_{j=1}^N \mathrm{ACov}\left[\sqrt{T} f_{ij,t}, \sqrt{T} s_{ij,t}\right]$, and ν_t measures the misspecification of the shrinkage target: $\nu_t = \sum_{i=1}^N \sum_{j=1}^N (\phi_{ij,t} - \sigma_{ij,t})^2$. Following their formulation and assumptions, $\sum_{i=1}^N \sum_{j=1}^N \mathrm{Var}\left[\sqrt{T}(f_{ij} - s_{ij})\right]$ converges to a positive limit, and so $\sum_{i=1}^N \sum_{j=1}^N \mathrm{Var}\left[f_{ij} - s_{ij}\right] = \mathrm{O}(1/T)$. Using this result and the $\sqrt{T}$ convergence in distribution of the elements of the sample covariance matrix, Ledoit and Wolf (2003) show that the optimal shrinkage constant is given by

$$\alpha_t^* = \frac{1}{T} \frac{\pi_t - \rho_t}{\nu_t} + O\left(\frac{1}{T^2}\right). \tag{9}$$

Since α^* is unobservable, it has to be estimated. Ledoit and Wolf (2004) propose a consistent estimator of α^* for the case where the shrinkage target is a matrix in which all pairwise correlations are equal to the same constant. This constant is the average value of all pairwise correlations from the sample correlation matrix. The covariance matrix resulting from combining this correlation matrix with the sample variances, known as the equicorrelated matrix, is the shrinkage target. The equicorrelated matrix is a sensible shrinkage target as it involves only a small number of free parameters (hence less estimation noise). Thus the elements of the sample covariance matrix, which incorporate a lot of estimation error and hence can take rather extreme values are 'shrunk' towards a much less noisy average.

[3] In their paper the formula appears without the subscript t. By adding it here we want to emphasize that these variables are changing over time.

Using the equicorrelated matrix as the shrinkage target F_t in Eq. (3) the forecast is given by

$$\hat{\Sigma}_{t+1|t}^{(ss)} = \Sigma_t^{SS}. \tag{10}$$

2.3 A RiskMetrics™ forecast

The RiskMetrics™ forecasting methodology is a modification of the sample covariance matrix, in which observations which are further in the past are given exponentially smaller weights, determined by a factor λ. For the generic (i, j), $i, j = 1, \ldots, N$ element of the EWMA covariance matrix Σ_t^{RM} we have

$$\sigma_{ij,t}^{RM} = (1 - \lambda) \sum_{s=1}^{t} \lambda^{s-1} \left(r_{i,s} - \bar{r}_i\right) \left(r_{j,s} - \bar{r}_j\right), \tag{11}$$

where $\bar{r}_i = \frac{1}{t} \sum_{s=1}^{t} r_{i,s}$. Again, the forecast is given by

$$\hat{\Sigma}_{t+1|t}^{(rm)} = \Sigma_t^{RM}. \tag{12}$$

Methods to choose the optimal λ are discussed in J.P. Morgan (1996). In this paper we set $\lambda = 0.97$, the value used by J.P. Morgan for monthly (co)volatility forecasts. Note that contrary to the sample covariance matrix, for which we use a rolling window scheme, in the RiskMetrics approach we use at each t all the available observations from the beginning of the observation period up to t. Since in the RiskMetrics approach the weights decrease exponentially, the observations which are further away in the past are given relatively smaller weights and hence do not influence the estimate as much as in the sample covariance matrix.

2.4 A simple realized covariance forecast

The realized covariance estimator was already defined in expression (1). Its univariate and multivariate properties have been studied among others, by Barndorff-Nielsen and Shephard (2004) and by Andersen et al. (2003). In the limit, when $M \to \infty$, Barndorff-Nielsen and Shephard (2004) have shown that realized covariance is an error-free measure for the integrated covariation of a very broad class of stochastic volatility models. In the empirical part we compute monthly realized covariance by using daily returns (see also French et al. (1987)). The simple forecast is defined by

$$\hat{\Sigma}_{t+1|t}^{(rc)} = \Sigma_t^{RC}. \tag{13}$$

Thus an investor who uses this strategy simply computes the realized covariance at the end of each month and then uses it as his best guess about the true covariance matrix of the next month. A nice feature of this method is that it only uses recent information which is of most value for the forecast, but unfortunately, it imposes a very simple and restrictive time dependence. Practically, Eq. (13) states that all variances and covariances follow a random walk process. However, as we shall see later, the estimated series of monthly variances and covariances show weak stationarity.

2.5 A shrinkage realized covariance forecast

Although the estimator discussed in the previous section is asymptotically error-free, in practice one cannot record observations continuously. A much more serious problem is the fact that at very high frequencies, the martingale assumption needed for the convergence of the realized covariances to the integrated covariation is no longer satisfied. At trade-by-trade frequencies, market microstructure affects the price process and results in microstructure noise induced autocorrelations in returns and hence biased variance estimates. Methods to account for this bias and correct the estimates have been developed by Hansen and Lunde (2006), Oomen (2005), Aït-Sahalia et al. (2005), Bandi and Russell (2005), Zhang et al. (2005), and Voev and Lunde (2007), among others. At low frequencies the impact of market microstructure noise can be significantly mitigated, but this comes at the price of higher variance of the estimator. Since we are using daily returns, market microstructure is not an issue. Thus we will suggest a possible way to reduce variance. Again as in Section 2.2, we will try to find a compromise between bias and variance applying the shrinkage methodology. The estimator looks very much like the one in expression (3). In this case we have

$$\Sigma_t^{SRC} = \hat{\alpha}_t^* F_t + (1 - \hat{\alpha}_t^*)\Sigma_t^{RC}, \tag{14}$$

where now F_t is the equicorrelated matrix, constructed from the realized covariance matrix Σ_t^{RC} in the same fashion as the equicorrelated matrix constructed from the sample covariance matrix, as explained in Section 2.2. Similarly to the previous section, the forecast is simply

$$\hat{\Sigma}_{t+1|t}^{(src)} = \Sigma_t^{SRC}. \tag{15}$$

Since the realized covariance is a consistent estimator, we can still apply formula (7) taking into account the different rate of convergence. In order to compute the estimates for the variances and covariances, we need a theory for the distribution of the realized covariance, which is developed in Barndorff-Nielsen and Shephard (2004), who provide asymptotic distribution results for the realized covariation matrix of continuous stochastic volatility semimartingales ($\mathcal{SVSM}^c$). Assuming that the log price process $\ln \mathbf{P} \in \mathcal{SVSM}^c$, we can decompose it as $\ln \mathbf{P} = a^* + m^*$, where a^* is a process with continuous finite variation paths and m^* is a local martingale. Furthermore, under the condition that m^* is a multivariate stochastic volatility process, it can be defined as $m^*(t) = \int_0^t \Theta(u)\mathrm{d}w(u)$, where Θ is the spot covolatility process and w is a vector standard Brownian motion. Then the spot covariance is defined as

$$\Sigma(t) = \Theta(t)\Theta(t)', \tag{16}$$

assuming that (for all $t < \infty$)

$$\int_0^t \Sigma_{kl}(u)\mathrm{d}u < \infty, \quad k, l = 1, \ldots, N, \tag{17}$$

where $\Sigma_{kl}(t)$ is the (k,l) element of the $\Sigma(t)$ process. With this notation, we will now interpret the 'true' covariance matrix as

$$\Sigma_{t+1} = \int_t^{t+1} \Sigma(u) \mathrm{d}u. \tag{18}$$

Thus the covariance matrix at time $t+1$ is the increment of the integrated covariance matrix of the continuous local martingale from time t to time $t+1$. The realized covariance as defined in expression (1) consistently estimates Σ_{t+1} as given in Eq. (18). Furthermore, Barndorff-Nielsen and Shephard (2004) show that under a set of regularity conditions the realized covariation matrix follows asymptotically, as $M \to \infty$, the normal law with $N \times N$ matrix of means $\int_t^{t+1} \Sigma(u)\mathrm{d}u$. The asymptotic covariance of

$$\sqrt{M}\left\{\Sigma_{t+1}^{RC} - \int_t^{t+1} \Sigma(u)\mathrm{d}u\right\}$$

is Ω_{t+1}, a $N^2 \times N^2$ array with elements

$$\Omega_{t+1} = \left\{\int_t^{t+1} \{\Sigma_{kk'}(u)\Sigma_{ll'}(u) + \Sigma_{kl'}(u)\Sigma_{lk'}(u)\}\,\mathrm{d}u\right\}_{k,k',l,l'=1,\ldots,N}.$$

Of course, this matrix is singular due to the equality of the covariances in the integrated covariance matrix. This can easily be avoided by considering only its unique lower triangular elements, but for our purposes it will be more convenient to work with the full matrix. The result above is not useful for inference, since the matrix Ω_{t+1} is not known. Barndorff-Nielsen and Shephard (2004) show that a consistent, positive semi-definite estimator is given by a random $N^2 \times N^2$ matrix:

$$H_{t+1} = \sum_{j=1}^{M} x_{j,t+1}x'_{j,t+1} - \frac{1}{2}\sum_{j=1}^{M-1}\left(x_{j,t+1}x'_{j+1,t+1} + x_{j+1,t+1}x'_{j,t+1}\right), \tag{19}$$

where $x_{j,t+1} = vec\left(\mathbf{r}_{t+\frac{j}{M}}\mathbf{r}'_{t+\frac{j}{M}}\right)$ and the vec operator stacks the columns of a matrix into a vector. It holds that $MH_{t+1} \xrightarrow{p} \Omega_{t+1}$ with $M \to \infty$.

With the knowledge of this matrix, we can combine the asymptotic results for the realized covariance, with the result in Eq. (7) to compute the estimates for π_t, ρ_t and ν_t.

For the equicorrelated matrix F we have that[4] $f_{ij} = \bar{r}\sqrt{\sigma_{ii}^{(RC)}\sigma_{jj}^{(RC)}}$, where $\bar{r}$ is the average value of all pairwise correlations, implied by the realized covariance matrix, and $\sigma_{ij}^{(RC)}$ is the (i,j) element of the realized covariance matrix. Thus Φ, the population equicorrelated matrix, has a typical element $\phi_{ij} = \bar{\varrho}\sqrt{\sigma_{ii}\sigma_{jj}}$, where σ_{ij} is the (i,j) of the true covariance matrix Σ and $\bar{\varrho}$ is the average correlation

[4] In the following exposition, the time index is suppressed for notational convenience.

implied by it. Substituting $\sigma_{ij}^{(RC)}$ for s_{ij} in Eq. (7) and multiplying by M gives for the optimal shrinkage intensity:

$$M\alpha^* = \frac{\sum_{i=1}^{N}\sum_{j=1}^{N}\left(\mathrm{Var}\left[\sqrt{M}\sigma_{ij}^{(RC)}\right] - \mathrm{Cov}\left[\sqrt{M}f_{ij}, \sqrt{M}\sigma_{ij}^{(RC)}\right]\right)}{\sum_{i=1}^{N}\sum_{j=1}^{N}\left(\mathrm{Var}\left[f_{ij} - \sigma_{ij}^{(RC)}\right] + (\phi_{ij} - \sigma_{ij})^2\right)}. \quad (20)$$

Note that this equation resembles expression (8). The only difference is the scaling by $\sqrt{M}$ instead of $\sqrt{T}$, which is due to the $\sqrt{M}$ convergence. In this case π_t, the first summand in the numerator, is simply the sum of all diagonal elements of Ω_t. By using the definition of the equicorrelated matrix, it can be shown that the second term, ρ_t, can be written as (suppressing the index t)

$$\rho = \sum_{i=1}^{N} \mathrm{AVar}\left[\sqrt{M}\sigma_{ii}^{(RC)}\right] + \sum_{i=1}^{N}\sum_{j=1, j\neq i}^{N} \mathrm{ACov}\left[\sqrt{M}\bar{r}\sqrt{\sigma_{ii}^{(RC)}\sigma_{jj}^{(RC)}}, \sqrt{M}\sigma_{ij}^{(RC)}\right]. \quad (21)$$

Applying the delta method the second term can be expressed as[5]

$$\frac{\bar{r}}{2}\left(\sqrt{\frac{\sigma_{jj}^{(RC)}}{\sigma_{ii}^{(RC)}}}\mathrm{ACov}\left[\sqrt{M}\sigma_{ii}^{(RC)}, \sqrt{M}\sigma_{ij}^{(RC)}\right] + \sqrt{\frac{\sigma_{ii}^{(RC)}}{\sigma_{jj}^{(RC)}}}\mathrm{ACov}\left[\sqrt{M}\sigma_{jj}^{(RC)}, \sqrt{M}\sigma_{ij}^{(RC)}\right]\right).$$

From this expression we see that ρ also involves summing properly scaled terms of the Ω matrix. In the denominator of Eq. (20), the first term is of order $O(1/M)$, and the second one is consistently estimated by $\hat{\nu} = \sum_{i=1}^{N}\sum_{j=1}^{N}\left(f_{ij} - \sigma_{ij}^{(RC)}\right)^2$.

Since we have a consistent estimator for Ω, we can now also estimate π and ρ. In particular, we have

$$\hat{\pi} = \sum_{i=1}^{N}\sum_{j=1}^{N} h_{ij,ij}$$

$$\hat{\rho} = \sum_{i=1}^{N} h_{ii,ii} + \frac{\bar{r}}{2}\sum_{i=1}^{N}\sum_{j=1}^{N}\sqrt{\frac{\sigma_{jj}^{(RC)}}{\sigma_{ii}^{(RC)}}}h_{ii,ij} + \sqrt{\frac{\sigma_{ii}^{(RC)}}{\sigma_{jj}^{(RC)}}}h_{jj,ij},$$

[5] cf. Ledoit and Wolf (2004).

where $h_{kl,k'l'}$ is the element of H which estimates the corresponding element of Ω. Thus we can estimate κ_t by $\hat{\kappa}_t = \frac{\hat{\pi}_t - \hat{\rho}_t}{\hat{\gamma}_t}$ and the estimator for the optimal shrinkage constant is

$$\hat{\alpha}_t^* = \max\left\{0, \min\left\{\frac{\hat{\kappa}_t}{M}, 1\right\}\right\}. \tag{22}$$

The estimated optimal shrinkage constants for our dataset range from 0.0205 to 0.2494 with a mean of 0.0562.

2.6 Dynamic realized covariance forecasts

This model is an alternative to the one in Section 2.4. The most popular models for time varying variances and covariances are the GARCH models. The most significant problem of these models is the large number of parameters in large-dimensional systems. The recent DCC models of Tse and Tsui (2002) and Engle (2002) propose a way to mitigate this problem by using the restriction that all correlations obey the same dynamics. Recently, Gourieroux et al. (2004) have suggested an interesting alternative—the WAR (Wishart autoregressive) model, which has certain advantages over the GARCH models, e.g. smaller number of parameters, easy construction of non-linear forecasts, simple verification of stationarity conditions, etc. Even quite parsimonious models, however, have a number of parameters of the order $N(N+1)/2$. With $N = 15$ this means more than 120 parameters, which would be infeasible for estimation. We therefore suggest a simple approach in which all variance and covariance series are modelled univariately as ARMA processes and individual forecasts are made, which are then combined into a forecast of the whole matrix. This approach can also be extended by including lags of squared returns, which can be interpreted as of ARCH-effects. A theoretical drawback of this model is that such a methodology does not guarantee the positive definiteness of the forecast matrix. It turns out that this problem could be quite severe, especially if we include functions of lagged returns in the specification. Hence we propose two possible solutions. First, if the above mentioned problem occurs relatively rarely, then in these cases we can define the forecast as in Section 2.4, which would ensure that all forecast matrices are positive definite. More precisely, instead of assuming a random walk process for the realized covariance series (as in Section 2.4) we now model each of them as ARMAX$(p, q, 1)$[6] processes as follows:

$$\sigma_{ij,t}^{(RC)} = \omega + \sum_{s=1}^{p} \varphi_s \sigma_{ij,t-s}^{(RC)} + \sum_{u=0}^{q} \theta_u \varepsilon_{ij,t-u} + \alpha r_{i,t-1} r_{j,t-1}, \tag{23}$$

with $\theta_0 = 1$ and $\varepsilon_{ij,t}$, a Gaussian white noise process. The model easily extends to an ARMAX(p, q, k) specification with k lags of crossproducts. The parameters φ_s, θ_u and α are estimated by maximum likelihood starting at $t = 100$ and the forecasts $\hat{\sigma}_{ij,t+1|t}^{(RC)}$ are collected in a matrix Σ_{t+1}^{DRC}. At time $t+1$ the new information

[6] The last parameter shows the number of lags of the X variable.

is taken into account and the procedure is repeated. The best model for each series is selected by minimizing the Akaike information criterion (AIC).

In this case the forecast is

$$\hat{\Sigma}^{(drc)}_{t+1|t} = \begin{cases} \Sigma^{DRC}_{t+1}, & \text{if } \Sigma^{DRC}_{t+1} \text{ is positive definite} \\ \Sigma^{RC}_{t}, & \text{otherwise.} \end{cases} \tag{24}$$

A more robust solution is to factorize the sequence of realized covariance matrices into their Cholesky decompositions, model the dynamics and forecast the Cholesky series, and then reconstruct the variance and covariance forecasts. This ensures the positive definiteness of the resulting forecast. In this case the Cholesky series are modelled like in Eq. (23), the forecasts are collected in a lower triangular matrix $\mathbf{C}_{t+1}$ and the covariance forecast is given by

$$\hat{\Sigma}^{(drc-Chol)}_{t+1|t} = \mathbf{C}_{t+1}\mathbf{C}'_{t+1}. \tag{25}$$

Analogously, we can use these two strategies to model dynamically the series of shrunk variance covariance matrices which defines the forecasts $\Sigma^{(dsrc)}_{t+1|t}$ and $\Sigma^{(dsrc-Chol)}_{t+1|t}$.

3 Data

The data we have used consists of 15 stocks from the current composition of the Dow Jones Industrial Average index from 1 January 1980 to 31 December 2002. The stocks are Alcoa (NYSE ticker symbol: AA), American Express Company (AXP), Boeing Company (BA), Caterpillar Inc. (CAT), Coca-Cola Company (KO), Eastman Kodak (EK), General Electric Company (GE), General Motors Corporation (GM), Hewlett-Packard Company (HPQ), International Business Machines (IBM), McDonald's Corporation (MCD), Philip Morris Companies Incorporated (MO), Procter & Gamble (PG), United Technologies Corporation (UTX) and Walt Disney Company (DIS). The reason that we have considered only 15 stocks is due to the fact that the realized covariance matrices are of full rank only if $M > N$, where M is the number of intra-period observations used to construct the realized covariance, in our case number of daily returns used to construct each monthly realized covariance. Usually there are 21 trading days per month, but some months have had fewer trading days (e.g. September 2001). With intradaily data this problem would not be of importance, since then we can easily have hundreds of observations within a day. Such datasets are already common, but they still do not cover large periods of time. Nevertheless, the dynamic properties of daily realized volatilities, covariances and correlations are studied by, e.g. Andersen et al. (2001a) and Andersen et al. (2001b). It has been shown that there is a long-range persistence, which allows for construction of good forecasts by means of ARFIMA processes.

All the stocks are traded on the NYSE and we take the daily closing prices and monthly closing prices to construct corresponding returns. The data is adjusted for splits and dividends. We find the typical properties of financial returns: negative skewness (with the exception of PG), leptokurtosis and non-normality. The average (across stocks) mean daily return is 0.05% and the average daily standard deviation is 1.9%. From the daily data log monthly returns are constructed by using the opening price of the first trading day of the month and the closing price of the last day. These returns are then used to construct rolling window sample covariance matrices, used in the first two forecasting models.

4 Results

In this section we present and discuss the results on the performance of the forecasting models described in Section 2.

In order to asses the forecasting performance, we employ Diebold–Mariano tests for each of the variance and covariance series. Then we measure the deviation of the forecast as a matrix from its target by using again the Frobenius norm, which gives an overall idea of the comparative performance of the models. Of course, if the individual series are well forecast, so will be the matrix. As a target or 'true' covariance matrix, we choose the realized covariance matrix. First, we present some graphical results. Out of the total of 120 variance and covariance forecast series, Figure 1 plots nine representative cases, for the sample covariance and the RiskMetrics™ model, against the realized series. The name, which appears above each block in the figure, represents either a variance series (e.g. EK), or a covariance one (e.g. GE,AA).

Both forecasts are quite close, and as can be seen, they cannot account properly for the variation in the series. As the tests show, however, the Riskmetrics™ fares better and is the best model among the sample based ones. It is already an acknowledged fact that financial returns have the property of volatility clustering. This feature is also clearly evident in the figure, where periods of low and high volatility can be easily distinguished, which suggests that variances and covariances tend to exhibit positive autocorrelation. Figure 2 shows the autocorrelation functions for the same nine series of realized (co)variances. The figure clearly shows that there is some positive serial dependence, which usually dies out quickly, suggesting stationarity of the series. Stationarity is also confirmed by running Augmented Dickey–Fuller (ADF) tests, which reject the presence of an unit root in all series at the 1% significance level.

The observed dependence patterns suggest the idea of modelling the variance and covariance series as well as their shrunk versions as ARMA processes. This resulted in a few cases in which the matrix forecast was not positive definite (16 out of 176 for the original series and 8 out of 176 for the shrunk series). Thus the forecast in expression (24) seems to be reasonable and as we shall see later, compares well to the sample covariance based models. In a GARCH framework, the conditional variance equation includes not only lags of the variance, but also lags of squared innovations (shocks). When mean returns are themselves unpredictable (the usual approach is to model the mean equation as an ARMA process), the shock is simply the return. This fact led us to include lags of squared returns (for the variance series) and cross-products (for the covariance series) as in the ARMAX(p, q, 1) model in Eq. (23). This added flexibility, however, comes at the price of a drastic increase of the non-positive definite forecasts (108 and 96 out of 176, respectively). Thus the forecast in Eq. (24) comes quite close to the simple realized and shrunk realized covariance models in Sections 2.4 and 2.5, respectively. A solution to this issue is to decompose the matrices into their lower triangular Cholesky factors, forecast

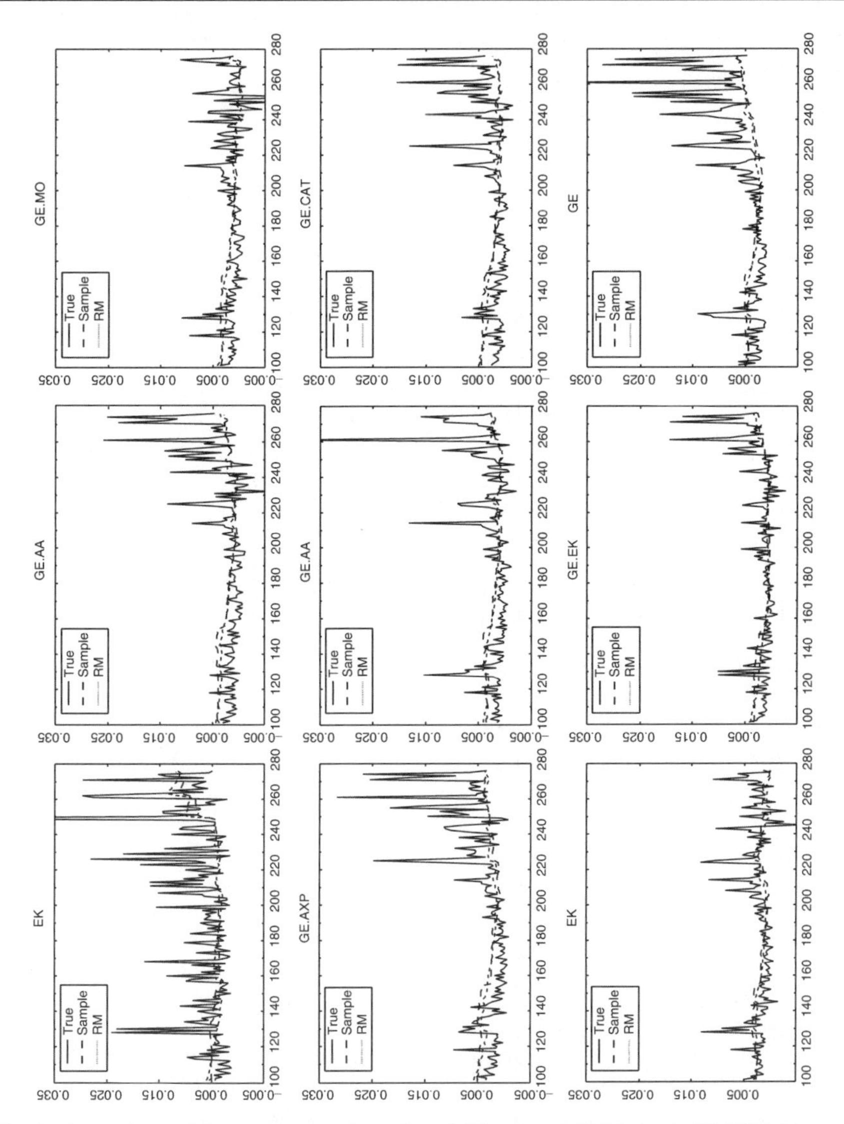

Fig. 1 Comparison of the sample covariance based (Sample) and Riskmetrics™ (RM) forecast against the realized covariance (True).

the Cholesky series, and then reconstruct the matrix. This leads to the forecasting formula in Eq. (25), which defines the $drc - Chol$ and $dsrc - Chol$ forecasting models for the simple realized and shrunk realized covariance case, respectively. A drawback of this approach is that the Cholesky series do not have an intuitive interpretation. They are simply used as a tool to constrain the forecasts to satisfy the complicated restrictions implied by the positive definiteness requirement. Another drawback is that the Cholesky decomposition involves nonlinear transformations of the original series. Thus, if one can adequately forecast the nonlinear transformation, this does not immediately mean that applying the inverse transformation

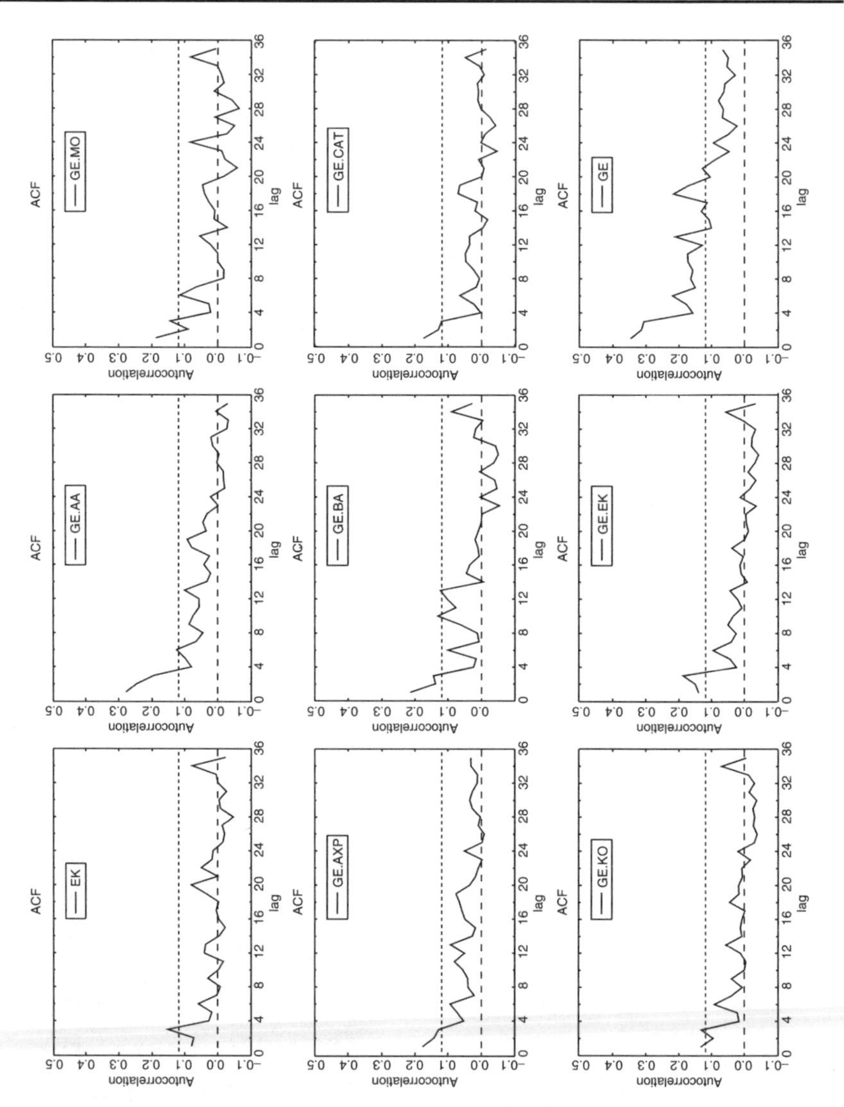

Fig. 2 Autocorrelation functions of the realized variance and covariance series.

to the forecast will produce a good forecast of the initial series. So there is a trade-off between the possibility of including more information in the forecast and obtaining positive definite matrices on the one hand, and the distortions caused by the non-linearity of the transformation on the other. It turns out that in our case the beneficial effects outweigh the negative ones. Figure 3 shows the $drc - Chol$ and the RiskMetrics™ forecast for the same nine variance and covariance series. From the figure it is evident that the dynamic forecasts track the true series much closer than the RiskMetrics™ forecasts, especially at the end of the period when the (co)volatilities were more volatile. The $dsrc - Chol$ forecast looks quite similar

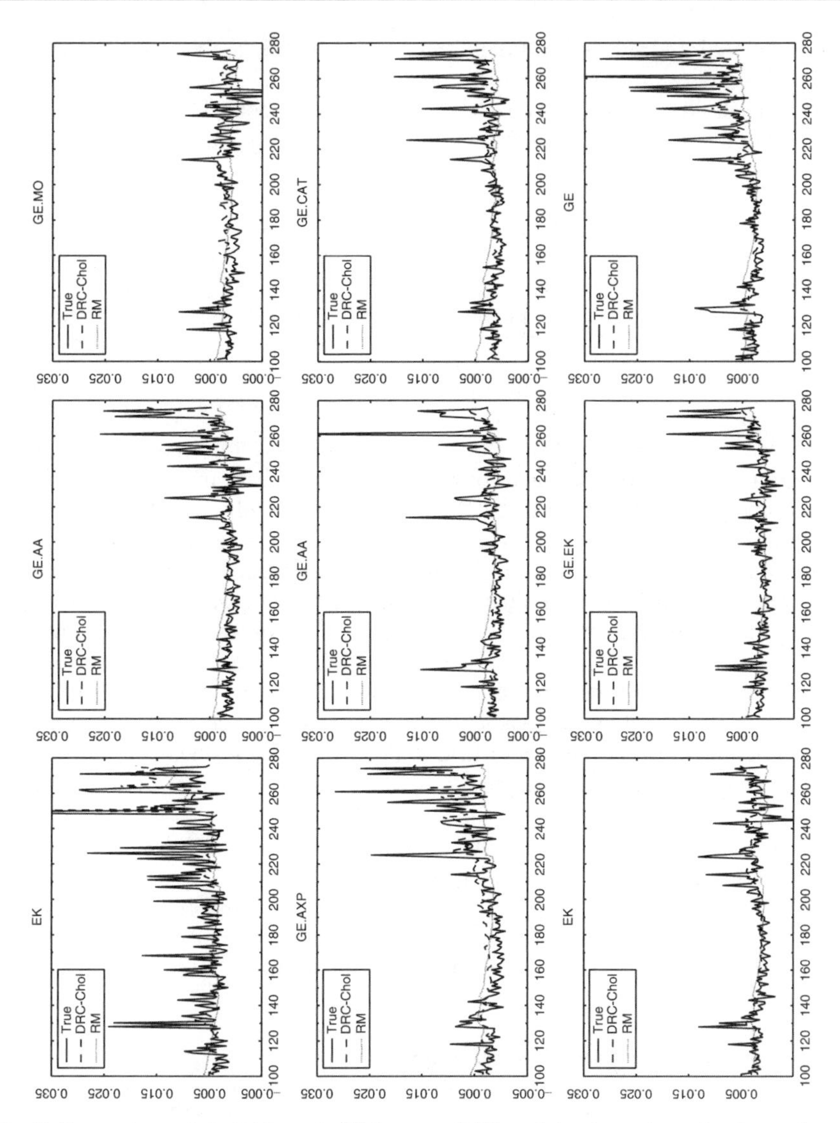

Fig. 3 Comparison of the Riskmetrics™ forecast (RM) and the dynamic realized covariance forecast based on Cholesky series (DRC-Chol) against the realized covariance (True).

to the $drc - Chol$ (due to the usually small shrinkage constants), but as we shall see later the forecasts are in fact somewhat better.

Turning to the statistical comparison of the forecasting methods, we first briefly present the Diebold–Mariano testing framework as in Harvey et al. (1997). Suppose a pair of l-step ahead forecasts h_1 and h_2, $h_1, h_2 \in H$ have produced errors (e_{1t}, e_{2t}), $t = 1, \ldots, T$. The null hypothesis of equality of forecasts is based on some function $g(e)$ of the forecast errors and has the form $\mathrm{E}\left[g(e_{1t}) - g(e_{2t})\right] = 0$. Defining the loss differential $d_t = g(e_{1t}) - g(e_{2t})$ and its average $\bar{d} = T^{-1} \sum_{t=1}^{T} d_t$, the authors note that 'the series d_t is likely to be autocorrelated. Indeed, for optimal l-steps ahead forecasts, the sequence of forecast errors follows a moving average process of order $(l-1)$. This result can be expected to hold approximately for any reasonably well-conceived set of forecasts.' Consequently, it can be shown that the variance of $\bar{d}$ is, asymptotically,

$$\mathrm{Var}\left[\bar{d}\,\right] \approx T^{-1}\left[\gamma_0 + 2\sum_{k=1}^{l-1}\gamma_k\right], \tag{26}$$

where γ_k is the kth autocovariance of d_t. The Diebold–Mariano test statistic is

$$S_1 = \left[\widehat{\mathrm{Var}}\left[\bar{d}\,\right]\right]^{-1/2}\bar{d}, \tag{27}$$

where $\widehat{\mathrm{Var}}\left[\bar{d}\,\right]$ is obtained from Eq. (26) by substituting for γ_0 and γ_k the sample variance and autocovariances of d_t, respectively. Tests are then based on the asymptotic normality of the test statistic. Noting that we only consider 1-step ahead forecasts in this paper, the series d_t should not be autocorrelated. As already noted above, this is expected to hold for any *reasonably* constructed forecasts. Actually, however, the sample based forecasts are not really *reasonable* in the sense that they do not account for the serial dependence of the process they are supposed to forecast. Thus, the degree of autocorrelation in the d_t series, when either h_1 or h_2 is a sample based forecast, will correspond to the degree of dependence in the series to be forecast. For this reason, ignoring autocovariances in the construction of the Diebold–Mariano tests will lead to an error in the test statistic. To correct for this we include in $\widehat{\mathrm{Var}}\left[\bar{d}\,\right]$ the first k significant autocorrelations for each of the 120 series.

Table 1 summarizes the results of the Diebold–Mariano tests carried out pairwise between all models for all 120 series.

The first entry in each cell of the table shows the number of series (out of 120) for which the model in the corresponding column outperforms the model in the corresponding row. The second entry corresponds to the number of significant

Table 1 Results from the Diebold–Mariano tests

	s	*ss*	*rm*	*rc*	*src*	*drc*	*dsrc*	*drc–Chol*	*dsrc–Chol*
s	–	**85/28**	106/50	14/1	16/1	47/20	89/37	93/49	100/55
ss	**20/0**	–	106/47	14/1	16/1	47/20	89/37	92/49	100/55
rm	14/0	14/0	–	7/1	11/1	37/7	73/29	85/33	89/37
rc	106/60	106/61	113/69	–	**105/86**	119/59	120/88	115/80	117/88
src	104/55	104/56	109/69	**0/0**	–	119/50	120/86	114/77	117/85
drc	73/12	73/12	83/26	1/0	1/0	–	**104/31**	98/47	103/58
dscr	31/3	31/3	47/8	0/0	0/0	**1/0**	–	69/28	83/35
drc (Chol)	27/8	28/8	35/10	5/1	6/1	22/7	51/12	–	91/19
dsrc (Chol)	20/7	20/7	31/8	3/1	3/1	17/6	37/11	29/3	–

Note: Due to the definition of the shrinkage target, the first numbers in the pairs highlighted in bold do not sum up to 120, since the variance series are unchanged in their respective 'shrunk' versions. Thus, in these cases there are only 105 series forecasts to be compared.

outperformances according to the Diebold–Mariano tests at the 5% significance level. Hence, the table is in a sense symmetric, as the number of times model h_1 outperforms model h_2 plus the number of times model h_2 outperforms model h_1 (given by the first number in each cell) sum up to 120—the total number of series. This is not the case, only for the pairs highlighted in bold, because the 15 variance series are unchanged in their respective 'shrunk' versions.[7] Thus, in these cases there are only 105 covariance series forecasts to be compared.

At first glance one can notice that the worst performing models are the rc and src models. Among the sample based forecasts the RiskMetrics™ is the one which delivers the best performances. The comparison between the sample and the shrinkage sample forecasts shows that shrinking has indeed improved upon the sample covariance matrix. This holds also for the realized covariance matrix. Here, the result is reinforced by the fact that shrinking also increases the probability of obtaining a positive definite forecast. In fact, the quite poor performance of the drc model is not due to the poor forecasting of the series themselves, but due to the large error, introduced by taking the previous realized covariance matrix, in case of a non-positive definite forecast (see Eq. (24)). Even though this only happens in 16 out of 176 cases, it is enough to distort the forecast considerably. The main result of this paper, however, arises from the comparison of the dynamic models with the sample based ones, which can be drawn by considering the last three columns of the table. For most of the series the dynamic models provide better forecasts, which results in smaller errors in the covariance matrix forecasts, as will be shown later. Despite the fact that the number of significant outperformances is not strikingly high (due to the small number of periods for evaluation), it is still clear that the dynamic models outperform decisively even the best model among the sample based ones. Furthermore, as noted earlier, the forecasts using the Cholesky decomposition appear to be better compared to those which model the variance and covariance series directly. This result comes mainly as a consequence of the considerable explanatory power of the lagged shocks in addition to the lagged (co)variances, which could not have been utilized had not we assured the positive definiteness of the forecasts.

In order to understand better the benefits from modelling the variance and covariance series dynamically, we shall consider an alternative (but closely related) measure of forecasting error. In Section 2.2 it was shown how the Frobenius norm can be used as a measure of distance between two matrices. Here we will utilize this concept again by considering the following definition of the forecast error in terms of a matrix forecast

$$e_t^{(h)} = \left\| \hat{\Sigma}_{t|t-1}^{(h)} - \Sigma_t^{RC} \right\|^2, \quad h \in H. \tag{28}$$

The root mean squared prediction errors (RMSPE) are collected in Table 2.

The ranking of the models according to this table is quite similar to the one following from Table 1. The only difference is that now the $dsrc$ model appears to be somewhat better than the $dsrc - Chol$, which is most probably due to chance, since as we saw earlier the latter model forecasts most of the series better. As a conclusion, we can state again that in general the dynamic models outperform the sample covariance based ones.

[7] By shrinking towards the equicorrelated matrix, the variances do not change.

Table 2 Root mean squared prediction errors based on the Frobenius norm

RMSPE^{s}	0.06021
RMSPE^{ss}	0.06016
RMSPE^{rm}	0.05887
RMSPE^{rc}	0.06835
RMSPE^{src}	0.06766
RMSPE^{drc}	0.06004
RMSPE^{dsrc}	0.05749
$\text{RMSPE}^{src-Chol}$	0.05854
$\text{RMSPE}^{dsrc-Chol}$	0.05799

5 Conclusion

Volatility forecasting is crucial for portfolio management, option pricing and other fields of financial economics. Starting with Engle (1982) a new class of econometric models was developed to account for the typical characteristics of financial returns volatility. This class of models grew rapidly and numerous extensions were proposed. In the late 1980s these models were extended to handle not only volatilities, but also covariance matrices. The main practical problem of these models is the large number of parameters to be estimated, if one decides to include more than a few assets in the specification. Partial solutions to this 'curse of dimensionality' were proposed, which impose restrictions on the system dynamics. Still, modelling and forecasting return covariance matrices remains a challenge. This paper proposes a methodology which is more flexible than the traditional sample covariance based models and at the same time is capable of handling a large number of assets. Although conceptually this methodology is more elaborate than the above mentioned traditional models, it is easily applicable in practice and actually requires shorter historical samples, but with a higher frequency. The gains come from the fact that with high-frequency observations, the latent volatility comes close to being observable. This enables the construction of realized variance and covariance series, which can be modelled and forecast on the basis of their dynamic properties. Additionally, we show that shrinking, which has been shown to improve upon the sample covariance matrix, can also be helpful in reducing the error in the realized covariance matrices. A practical drawback which appears in this framework is that the so constructed forecasts are not always positive definite. One possible solution to this is to use the Cholesky decomposition as a method of incorporating the positive definiteness requirement in the forecast.

The paper shows that on the monthly frequency this approach produces better forecasts based on results from Diebold–Mariano tests. The possible gains from a better forecast are, e.g. construction of mean-variance efficient portfolios. Providing a more accurate forecast of future asset comovements will result in better balanced portfolios. These gains will be most probably higher and more pronounced if intradaily returns are used for the construction of daily realized covariance matrices, which remains a possible avenue for further research. It has been shown (e.g. by Andersen et al. (2001a)) that realized daily volatilities and correlations exhibit high persistence. Since by incorporating intra-daily information these realized measures are also quite precise, this serial dependence can be exploited for volatility forecasting. A possible extension of the methodological framework suggested in the paper could be modelling the realized series in a vector ARMA system, in order to analyze volatility spillovers across stocks, industries or markets, which however would again involve a large number of parameters.

A closely related area of research is concerned with the methods for evaluation of covariance matrix forecasts. In this paper we have used purely statistical evaluation tools based on a symmetric loss function. An asymmetric measure in this case may have more economic meaning, since it is quite plausible to assume that if a portfolio variance has been overestimated, the consequences are less adverse than if it has been underestimated. In a multivariate context Byström (2002) uses as an evaluation measure of forecasting performance the profits generated by a simulated trading of portfolio of rainbow options. The prices of such options depend on the correlation between the underlying assets. Thus the agents who forecast the correlations more precisely should have higher profits on average.

Further, the models presented in this paper can be extended by introducing the possibility of asymmetric reaction of (co)volatilities to previous shocks (leverage). This can be achieved by introducing some kind of asymmetry in Eq. (23), e.g. by including products of absolute shocks or products of indicator functions for positivity of the shocks.

Acknowledgements Financial support from the German Science Foundation, Research Group 'Preis-, Liquiditäts- und Kreditrisiken: Messung und Verteilung' is gratefully acknowledged. I am thankful to the referee whose comments significantly improved the overall quality of the paper. I would like to thank Winfried Pohlmeier, Michael Lechner, Jens Jackwerth and Günter Franke for helpful comments. All remaining errors are mine.

References

Aït-Sahalia Y, Mykland PA, Zhang L (2005) How often to sample a continuous-time process in the presence of market microstructure noise. Rev Financ Stud 18(2):351–416

Andersen T, Bollerslev T, Christoffersen PF, Diebold FX (2006) Volatility forecasting. In: Elliott G, Granger C, Timmermann A (eds) Handbook of economic forecasting, 1st edition, chap 15, Elsevier

Andersen TG, Bollerslev T (1998) Answering the skeptics: yes, standard volatility models do provide accurate forecasts. Int Econ Rev 39:885–905

Andersen TG, Bollerslev T, Diebold FX, Ebens H (2001a) The distribution of stock return volatility. J Financ Econ 61:43–76

Andersen TG, Bollerslev T, Diebold FX, Labys P (2001b) The distribution of exchange rate volatility. J Am Stat Assoc 96:42–55

Andersen TG, Bollerslev T, Diebold FX, Labys P (2003) Modeling and forecasting realized volatility. Econometrica 71:579–625

Bandi FM, Russell JR (2005) Microstructure noise, realized volatility, and optimal sampling. Working paper, Graduate School of Business, The University of Chicago

Barndorff-Nielsen OE, Shephard N (2004) Econometric analysis of realised covariation: high frequency based covariance, regression and correlation in financial economics. Econometrica 72:885–925

Bauwens L, Laurent S, Rombouts J (2006) Multivariate GARCH models: a survey. J Appl Econ 21:79–109

Black F, Litterman R (1992) Global portfolio optimization. Financ Anal J 48(5):28–43

Byström H (2002) Using simulated currency rainbow options to evaluate covariance matrix forecasts. J Int Financ Mark Inst Money 12:216–230

Diebold FX, Mariano RS (1995) Comparing predictive accuracy. J Bus Econ Stat 13(3):253–263

Engle R (1982) Autoregressive conditional heteroscedasticity with estimates of the variance of United Kingdom inflation. Econometrica 50:987–1007
Engle R (2002) Dynamic conditional correlation: a simple class of multivariate generalized autoregressive conditional heteroscedasticity models. J Bus Econ Stat 20:339–350
French KR, Schwert GW, Stambaugh RF (1987) Expected stock returns and volatility. J Financ Econ 19:3–29
Gourieroux C, Jasiak J, Sufana R (2004) The wishart autoregressive process of multivariate stochastic volatility. Working Paper, University of Toronto
Hansen PR, Lunde A (2006) Realized variance and market microstructure noise. J Bus Econ Stat 24:127–218
Harvey D, Leyborne S, Newbold P (1997) Testing the equality of prediction mean squared errors. Int J Forecast 13:281–291
Ledoit O, Wolf M (2003) Improved estimation of the covariance matrix of stock returns with an application to portfolio selection. J Empir Finance 10(5):603–621
Ledoit O, Wolf M (2004) Honey, I shrunk the sample covariance matrix. J Portf Manage 31:110–119
Michaud RO (1989) The Markowitz optimization enigma: is 'optimized' optimal? Financ Anal J 45(1):31–42
Oomen RCA (2005) Properties of bias-corrected realized variance under alternative sampling schemes. J Financ Econ 3:555–577
Tse Y, Tsui A (2002) A multivariate generalized auto-regressive conditional heteroscedasticity model with time-varying correlations. J Bus Econ Stat 20:351–362
Voev V, Lunde A (2007) Integrated covariance estimation using high-frequency data in the presence of noise. J Financ Econ 5:68–104
Zhang L, Mykland PA, Aït-Sahalia Y (2005) A tale of two time scales: determining integrated volatility with noisy high frequency data. J Am Stat Assoc 100:1394–1411

Printing: Krips bv, Meppel, The Netherlands
Binding: Stürtz, Würzburg, Germany